# Renewable Energy and Green Metallurgy Technology

# Renewable Energy and Green Metallurgy Technology

Editors

**Xin Yao**

**Huaqing Xie**

MDPI

Basel • Beijing • Wuhan • Barcelona • Belgrade • Novi Sad • Cluj • Manchester

*Editors*
Xin Yao
North China University of Science and Technology
Tangshan
China

Huaqing Xie
Northeastern University
Shenyang
China

*Editorial Office*
MDPI
St. Alban-Anlage 66
4052 Basel, Switzerland

This is a reprint of articles from the Special Issue published online in the open access journal *Processes* (ISSN 2227-9717) (available at: https://www.mdpi.com/journal/processes/special_issues/3F2E903800).

For citation purposes, cite each article independently as indicated on the article page online and as indicated below:

| Lastname, A.A.; Lastname, B.B. Article Title. *Journal Name* **Year**, *Volume Number*, Page Range. |
| --- |

**ISBN 978-3-7258-0639-3 (Hbk)**
**ISBN 978-3-7258-0640-9 (PDF)**
**doi.org/10.3390/books978-3-7258-0640-9**

# Contents

# Preface

Iron and steel enterprises are the pillar industries of the national economy. But at present, the iron and steel industry is an energy-intensive and CO2-intensive industry that accounts for approximately 5% of the world's energy consumption. Energy conservation and emission reductions in the iron and steel industry will play a significant role in achieving sustainable development. Mastering the principle of advanced metallurgical technology, using renewable energy to replace fossil fuels and recycling the waste heat of steel production process could provide a guarantee for the green and high-quality development of the metallurgical field.

The Special Issue "Renewable Energy and Green Metallurgy Technology" focused on green metallurgy production was established by Dr. Xin Yao from North China University of Science and Technology and Dr. Huaqing Xie from Northeastern University. It summarizes the advanced technology of iron and steel enterprises, and mainly presents the latest research results for six research directions, such as biomass energy and solar energy as replacements for fossil fuels, resource utilization of metallurgical slag, low-carbon smelting technology in steel, CO2 resource utilization, strengthening mechanisms and smelting processes of non-quenched and tempered steel for automobiles, high-nitrogen steel smelting technology. Thirteen papers were finally accepted and published in the Special Issue. The papers in the Special Issue proposed a new method for predicting defects in non-oriented silicon steel billets, established a heat transfer model during the process of centrifugal fiber formation, presented the solid-state decarburization method, illustrated recovery waste heat from the iron and steel process, and described the mechanism of iron and steel production.

The publication of the Special Issue has been strongly supported by MDPI. We would like to thank the academic editors and publishing editors for proofreading the text and the illustrations. MDPI offers generous writing timelines for guest editors, allowing the book to be published on schedule. We would like to express our heartfelt thanks.

The Special Issue provides a detailed description of green metallurgy technology, which can be used as a reference for scholars and engineers in metallurgy, energy or related fields in universities. We hope that this Special Issue can be used for reference by scholars and engineering and technical personnel engaged in steel production, energy recovery and other related directions.

**Xin Yao and Huaqing Xie**
*Editors*

processes

*Editorial*

# Renewable Energy and Green Metallurgy Technology

**Xin Yao [1,2,*] and Huaqing Xie [3]**

1 School of Metallurgy and Energy, North China University of Science and Technology, Tangshan 063009, China
2 Tangshan Chuangyuan Fangda Electric Co., Ltd., Tangshan 063000, China
3 School of Metallurgy, Northeastern University, Shenyang 110819, China; xiehq@smm.neu.edu.cn
* Correspondence: yaoxin_0129@163.com

Iron and steel are regarded as the foundation for national development, but their processing consumes huge amounts of fossil fuel and produces large amounts of carbon dioxide gas, which is not conducive to the sustainable development of society [1,2]. The traditional metallurgy process consists of coking, ironmaking in a blast furnace, steelmaking in a converter, refining, continuous casting, steel rolling, etc. The utilization of renewable energy to replace fossil fuels, the recovery of waste heat from the metallurgy process, and the development green metallurgy technology could achieve the goal of green and high-quality iron and steel production. The Special Issue titled *"Renewable Energy and Green Metallurgy Technology"* focused on steel production was established by Dr. Xin Yao and Dr. Huaqing Xie. The Special Issue was published with the support of the North China University of Science and Technology, Tangshan Chuangyuan Fangda Electric Co., Ltd, and Northeastern University. This Special Issue is mainly focused on six selected topics of different aspects of iron and steel production, expanding on biomass energy and solar energy as replacements for fossil fuels, resource utilization of metallurgical slag, low-carbon smelting technology in steel, $CO_2$ resource utilization, the strengthening mechanisms and smelting processes of non-quenched and tempered steel for automobiles, and high-nitrogen steel smelting technology, aiming to master the mechanism of green metallurgy production and decrease fossil fuel consumption and carbon dioxide emissions from the metallurgy process.

Topics concerning the mechanism of green metallurgy production and the decrease in fossil fuel consumption and carbon dioxide emissions from the metallurgy process are of great interest, and the main relevant areas of research over the last 10 years are ordered as follows.

*Biomass energy and solar energy as replacements for fossil fuels.* Biomass energy is considered a green energy [3], but the utilization of biomass to obtain valuable high-end products might consume energy. Solar energy, regarded as a renewable energy, could provide energy for chemical reactions [4]. Combining biomass and solar energy to replace fossil fuels could decrease the energy consumed during the metallurgy process. Naveen S summarized the utilization of solar energy to produce biodiesel, bioethanol, biohydrogen, and biomethane [5]. It was concluded that solar energy could promote the energy carrier for biomass technology and obtain high-end products. The high-end products could be utilized during the metallurgy process, thus decreasing fossil fuel consumption and carbon dioxide emissions.

*Resource utilization of metallurgical slag.* During the metallurgy process, metallurgical slag of high temperature is discharged as a by-product [6], such as blast furnace slag, steel slag, etc. In addition, metallurgical slag could be utilized as a raw material in cement. Sun investigated the thermodynamic characteristics of biomass steam gasification within metallurgical slags [7]. The results implied that blast furnace slag and steel slag are high-quality heat carriers for the process of biomass steam gasification, and the production of syngas could be utilized in the metallurgy process, decreasing the consumption of fossil fuels.

**Citation:** Yao, X.; Xie, H. Renewable Energy and Green Metallurgy Technology. *Processes* **2024**, *12*, 340. https://doi.org/10.3390/pr12020340

Received: 23 January 2024
Accepted: 31 January 2024
Published: 5 February 2024

*Low-carbon smelting technology in steel.* Hydrogen smelting is regarded as an outstanding method for low-carbon ironmaking. Chen investigated low-carbon ironmaking using hydrogen-rich fuel [8]. The results implied that the injection of hydrogen-rich fuel could save coke and strengthen blast furnace smelting, which could provide the groundwork for low-carbon ironmaking.

*$CO_2$ resource utilization.* The huge amount of $CO_2$ discharged from the metallurgy process disturbs the environmental balance. CCUS (carbon capture, utilization, and storage) could achieve the goal of low-carbon steelmaking. McLaughlin reviewed CCUS in the industry [9]. The results implied that CCUS could reduce the negative impacts on the climate and the stored $CO_2$ could be utilized as fertilizer raw materials, fuel cells, chemicals, etc.

*The strengthening mechanisms and smelting processes of non-quenched and tempered steel for automobiles.* Non-quenched and tempered steel could be utilized in the automobile field, and the strengthening mechanisms and smelting processes of steel are especially significant. Jiang investigated the microalloying behavior of non-quenched and tempered steel [10]. The results implied that the yield strength of V-Ti-Nb-N steel could achieve 509.7 MPa and the elongation was 19.2%. These results could provide the groundwork for steel use in automobiles.

*High-nitrogen steel smelting technology.* High-nitrogen steel has the advantages of high strength, good toughness, and corrosion resistance compared with traditional steel [11]. Specifically, high-nitrogen steel with excellent corrosion resistance shows unique advantages in the field of marine equipment. Svyazhin investigated the mechanisms of nitrogen on steel properties and obtained the kinetic characterizations and thermodynamic characterizations of alloying with nitrogen [12]. These results could provide the groundwork for high-nitrogen steel applications.

The Special Issue titled "*Renewable Energy and Green Metallurgy Technology*" mainly focuses on six selected topics to collect research on the advanced technologies for iron and steel processing, achieving green metallurgy production. Centered around research into the selected topics, thirteen papers were finally accepted and published in the Special Issue. The contributions are listed below:

1. Huo, W.; Zhang, C.; Zhang, Y.; Li, X. Numerical Simulation of Mold Slag Entrapment Behavior in Nonoriented Silicon Steel Production Process. *Processes* **2024**, *12*, 167. https://doi.org/10.3390/pr12010167.
2. Liu, C.; Wang, W.; Qi, X.; Wang, B.; Chen, W.; Zhao, K.; Zhen, J.; Zhang, Q. Numerical Simulation of Heat Transfer of Roller Slag in Centrifugal Preparation of Inorganic Fiber. *Processes* **2023**, *11*, 3225. https://doi.org/10.3390/pr11113225.
3. Wen, L.; Ai, L.; Hong, L.; Zhou, Y.; Zhu, G.; Sun, C. Diffusion Behavior of Carbon and Silicon in the Process of Preparing Silicon Steel Using Solid-State Decarburization. *Processes* **2023**, *11*, 3176. https://doi.org/10.3390/pr11113176.
4. Ding, Z.; Xue, Y.; Zhang, L.; Li, C.; Wang, S.; Ni, G. Study on Mushy Zone Coefficient in Solidification Heat Transfer Mathematical Model of Thin Slab with High Casting Speed. *Processes* **2023**, *11*, 3108. https://doi.org/10.3390/pr11113108.
5. Li, C.; Zhang, Y.; Xue, Y.; Zhang, K.; Wang, S.; Sun, H.; Xie, H. The Kinetic Mechanism of the Thermal Decomposition Reaction of Small Particles of Limestone at Steelmaking Temperatures. *Processes* **2023**, *11*, 2712. https://doi.org/10.3390/pr11092712.
6. Zhou, M.; Ai, L.; Hong, L.; Sun, C.; Tong, S. Promoting Effect of Microwave Field on Gas Phase Diffusion Limited Magnetite Reduction in Carbon Monoxide. *Processes* **2023**, *11*, 2709. https://doi.org/10.3390/pr11092709.
7. Li, H.; Xue, H.; Zhang, J.; Zhang, G. Study on Efficient Removal Method of Fine Particulate Dust in Green Metallurgy Process. *Processes* **2023**, *11*, 2573. https://doi.org/10.3390/pr11092573.
8. Meng, L.; Cui, X.; Liu, R.; Zheng, Z.; Shao, H.; Liu, J.; Peng, Y.; Zheng, L. Research on Metallurgical Saw Blade Surface Defect Detection Algorithm Based on SC-YOLOv5. *Processes* **2023**, *11*, 2564. https://doi.org/10.3390/pr11092564.

9. Guo, H.; Tan, Z.; Li, H.; Long, Y.; Ji, A.; Liu, L. Dynamic Characteristics Analysis of Metallurgical Waste Heat Radiative Drying of Thin Layers of Sewage Sludge. *Processes* **2023**, *11*, 2535. https://doi.org/10.3390/pr11092535.
10. Ding, Z.; Liu, Y.; Yao, X.; Xue, Y.; Li, C.; Li, Z.; Wang, S.; Wu, J. Thermodynamic Analysis of Hydrogen Production from Bio-Oil Steam Reforming Utilizing Waste Heat of Steel Slag. *Processes* **2023**, *11*, 2342. https://doi.org/10.3390/pr11082342.
11. Hui, X.; Qi, M.; Wang, W.; Yang, S.; Zhang, C. A Comprehensive Model for Evaluating Titanium Industry Security in China. *Processes* **2023**, *11*, 2286. https://doi.org/10.3390/pr11082286.
12. Tong, S.; Li, C.; Ai, L.; Wang, S.; Zhang, S. Behavior of Carbothermal Dephosphorization of Phosphorus-Containing Converter Slag and Its Resource Utilization. *Processes* **2023**, *11*, 1943. https://doi.org/10.3390/pr11071943.
13. Sun, X.; Ren, J.; Wang, S.; Zhao, D. Effect of Powder Formulation and Energy Density on the Nitrogen Content, Microstructure, and Mechanical Properties of SLMed High-Nitrogen Steel. *Processes* **2023**, *11*, 1937. https://doi.org/10.3390/pr11071937.

Focusing on the common surface defects during the continuous casting process of non-oriented silicon steel, contribution 1 proposed a new method for predicting defects in non-oriented silicon steel billets. Based on the actual slab crystallizer parameters, a three-dimensional model of the slab crystallizer was established using Fluent software. The LES+VOF coupling algorithm was combined to conduct in-depth research on the steel flow field and the unique slag rolling behavior in the crystallizer. Combined with its unique production process characteristics, corresponding operating processes were established, and a unique production process suitable for the current stage of non-oriented silicon steel was obtained. These results could be of great benefit to improve the quality of non-oriented silicon steel products further.

Centered around the heat transfer in the process of centrifugal fiber formation, contribution 2 conducted a numerical simulation study and established a heat transfer model using the Fluent module in ANSYS software. The effects of the internal circulating water, slag temperature, the width of the slag film, and the boundary layer thickness were investigated. The results implied that the average temperature of the water outlet was about 6 °C higher than that of the water inlet. When the slag temperature increased by 1 °C, the roll surface temperature in that part increased by 0.9064 °C. The conditions involving a slag width from 11 mm to 17 mm and a slag thickness (the thickness of the boundary layer) less than 1 mm were advantageous for fiber production.

Aimed toward the preparation process of silicon steel, contribution 3 implied that the use of gas–solid reaction decarburization could avoid the inclusion generation and bubble generation and improve steel cleanliness. On this basis, the feasibility of preparing silicon steel via the solid-state decarburization method and its decarburization effect were discussed in an Ar-$H_2O$-$H_2$ atmosphere. The results implied that an increase in Si content could promote carbon diffusion to the surface. The self-diffusion coefficient of Si increased with the increasing Si content, and the morphology of the oxide layer was also different. Combined with a molecular dynamics study, temperature had an effect on the crystal structure of atoms.

Researching the solidification processes of iron and steel, contribution 4 took a thin slab mold with high casting speed and discussed the effects of the mushy zone coefficient on the thickness of the solidified shell of molten steel in the mold. Through the numerical simulation of the temperature field using Fluent software, it was found that the size of the mushy zone coefficient had a significant influence on the formation thickness of the solidified shell. In order to make the solidified shell thickness of the thin slab at high casting speed close to the real solidified shell thickness, the mushy zone coefficient should be increased. However, too large a mushy zone coefficient will have adverse effects on the calculation of the solidified thickness, so the mushy zone coefficient should be controlled within a reasonable range.

Focusing on high temperature heat waste during the steelmaking process, contribution 5 proposed to study the decomposition behavior of small limestone particles at steelmaking temperature by using the double extrapolation method. The results implied that with the increasing heating rate, the decomposition temperature increased. With the decreasing limestone particle size, the lower activation energy of decomposition decreased. The thermal decomposition reactions of small limestone particles in different $CO_2$ partial pressures were modeled by stochastic nucleation. With the $CO_2$ partial pressure increasing from 25% to 100%, the activation energy and number of reaction stages of decomposition reaction increased.

Aimed toward increasing the reduction rate of the early stage of CO reduction, contribution 6 investigated the effects of increasing temperature and applying microwaves on gas phase diffusion during CO reduction of magnet powder. The results implied that the formation of porous iron productions in the reduction process was dependent on the original gas composition, temperature, and mineral composition. The reduction process (oxygen loss rate) and the local and current partial pressure of $CO_2$ were also key influencing factors. The reduction process under the control of gas diffusion seriously delayed and restricted the occurrence of bubbling. The reaction diffusion of CO reduction was weak due to the low reduction rate at high temperature, which was conducive to sintering diffusion and led to sintering of the material layer. The microwave had a certain effect on improving sintering. The key to increase the CO reduction rate was to improve the CO diffusion conditions.

Focusing on fine particulate dust discharging during the converter steelmaking process, contribution 7 investigated the agglomeration effect and gravity dust removal efficiency of fine dust in converter flue gas using konjac gum, xanthan gum, and their mixed coagulation solutions. The results implied that the mixed solution of 1 g/L mixed glue and 0.5 g/L SDS had a significant influence on increasing the particle size of fine dust particles. Furthermore, the removal rates of PM2.5 and PM10 were 51.46% and 53.13%, respectively. These results could provide effective technological support for efficient removal of the fine particles during the metallurgy process.

Aimed toward solving poor real-time detection, a high false detection rate, and the difficult deployment of detection models, contribution 8 proposed a surface defect detection algorithm of SC-YOLOv5 based on a metallurgical saw blade. The results implied that the accuracy of the improved YOLOv5 model was 88.5%, and the number of parameters were 31.1 M. Compared to the original model, the calculation amount was reduced by 56.36% and the accuracy was increased by 2.1%. The improved algorithm ensured a high detection rate and greatly reduced the complexity and the extent of parameter calculations. It satisfied the requirements of enterprise mobile terminal deployment and provided an outstanding direction for enterprise applications.

Researching the limited methods and low utilization rate of low-grade metallurgical heat waste, contribution 9 simulated the application of low-grade heat waste from the metallurgy process for sludge drying. The seven drying models were analyzed. The kinetic parameters and effective coefficients of diffusion for the three stages of sludge drying were obtained, containing an increasing rate, a constant rate, and a decreasing rate. The relationship between temperature and the effective diffusion coefficient was established using the Arrhenius equation. The activation energies of the three stages of the sludge drying process were calculated to be 29.772 $kJ \cdot mol^{-1}$, 37.129 $kJ \cdot mol^{-1}$, and 39.202 $kJ \cdot mol^{-1}$, respectively.

Centered around the heat waste of steel slag during the metallurgy process, contribution 10 proposed utilizing steam reforming of bio-oil for the recovery of heat waste from steel slag. The equilibrium productions were obtained through thermodynamic analysis. The results implied that steel slag was an outstanding heat carrier to provide heat for the bio-oil steam reforming reaction. At the optimal temperature of 706 °C and the S/C of 6, the hydrogen yield was 109.13 mol/kg and the hydrogen component could achieve 70.21%. The obtained syngas containing hydrogen, carbon monoxide, and methane from the steam reforming of bio-oil could be utilized for the metallurgy process, reducing energy consumption.

Ti is an important deoxidizer and alloying agent during steelmaking production. Researching the development and structure of the Ti industry, contribution 11 established a comprehensive evaluation system, encompassing aspects of availability, economics, and sustainability. The safety level of the Ti industry chain in China from 2010 to 2020 was assessed through the entropy weight technique for order preference and the gray correlation method was used to determine the negative impact on the safety level of the Ti industry chain. The dimension layers and index system were analyzed in terms of coupling degree and sensitivity.

Aimed toward resolving the problem of phosphorus enrichment in converter slag, contribution 12 proposed gasification dephosphorization of converter slag. Firstly, combined with thermodynamic calculations and experiments, the product of gasification dephosphorization was identified as $P_2$, and the migration behavior of carbothermal phosphorus reduction was clearly analyzed. Then, the production practice showed that circulating steelmaking with converter slag gasification dephosphorization had multiple purposes of environmental protection, saving energy, and efficient dephosphorization. Finally, the prospect of resource utilization of phosphorus-containing slag was proposed. To ensure the full reduction of $P_2O_5$ in converter slag, the carbothermal reduction conditions and optimal gasification dephosphorization parameters should be designed appropriately.

Focusing on fabrication techniques of high-nitrogen steel using selective laser melting, contribution 13 investigated the effects of energy density and powder formulations on the microstructural features and nitrogen content of the steel. The results implied that the samples made with elemental mixed powder (EMP) exhibited more non-fusion flaws and poorer density, only reaching a density of 92.36%. Meanwhile, alloy mixed powder (AMP)-prepared samples showed higher structural integrity, reaching a density of 97.21%. With the energy density increasing from 83.3 $J/mm^3$ to 125 $J/mm^3$, the density of EMP samples increased from 88.29% to 92.36%. When energy density, scanning speed, and layer thickness, respectively, reached 104.2 $J/mm^3$, 1000 mm/s, and 30 μm, the optimal mechanical property for AMP was obtained. Under the optimal condition, the ultimate tensile strength, elongation, and yield strength of AMP were 1189.2 MPa, 30.66%, and 958.8 MPa, respectively.

The Special Issue titled "*Renewable Energy and Green Metallurgy Technology*" summarizes the recent findings on advanced technology used in the traditional metallurgy process, including special steel smelting, waste heat recovery, fine particulate dust removing, the utilization of thin slabs with high casting speed, etc. These results could provide a technological guarantee for the development of renewable energy and green metallurgy. However, other emerging technologies were proposed for green metallurgy production. Hydrogen is regarded as an environmentally friendly energy utilized in iron and steel production. The emerging technology named hydrogen metallurgy technology applies hydrogen as a reducing agent instead of carbon to reduce $CO_2$ emissions [13]. A large amount of carbon dioxide is discharged from the metallurgy process. The utilization of emerging technology to achieve the goal of CCUS becomes a great challenge for green iron and steel production [9]. In addition, compared with blast converter steelmaking, electric furnace steelmaking has the advantages of lower energy costs, higher production efficiency, higher product quality, and better environmental performance [14]. Mastering the detailed mechanisms of hydrogen metallurgy technology, CCUS, and electric furnace steelmaking will become the promising direction for iron and steel production in the future.

**Author Contributions:** Conceptualization, X.Y. and H.X.; methodology, X.Y.; investigation, X.Y.; writing—original draft preparation, X.Y. and H.X.; writing—review and editing, X.Y. and H.X.; visualization, X.Y.; supervision, H.X. As the Guest Editors of the Special Issue "Renewable Energy and Green Metallurgy Technology", we would like to express our deep appreciation to the authors, whose valuable research was published in the Special Issue. All authors have read and agreed to the published version of the manuscript.

**Funding:** This work is supported by the Hebei scientific research project of the college and university (BJK2024058) and the Applied Basic Research Project of Tangshan (22130233H).

**Conflicts of Interest:** The authors declare no conflicts of interest.

## References

1. Duan, W.; Yu, Q.; Wang, Z.; Liu, J.; Qin, Q. Life cycle and economic assessment of multi-stage blast furnace slag waste heat recovery system. *Energy* **2018**, *142*, 486–495. [CrossRef]
2. Yao, X.; Liu, Y.; Yu, Q.; Wang, S. Energy consumption of two-stage system of biomass pyrolysis and bio-oil reforming to recover waste heat from granulated BF slag. *Energy* **2023**, *273*, 127204. [CrossRef]
3. Moud, P.H.; Kantarelis, E.; Andersson, K.J.; Engvall, K. Biomass pyrolysis gas conditioning over an iron-based catalyst for mild deoxygenation and hydrogen production. *Fuel* **2018**, *211*, 149–158. [CrossRef]
4. Huang, X.; Lv, Z.; Zhao, B.; Zhang, H.; Yao, X.; Shuai, Y. Optimization of operating parameters for methane steam reforming thermochemical process using Response Surface Methodology. *Int. J. Hydrogen Energy* **2022**, *47*, 28313–28321. [CrossRef]
5. Naveen, S.; Aravind, S.; Yamini, B.; Vasudhareni, R.; Gopinath, K.P.; Arun, J.; Pugazhendhi, A. A review on solar energy intensified biomass valorization and value-added products production: Practicability, challenges, techno economic and lifecycle assessment. *J. Clean. Prod.* **2023**, *405*, 137028. [CrossRef]
6. Ding, Z.; Xue, Y.; Zhang, L.; Li, C.; Wang, S.; Ni, G. Study on Mushy Zone Coefficient in Solidification Heat Transfer Mathematical Model of Thin Slab with High Casting Speed. *Processes* **2023**, *11*, 3108. [CrossRef]
7. Sun, Y.; Zhang, Z.; Liu, L.; Wang, X. Integration of biomass/steam gasification with heat recovery from hot slags: Thermodynamic characteristics. *Int. J. Hydrogen Energy* **2016**, *41*, 5916–5926. [CrossRef]
8. Chen, Y.; Yu, Y.; Gao, Y.; Wang, J.; Xue, Q.; Zuo, H. Effect of Mixed Charging of Nut Coke and Sinter on Hydrogen-Rich Smelting Process of Blast Furnace. *J. Sustain. Metall.* **2023**, *9*, 280–293. [CrossRef]
9. McLaughlin, H.; Littlefield, A.A.; Menefee, M.; Kinzer, A.; Hull, T.; Sovacool, B.K.; Bazilian, M.D.; Kim, J.; Griffiths, S. Carbon capture utilization and storage in review: Sociotechnical implications for a carbon reliant world. *Renew. Sustain. Energy Rev.* **2023**, *177*, 113215. [CrossRef]
10. Jiang, B.; Fang, W.; Chen, R.; Guo, D.; Huang, Y.; Zhang, C.; Liu, Y. Mechanical properties and microstructural characterization of medium carbon non-quenched and tempered steel: Microalloying behavior. *Mater. Sci. Eng. A* **2019**, *748*, 180–188. [CrossRef]
11. Sun, X.; Ren, J.; Wang, S.; Zhao, D. Effect of Powder Formulation and Energy Density on the Nitrogen Content, Microstructure, and Mechanical Properties of SLMed High-Nitrogen Steel. *Processes* **2023**, *11*, 1937. [CrossRef]
12. Svyazhin, A.; Kaputkina, L.; Smarygina, I.; Kaputkin, D. Nitrogen Steels and High-Nitrogen Steels: Industrial Technologies and Properties. *Steel Res. Int.* **2022**, *93*, 2200160. [CrossRef]
13. Tang, J.; Chu, M.-S.; Li, F.; Feng, C.; Liu, Z.-G.; Zhou, Y.-S. Development and progress on hydrogen metallurgy. *Int. J. Miner. Metall. Mater.* **2020**, *27*, 713–723. [CrossRef]
14. Shen, J.; Yang, L.; Yang, Q.; Zhao, X.; Liu, G.; Zheng, M. Polychlorinated Biphenyl Emissions from Steelmaking Electric Arc Furnaces. *Bull. Environ. Contam. Toxicol.* **2021**, *106*, 670–675. [CrossRef] [PubMed]

*processes*

MDPI

*Article*

# Numerical Simulation of Mold Slag Entrapment Behavior in Nonoriented Silicon Steel Production Process

**Wenjie Huo [1,2], Caijun Zhang [1,2,*], Yanchao Zhang [3,*] and Xuekai Li [3]**

1 Hebei High Quality Steel Continuous Casting Engineering Technology Research Center, College of Metallurgy and Energy, North China University of Science and Technology, Tangshan 063000, China; huowenjie@ncst.edu.cn
2 Department of Development Planning and Discipline Construction, North China University of Science and Technology, Tangshan 063210, China
3 Hebei Vocational University of Technology and Engineering, Xingtai 054000, China; lxk15833283430@126.com
* Correspondence: zhangcaijun@126.com (C.Z.); zhangyanchao@163.com (Y.Z.)

**Abstract:** This paper is based on the surface defects of casting billets in the production process of nonoriented silicon steel plates at a steel plant in North China. Taking the parameters of a slab mold in the nonoriented silicon steel production process as a prototype, the flow field characteristics of the mold under the same section, different drawing speed and immersion depth were systematically studied by using a LES (large eddy simulation) and VOF (volume of fluid) coupling algorithm. The results show that under the current conditions, when the critical slag entrapment speed increases from 1.0 m/min to 1.2 m/min, the nozzle insertion depth increases linearly with the critical slag entrapment speed, while when the nozzle insertion depth exceeds 130 mm, the increasing effect of further increasing the nozzle insertion depth on the critical slag entrapment speed begins to decrease. When the drawing speed of continuous casting is kept constant at 1.4 m/min, the abnormal fluctuation height of the steel slag interface is significantly improved when the angle of the water nozzle is increased from 15° to 20°, and the proportion of slag entrapment is also reduced from 0.376% to 0.015%. When the nozzle angle is 25°, the slag entrapment ratio is reduced to 0%, and the steel slag interface also ensures a certain activity. The numerical simulation results were applied to the industrial site, and the slag inclusion rate and crack rate of the billet in the continuous casting process of nonoriented silicon steel were obviously improved after the optimization process.

**Keywords:** nonoriented silicon steel; numerical simulation; slag entrapment behavior; slag drawing speed; immersion depth; nozzle angle

**Citation:** Huo, W.; Zhang, C.; Zhang, Y.; Li, X. Numerical Simulation of Mold Slag Entrapment Behavior in Nonoriented Silicon Steel Production Process. *Processes* **2024**, *12*, 167. https://doi.org/10.3390/pr12010167

Academic Editor: Sara Liparoti

Received: 24 November 2023
Revised: 22 December 2023
Accepted: 28 December 2023
Published: 10 January 2024



## 1. Introduction

Nonoriented silicon steel is an important Fe–Si binary alloy, and surface defects of continuous casting billets are a common quality issue in the production process of nonoriented silicon steel in various steel-making plants. After tracking and analyzing the surface defects that occur during the production process, it can be found that the main reason for the occurrence of casting defects in the continuous casting production process is the occurrence of slag curling behavior in the mold [1–8]. In the continuous casting production process, the behavior of slag entrapment in the mold is an extremely complex process, and many scholars have conducted extensive research on it. Lu Qiaotong et al. [9] studied the slag enrolling behavior by using the experimental method of cold-water simulation. The research showed that there are two kinds of slag enrolling behavior on the steel slag interface: one is shear slag enrolling caused by the upper circulation in the mold, and the other is slag enrolling formed when the steel slag interface breaks, when the vortex slag enrolling near the nozzle and argon bubble impact. The slag entrapment behavior at the steel slag interface under various working conditions is described. Li Baokuan et al. [10], using a combination of water simulation and numerical simulation, conducted a systematic study on the casting blank defects caused by swirling slag on the steel slag interface and the deflection of molten steel in the mold. The results show that the drift of molten steel in the mold is the main cause of the swirling slag at the steel slag interface, and the drawing speed and strength of the drift determine the probability of the slag behavior. LIUZ et al. [11], using water, oil and air as continuous phases and argon as the discrete phase, used numerical simulation methods to simulate and calculate the slag enrolling behavior in the mold, respectively, and obtained the influence of the slag eye and liquid surface exposure on the slag interface under different operating parameters. Qian Zhongdong et al. [12] conducted an in-depth study on the vortex flow in the mold. The research shows that the maintenance time of "turbulent vortex" in the mold can be increased by increasing the dip angle and immersion depth. The asymmetry of the flow field of molten steel in the mold is the main cause of the "bias vortex". The "turbulent vortex" and "bias vortex" in the mold can be effectively improved by using the plate-type vortex damper, and the slag enrolling behavior at the steel slag interface can be further improved. It can be seen from the previous studies that a lot of studies have been done on the mechanism of slag entrapment in the mold, the formation conditions of slag entrapment and the influencing factors through numerical and physical simulation. However, there are few studies on the critical slagging speed and evaluation criteria of whether slagging occurs.

Therefore, this paper is based on the surface defects of casting billets in the production process of nonoriented silicon steel plates at a steel plant in North China. First, the surface defects of nonoriented silicon steel plates after cold rolling were detected using a scanning electron microscope and energy dispersive spectrometer, and the inclusion species and sources at the crack were analyzed in combination with large-scale electrolysis and micro-inclusion detection methods. It was found that more than 60% of the inclusions at the crack contained K and Na elements, so it could be concluded that the surface inclusion was mainly caused by the involvement of mold protection slag. In order to control the occurrence of slag inclusion defects, a three-dimensional model of the slab mold was established by using a numerical simulation software based on the actual parameters of slab mold, and the characteristics of the steel flow field in the mold under different operating parameters were deeply studied by combining the LES/VOF coupling algorithm.

## 2. Construction of Mathematical Model of Computational Domain

### 2.1. Construction of Model Computing Domain and Division of Hexahedral Grid

Based on the mold size used in the production process of nonoriented silicon steel plates at a steel plant in North China, the mold model established using solidworks software is mainly divided into steel liquid phase and protective slag phase. As shown in the Figure 1, due to the axisymmetric patterns of the mold and water port, the 4/1 part of the original mold was taken as the fluid calculation domain to improve the calculation efficiency. In order to improve the calculation accuracy, ICEM15.0 software was used to divide the whole calculation domain into hexahedral grids, and mesh encryption was carried out at the steel slag junction and water outlet, so as to accurately capture the two-phase change law at the steel slag interface. As shown in the figure below, the grid aspect ratio is 12, the number of grids is about 500,000, and the overall grid quality can reach more than 0.45, meeting the computing requirements.

### 2.2. Basic Assumptions and Boundary Conditions

Due to the limitations of the numerical simulation method and the complexity of the flow of molten steel in the mold, the following assumptions should be made based on the actual flow process of molten steel in the mold:

(1) Assume that the liquid steel fluid in the calculation domain is an incompressible fluid.
(2) The solidification fusion and phase change process of the molten steel fluid in the mold are not considered.

(3) The solidification and phase change process of the upper protective slag of the liquid steel fluid is not considered.
(4) Ignore the influence of mold vibration and taper on the flow field of molten steel in the mold.

Boundary conditions:

(1) Inlet: Select the inlet of the nozzle as the inlet of the calculation domain, calculate the inlet speed of the nozzle according to the law of drawing speed and conservation of mass on site, and set the inlet of the calculation domain as the speed inlet.
(2) Mold outlet: The bottom of the mold is set as the outlet of the calculation domain, and the static pressure of the overall steel is set as the pressure outlet.
(3) Wall surface: The wall surface of the fluid domain is defined as the nonslip boundary condition. In the calculation process, the standard wall function is adopted for the near-wall surface [13–15].

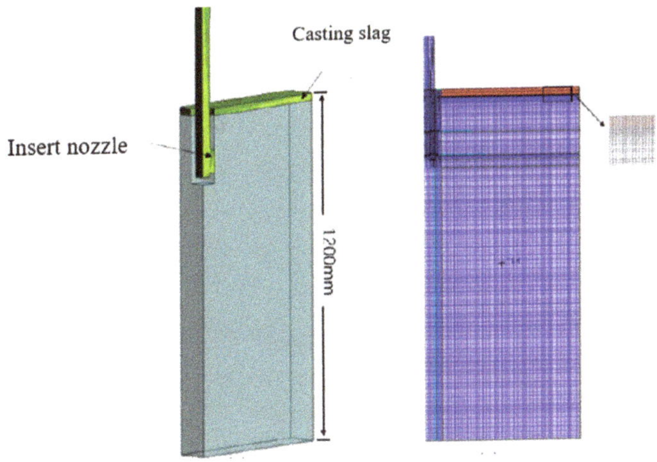


**Figure 1.** The calculation domain of mold is established, and the hexahedral mesh is divided.

### *2.3. Calculation Strategy and Simulation Parameters*

Through the above analysis, it can be seen that this simulation is mainly a two-phase flow simulation of steel liquid phase and protective slag phase. In order to avoid divergence in the calculation process, the variable density parameter in the working condition is first set to 0 $kg/m^2$, and the working temperature is set to a temperature value similar to the inlet temperature and liquid steel temperature. Then, each parameter of the subrelaxation factor in the calculation control is reduced by 0.2, respectively. Finally, the initial turbulent kinetic energy and specific dissipation rate in the initialization are set to 0.01 before initialization. In the calculation process, the weighted average velocity of the interface between liquid protection slag and liquid steel is monitored. The convergence standard of other RMS residuals is $10^{-3}$, except that the residual and kinetic energy are $10^{-6}$. Combined with grid parameters, the initial step of the calculation is set to 0.0001, and one step is iterated 20 times. When the Courante number in the calculation process is stable, the time step is gradually increased to improve the calculation efficiency.

The physical property parameters involved in the simulation calculation are shown in the Table 1:

**Table 1.** Simulation parameters of various physical quantities.

| Argument | Numerical Value | Argument | Numerical Value |
|---|---|---|---|
| Mold (mm) | 950 × 30 | Density of molten steel/($kg \cdot m^{-3}$) | 7100 |
| Nozzle insertion depth (mm) | 110, 120, 130, 150 | Slag density/($kg \cdot m^{-3}$) | 3100 |
| Drawing speed of continuous casting (m/min) | 1.2, 1.3, 1.4 | Viscosity of molten steel/($kg \cdot s^{-1} \cdot m^{-1}$) | 0.0062 |
| Inner diameter of the nozzle (mm) | 72 | Viscosity of slag/($kg \cdot s^{-1} \cdot m^{-1}$) | 0.14 |
| Calculated area length (mm) | 1200 | Interfacial tension of steel slag/($N \cdot m^{-1}$) | 1.4 |

### *2.4. Model Algorithm in Simulation Computation*

This paper mainly simulates the flow field and liquid surface fluctuation behavior of instantaneous turbulent flow in the mold. The large eddy simulation (LES) model of FLUENT was used to describe the turbulent flow in the mold. Meanwhile, the fluid volume (VOF) model was used to trace the interface between liquid steel and liquid protective slag in the mold in order to clarify the fluctuation of the liquid surface of the mold. Liquid steel flow can be regarded as unsteady and incompressible turbulent flow, and its governing equation is as follows [16–18]:

$$\frac{\partial \rho \mu_i}{\partial x_j} = 0 \tag{1}$$

Momentum equation:

$$\frac{\partial \mu_i}{\partial t} + \frac{\partial \mu_i \mu_j}{\partial x_j} = -\frac{1}{\rho}\frac{\partial \rho}{\partial x_i} + v\nabla^2 \mu_i \tag{2}$$

In the formula:
$\rho$—fluid density, $kg/m^3$;
$\mu$—speed, m/s;
$x$—nodal coordinates;
$v$—kinematic viscosity of fluid, $m^2/s$;
$p$—stress, Pa;
$i, j$—axis direction.

Large eddy simulation (LES) is a computational method used to simulate fluid dynamics, which can simulate a much finer turbulent structure than the Reynolds mean method (RANS). LES uses a filter to filter out small-scale turbulent structures, leaving large-scale structures intact. In LES, the flow field is divided into large-scale and small-scale parts, in which the large-scale part is simulated and the small-scale part is processed by the model. The transport equation of LES and turbulence equation are as follows [19–21]:

Transmission equation of LES model:

$$\frac{\partial}{\partial t}(\rho \mu_i) + \frac{\partial}{\partial x_j}(\rho \mu_i \mu_j) = -\frac{\partial p}{\partial x_i} + \frac{\partial}{\partial x_j}\left[(\mu + \mu_i)\left(\frac{\partial \mu_i}{\partial u_j} + \frac{\partial \mu_j}{\partial u_i}\right)\right] \tag{3}$$

The accuracy of large eddy simulation mainly depends on the selection of sublattice scale, which determines the role of unsolvable turbulence scale. In this study, the Smagorinsky–Lilly model was used to calculate turbulent viscosity.

$$\mu_t = \rho L_s^2 |\overline{S}| \tag{4}$$

In the formula:
$L_s$—mixing length of sublattice, m;

$S$—average deformation rate tensor, $s^{-1}$.

$$|\overline{S}| = \sqrt{2\overline{S}_{ij}\overline{S}_{ij}} \tag{5}$$

$$L_s = \min\left(kd, C_s V^{\frac{1}{3}}\right) \tag{6}$$

In the formula:
$k$—Karman constant;
$d$—the closest distance to the wall, m;
$C_s$—Smagorinsky constant.

The volume of fluid (VOF) model is a numerical method used in computational fluid dynamics (CFD) to simulate two-phase flows (e.g., gas–liquid and liquid–liquid). It is based on the volume fraction of the fluid, that is, the ratio of the volume of the fluid at a certain position to the total volume, to describe the distribution and interface morphology of two fluids. In the VOF model, fluid volume fraction is used as a scalar field to track the distribution of two fluids in the flow field. The position and shape of the interface between the two fluids can be determined by tracing the motion trajectories of fluids with different volume fractions while solving the fluid dynamics equation.

The advantage of the VOF model is that it can accurately simulate the interaction between two fluids, including the interface morphology, interface strength, physical quantities at the interface, etc. For the q phase, the following continuity equation [22–24] should be satisfied:

$$\frac{1}{\rho_q}\left[\frac{\partial}{\partial t}(a_q\rho_q) + \nabla\left(a_q\rho_q\vec{\mu}_q\right)\right] = 0 \tag{7}$$

In the formula:
$a_q$—the volume fraction of phase $q$;
$\rho_q$—the density of phase $q$, $kg/m^3$.

All phase volume fractions in the calculation process meet the following requirements:

$$\sum_{q=1}^{n} a_q = 1 \tag{8}$$

When there are multiple phases in the calculation domain, the fluid density and viscosity in the calculation domain are calculated by the following formulas [25–27]:

$$\rho = \sum_{q=1}^{n} a_q\rho_q \tag{9}$$

$$\mu = \sum_{q=1}^{n} a_q\mu_q \tag{10}$$

## 3. Influence of Different Operating Parameters on Flow Field of Mold

The types and sources of inclusions at the crack were analyzed by means of large sample electrolysis and microscopic inclusions. It was found that more than 60% of inclusions at the crack contained K and Na elements, so it could be concluded that the surface inclusions were mainly caused by the involvement of mold protective slag. In order to control the occurrence of slag inclusion defects, the flow field characteristics of a section, different drawing speed and immersion depth were systematically studied using a numerical simulation method combined with the actual operating parameters of the mold under the current production situation of nonoriented silicon steel.

### 3.1. The Formulation of the Simulation Scheme

The Table 2 shows the simulation plan formulated in combination with the actual production operation parameters. In the actual production process, combined with the production characteristics of nonoriented silicon steel, four kinds of immersion depths and three kinds of matching pulling speeds are formulated for the mold with a section size of 950 mm × 230 mm. Therefore, a total of eight groups of simulation plans are formulated in combination with the actual situation on site. In the simulation process, the distribution characteristics of the steel flow field in the mold under different operating parameters, the slag interfacial slag entrapment behavior, the steel slag interfacial velocity distribution and the fluctuation height are used as the evaluation criteria to calculate the critical slag entrapment speed under different operating parameters, which provides theoretical guidance for the actual production.

**Table 2.** Operating parameters used for simulation.

| Nozzle Insertion Depth/mm | Drawing Speed of Continuous Casting/(m/min) |
|---|---|
| 120 | 1.0 |
| 120 | 1.1 |
| 120 | 1.2 |
| 120 | 1.3 |
| 110 | 1.2 |
| 120 | 1.2 |
| 130 | 1.2 |

### 3.2. Simulation of Mold Flow Field Characteristics under Different Drawing Speeds

The Figure 2 shows the distribution of the molten steel flow field in the mold at the same immersion depth (120 mm) and at the same time under the conditions of different pulling speeds. It can be seen from the figure that the liquid steel first impacts the narrow edge of the mold after entering the mold from the water outlet. After the impact to the narrow edge, part of the liquid steel codified along the narrow edge of the mold to the direction of the liquid steel surface to form an upper circulation, and the other part of the liquid steel flowed down along the narrow edge to form a small range of lower circulation at the bottom of the mold. When the pulling speed is 1.0 m/min, the strength of the upper circulation is small, and the vortex center is close to the side of the water outlet. With the continuous increase of the pulling speed, the position of the jet impact on the narrow side of the mold continues to move upward, and the strength of the upper circulation continues to increase. At this time, the vortex center of the upper circulation continues to move toward the narrow side of the mold, and the eddy strength also continues to increase, resulting in a significant increase in the fluid velocity at the steel slag interface, aggravating the fluctuation strength of the steel slag interface.

In order to describe more quantitatively the velocity variation law of steel slag interfaces and the fluctuation strength of the steel slag under the conditions of different drawing speeds, the central line velocity of the steel slag interface at the same time and at different drawing speeds is extracted. The extraction results are shown in the Figure 3. As can be seen from the figure, as the drawing speed of the mold increases from 1.0 m/min to 1.3 m/min, the maximum surface velocity at the steel slag interface increases from 0.13 m/s to 0.24 m/s, and the velocity curve presents a curve characteristic of being high in the middle and low at both ends. The overall velocity in the central region of the liquid surface of the steel slag is greatly increased, but a small range of sudden increases in the liquid surface velocity of the steel slag near the narrow edge of the mold indicates that a small range of eddy current is formed here, and with the increase of the pulling speed, the intensity of the eddy current is constantly enhanced, after which it is easy to generate convection in the central region of the slag interface, resulting in slag entrapment.

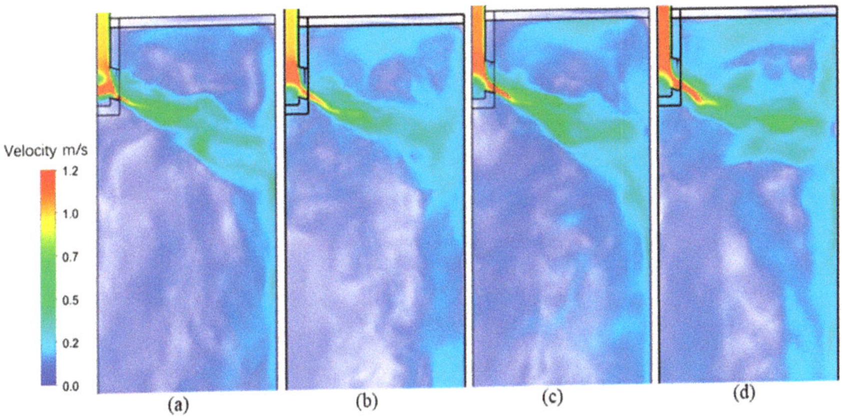


**Figure 2.** Cloud chart of flow field distribution of molten steel in the mold under different casting speed conditions. (**a**)—1.0 m/min; (**b**)—1.1 m/min; (**c**)—1.2 m/min; (**d**)—1.3 m/min.

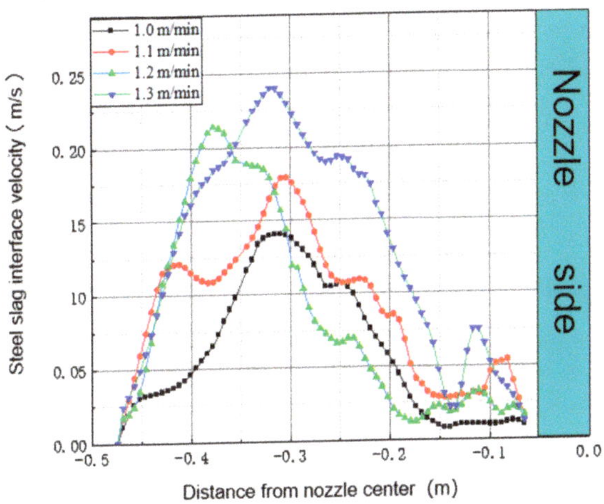


**Figure 3.** Velocity distribution in the central area of the steel slag interface at different drawing speeds.

During the model calculation process, the slag entrapment situation was extracted and statistically analyzed under different pulling speed conditions and at different times (40 s, 45 s and 50 s). The results are shown in Figure 4. It can be seen from the figure that the steel slag interface of the mold is relatively stable and no slag entrapment occurs when the drawing speed is 1.0 m/min and 1.1 m/min. When the pulling speed increases to 1.2 m/min, abnormal fluctuations appear on the side near the narrow edge of the mold, resulting in convective flow between the upper reflux liquid steel and liquid steel near the water mouth of the mold. When the convective intensity increases to a certain level, a downward convex vortex will be generated. Coupled with the effect of negative pressure in the flow field, the enrolling strength of the vortex becomes larger and larger, causing the protection slag from the slag phase into the liquid steel. When the pulling speed is further increased to 1.3 m/min, the occurrence probability of vortex slag-rolling behavior is further enhanced, the amount of protective slag involved near the nozzle is significantly increased, the abnormal fluctuation of the liquid surface of steel slag near the narrow edge is intensified, and the interface of steel slag suddenly rises at a certain position and then goes deep into the liquid steel to form a certain range of fluctuating slag rolling when it suddenly falls.

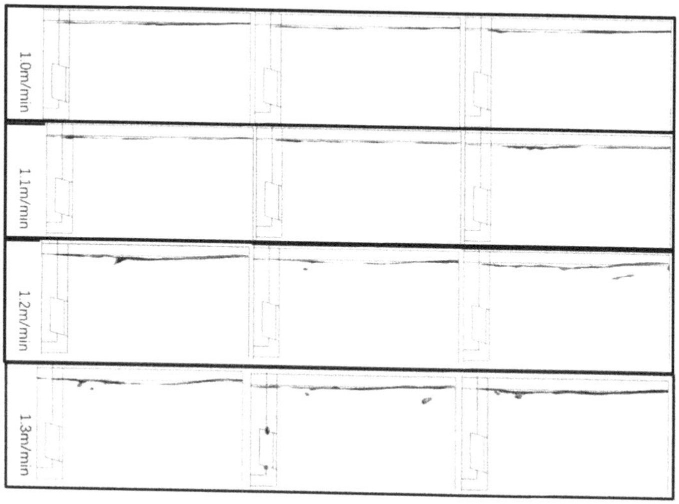


**Figure 4.** Distribution of slag phase at the interface of steel slag under different casting speeds.

### 3.3. *Analysis of Flow Field Characteristics of Mold under Different Insertion Depths*

In the above analysis, it can be found that under the same conditions, when the drawing speed is increased to 1.2 m/min, the slag interface will have a slag entrapment trend. Therefore, this section analyzes the flow field characteristics of the mold under the conditions of the same drawing speed (1.2 m/min), same time and different nozzle insertion depths.

The Figure 5 shows the distribution cloud of the liquid steel flow field under the condition of different inlet insertion depths. You can see from the picture that under the same pulling speed, the increase of nozzle insertion depth makes the position of the narrow edge of the mold move down continuously. After measurement, as the depth of nozzle insertion increases from 110 mm to 150 mm, the distance between the impact point of the liquid steel main strand on the narrow surface and the interface of the mold steel slag is 212 mm, 221 mm, 261 mm and 282 mm, respectively. As the position of the impact point continues to move down, the impact distance of the main stream in the mold continues to increase, which indirectly leads to the energy of the main stream impacting on the narrow edge of the mold, which decreases with the increase in the depth of the nozzle. Therefore, with the increase of the nozzle insertion depth, the energy of the upper reflux formed by the upward return after the jet impinges on the narrow edge of the mold continues to decrease, resulting in the downward movement of the vortex center of the upper reflux and the decreasing strength of the upper reflux, thus reducing the fluctuation of the steel slag interface. Especially when the immersion depth increases to 150 mm, the upward return of the liquid steel is less, and the upper reflux strength is lower. The flow rate of molten steel in the lower reflux zone is very large, which easily leads to uneven melting of the protective slag.

The Figure 6 shows the velocity distribution of the center line of the steel slag interface at the same time and under different immersion depths. It can be seen from the figure that the interface velocity curve of steel slag as a whole presents a change characteristic of first increasing and then decreasing. The maximum velocity area is mainly distributed at 150 mm from the narrow side of the mold, and the minimum velocity area is mainly distributed at the water mouth side. During the process of increasing the insertion depth of the nozzle from 110 mm to 130 mm, the centerline velocity of the steel slag interface showed a slight decrease overall with the increase of insertion depth, but the velocity distribution at the steel slag interface became extremely uneven. When the depth of the water inlet is 120 mm, the maximum velocity difference at the steel slag interface is

0.19 m/s, which can easily cause the phenomenon of slag entrapment at the steel slag interface. When the insertion depth of the water inlet is further increased to 150 mm, the centerline velocity of the steel slag interface significantly decreases, the overall velocity distribution of the steel slag interface becomes more uniform, and the velocity curve becomes smoother. Therefore, making changes within this immersion depth range can reduce the occurrence of slag entrapment.

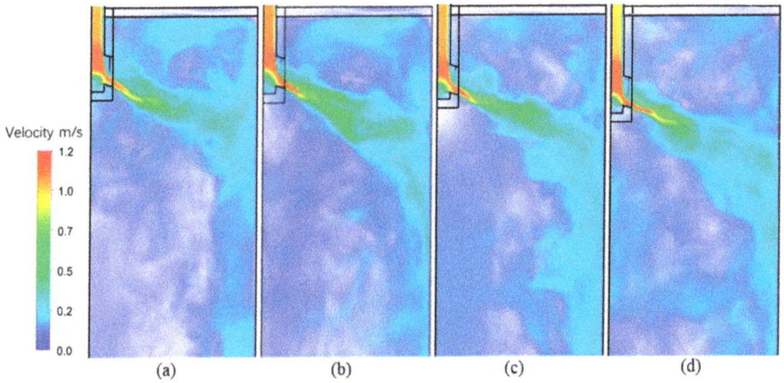


**Figure 5.** Cloud chart of flow field distribution of liquid steel under different nozzle immersion depth conditions. (**a**)—110 mm; (**b**)—120 mm; (**c**)—130 mm; (**d**)—150 mm.

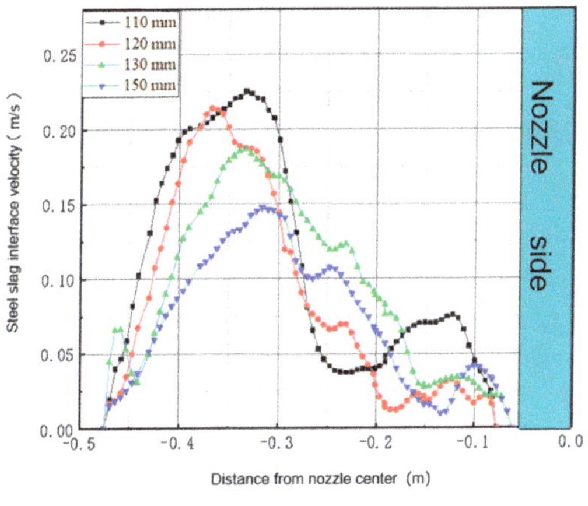


**Figure 6.** Velocity distribution in the central area of the steel slag interface under different immersion depths.

As shown in Figure 7, the fluctuation height distribution diagram in the central region of the steel slag interface at the same time shows that the flow field distribution characteristics of the upper and lower circulation in the mold result in the fluctuation of the steel slag interface showing an approximate "S" shape. The side near the narrow edge of the mold fluctuates upward in the direction of air, forming a wave peak, while the side near the water outlet fluctuates downward in the direction of liquid steel, forming a trough. With the increasing nozzle insertion depth, the wave curve of the steel slag interface becomes more gentle, the wave trough near the narrow edge gradually decreases, and the wave height difference between the two sides of the steel slag interface also gradually decreases. When the nozzle insertion depth is changed in the range of 110–120 mm, the slag interface fluctuates more violently, and the wave near the narrow side of the mold reaches about 10 mm, which is very easy to cause the slag interface rolling phenomenon. However, when the immersion depth is changed in the range of 120–150 mm, the interface fluctuation of steel slag is significantly improved, and the fluctuation curve is gradually smoothed. Operation in this range can greatly reduce the occurrence of slag entrapment.

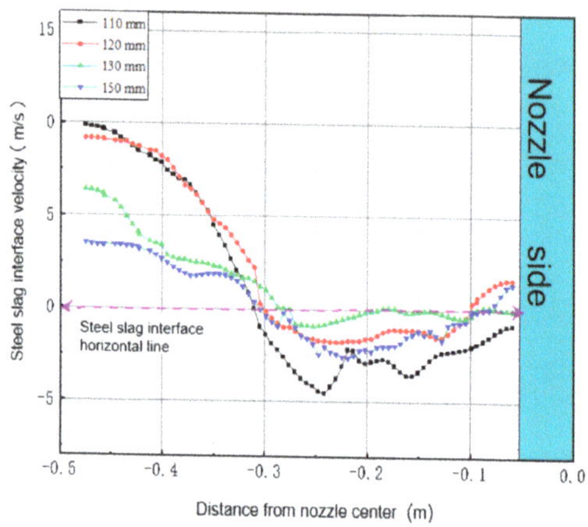


**Figure 7.** Distribution of fluctuation height in the central area of the steel slag interface.

On the basis of the above analysis, in order to more intuitively reflect the slag entrapment condition of the mold steel slag interface under the same drawing speed and different immersion depth, the slag entrapment conditions at the steel slag interface at different drawing speeds and different times (40 s, 45 s and 50 s) were extracted. The result is shown in Figure 8. As can be seen from the figure, under the same pulling speed, when the water nozzle insertion depth is operated within the range of 110–120 mm, there is an obvious slag entrapment phenomenon at the steel slag interface. It can also be seen that due to the distribution characteristics of the upper circulation in the mold and the fluctuation characteristics of the steel slag interface, the protective slag layer on the steel slag interface near the nozzle is thicker, resulting in the formation of a relatively obvious vortex slag at 1/4 of the mold. This is consistent with the above analysis results, indicating that the unreasonable distribution of steel slag interfacial velocity and fluctuation will greatly improve the occurrence of slag entrapment.

When the nozzle insertion depth is greater than 120 mm, there is no obvious phenomenon of slag droplets being sucked into the liquid steel in the figure, the slag entrapment situation at the steel slag interface is significantly improved, and the steel slag interface becomes more gentle. When the immersion depth is 130 mm, although there is a certain fluctuation in the steel slag interface, the fluctuation is relatively gentle, and the distribution of protective slag on the steel slag interface is relatively uniform, and no obvious slag entrapment occurs. When the water nozzle insertion depth is 150 mm, the steel slag interface becomes more gentle, and the interface fluctuation is basically no longer visible under this condition. Although this effectively prevents the occurrence of the slag entrapment phenomenon, the interface of the steel slag is too smooth, and there is no certain activity that will affect the melting of the protective slag and the inflow of the narrow side, which is not conducive to the heat transfer and solidification of the liquid steel in the mold. If the water nozzle is inserted too deep, it will lead to the crystallization of nonmetallic inclusions with a floating distance that is too long, affecting the removal of inclusions. Therefore, under the current conditions, when the pulling speed is 1.2 m/min, the insertion depth of the water nozzle is controlled at about 130 mm, which can not only reduce the occurrence of the slag entrapment phenomenon, but also ensure that there is a certain degree of activity at the slag interface, which has a certain positive effect on improving the quality of the casting billet.

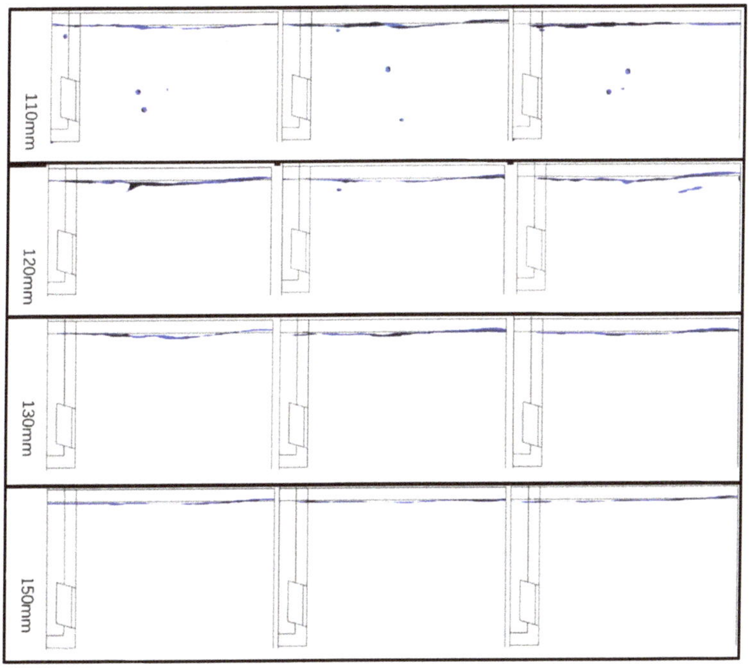


**Figure 8.** Slag entrapment behavior in molds at different nozzle immersion depths.

*3.4. Determination of Critical Slag Drawing Speed*

It can be seen from the above analysis results that under different processing conditions, the flow velocity distribution, fluctuation height and shape of steel slag interface are indicative of whether slag entrapment occurs in the mold. Therefore, in order to further establish the critical slag entrapment speed under the experimental conditions using the method of numerical simulation, this provides theoretical support for the actual production. Firstly, the interface characterization parameters of steel slag under different process parameters during the numerical simulation calculation were extracted and counted. The statistical results are shown in the table below.

As can be seen from the data in the Table 3, when the nozzle insertion depth remains unchanged at 120 mm, the maximum surface flow rate and wave height of the steel slag interface increase from 1.1 m/min to 1.2 m/min, and the proportion of slag entrapment liquid rapidly increases to 0.175%. When the pulling speed exceeds 1.2 m/min, the maximum surface flow rate and wave height increase. Although the maximum surface flow rate and wave height of the slag interface decreased, the proportion of slag entrapment liquid increased exponentially, indicating that when the pulling speed exceeded 1.2 m/min, more slag droplets in the mold began to be involved in the liquid steel, which then affected the quality of the casting billet. However, when the pulling speed remains unchanged at 1.2 m/min, the insertion depth of the water nozzle is continuously increased until it is increased to 130 mm, the interfacial flow rate and fluctuation height of the steel slag are greatly reduced, and the slag entrapment proportion is also reduced to 0%.

Based on the above statistical data, the distribution diagram of critical slag drawing speeds under different nozzle insertion depths is shown in the following Figure 9. As can be seen from the figure, when the critical slag entrapment speed increases from 1.0 m/min to 1.2 m/min, the nozzle insertion depth increases linearly with the critical slag entrapment speed; however, when the nozzle insertion depth exceeds 130 mm, the proportion of the increase in the critical slag entrapment speed begins to decrease, indicating that when the nozzle insertion depth exceeds 130 mm, the critical slag entrapment speed decreases. A further increase of the nozzle insertion depth has a lower effect on the critical slag drawing speed. Therefore, if you want to further increase the critical slag entrapment speed to 1.3 m/min, it is necessary to increase the nozzle insertion depth to 150 mm to reduce the occurrence of slag entrapment.

**Table 3.** Characterization parameters of steel slag interface under different process parameters.

| Pulling Speed/(m/min) | Immersion Depth/mm | Maximum Surface Velocity/(m/s) | Maximum Fluctuation Height/mm | Slag Proportion |
|---|---|---|---|---|
| 1.0 | 120 | 0.151 | 3.58 | 0% |
| 1.1 | 120 | 0.178 | 5.42 | 0.0013% |
| 1.2 | 120 | 0.214 | 9.19 | 0.175% |
| 1.3 | 120 | 0.240 | 10.18 | 0.415% |
| 1.2 | 110 | 0.225 | 9.85 | 0.308% |
| 1.2 | 120 | 0.214 | 9.19 | 0.114% |
| 1.2 | 130 | 0.188 | 6.42 | 0% |
| 1.2 | 150 | 0.146 | 3.53 | 0% |

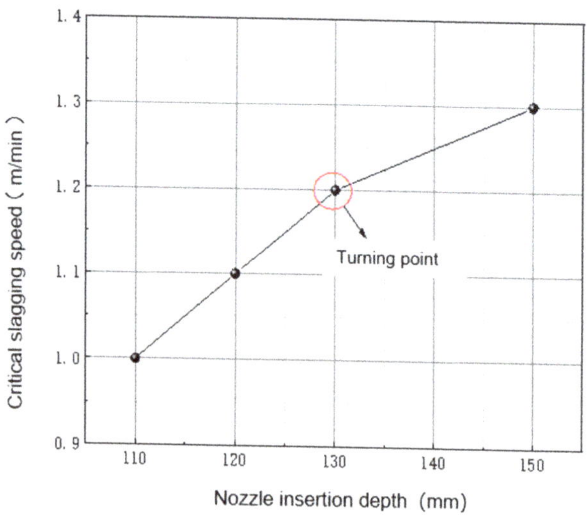


**Figure 9.** Maximum pulling speed under different nozzle immersion depths.

## 4. Influence of Different Nozzle Angles on Slag Enrolling Behavior in Mold

It can be seen from the above analysis that under the condition of the existing production equipment, the optimal critical slag drawing speed can be reached by simply changing the production operation parameter (nozzle insertion depth), which is 1.2 m/min. In order to further improve the production efficiency without changing the existing production process and further increase the critical slag entrapment speed to 1.4 m/min, it is necessary to optimize the dip angle of the submerged nozzle. Therefore, according to the actual production requirements of steel mills, four kinds of immersion spout with inclination angles of 15°, 18°, 20°, 22° and 25° were selected as prototypes, and the LES/VOF model was used to further simulate slag entrapment behavior in the mold under different spout inclination angles.

The Figure 10 shows the contour simulation results of the steel slag interface in the mold under different dip angles at the same time. As can be seen from the figure, as the angle of the nozzle increases from 15° to 25°, the fluctuation amplitude of the steel slag interface is gradually gentler, and the fluctuation height difference between the two sides of the steel slag interface also gradually decreases. When the angle of the nozzle is changed in the range of 15°–20°, the maximum velocity area on the slag interface is mainly distributed in the range of 200 mm–400 mm from the nozzle side, and the velocity distribution is very uneven, while the distribution in the region with large velocity is more concentrated, which makes it easy to produce downward vortices on the slag interface and slag entrapment. However, when the nozzle inclination continues to increase to 22° and 25°, the velocity distribution of the slag interface becomes more uniform, and the velocity difference between the maximum and minimum velocity regions gradually decreases. Changes within the nozzle inclination range can not only reduce the occurrence of the slag entrapment phenomenon, but can also ensure that the slag interface has a certain activity, which is conducive to the melting of the protective slag.

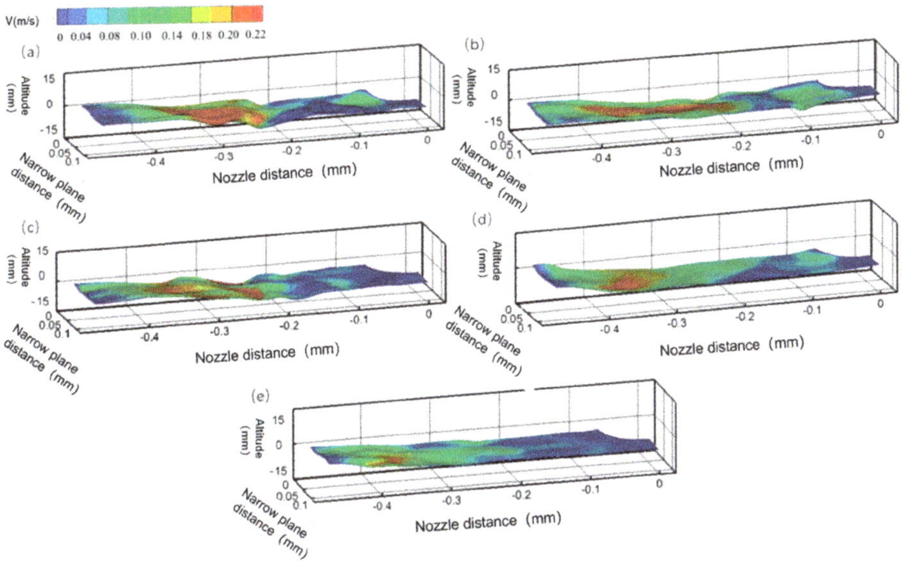


**Figure 10.** Outline of steel slag interface under different nozzle angles. (**a**)—15°; (**b**)—18°; (**c**)—20°; (**d**)—22°; (**e**)—25°.

In order to more intuitively understand the slag enrolling situation at the steel slag interface in the mold under different nozzle inclination angles, the slag enrolling behavior in the mold under different nozzle inclination angles was simulated, and the results were shown in Figure 11: It can be seen from the figure that under the same pulling speed, when the nozzle angle is operated in the range of 15°–20°, the distribution of the slag layer is unreasonable due to the abnormal fluctuation of the slag interface; most of the protective slag is deposited on the narrow side of the mold, and the slag layer in the middle part of the slag interface is thin. However, the uneven velocity distribution on the steel slag interface directly leads to obvious slag entrapment behavior in the range of 200 mm–400 mm from the nozzle side, and the slag droplets are gradually sucked into the bottom of the mold. With the increasing angle of the nozzle, the slag entrapment condition at the steel slag interface is gradually improved. When the angle of the nozzle is increased to 25°, the distribution of protective slag on the steel slag interface is more uniform, and no obvious slag drops are involved in the deep steel.

In order to further study the characteristic state of the steel slag interface when slag entrapment behavior occurs, the interface characterization parameters of steel slag under different nozzle inclination angles were extracted. The results are shown in the Table 4. It can be seen from the data in the table that when the pulling speed is kept constant at 1.4 m/min, the abnormal fluctuation height of the steel slag interface is significantly improved when the dip angle of the water nozzle is increased from 15° to 20°, which is reduced from the initial 15.2 mm to 7.4 mm. The proportion of slag was also reduced from

0.376% to 0.015%. In the process of increasing the nozzle angle from 20° to 25°, although the maximum surface flow rate and fluctuation height decrease, the slag proportion decreases further. When the angle of the nozzle is 25°, the slag entrapment ratio is reduced to 0%, and the steel slag interface also ensures a certain activity.

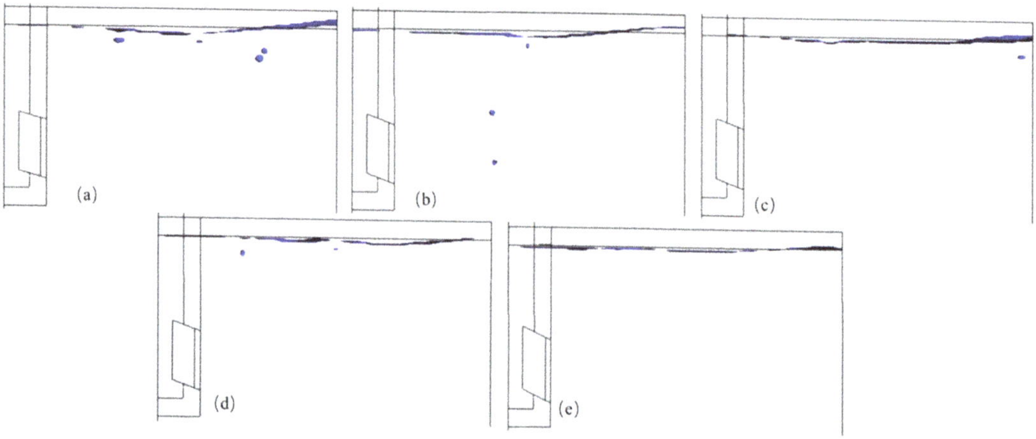


**Figure 11.** Slag entrapment behavior in mold under different nozzle angles. (**a**)—15°; (**b**)—18°; (**c**)—20°; (**d**)—22°; (**e**)—25°.

**Table 4.** Characterization parameters of steel slag interface under different inclination angles.

| Nozzle Angle/° | Maximum Surface Velocity/(m/s) | Maximum Fluctuation Height/mm | Slag Proportion |
|---|---|---|---|
| 15 | 0.287 | 15.2 | 0.376% |
| 18 | 0.259 | 12.2 | 0.209% |
| 20 | 0.235 | 7.4 | 0.015% |
| 22 | 0.215 | 6.8 | 0.0032% |
| 25 | 0.176 | 6.1 | 0% |

## 5. Industrial Verification Effect

In order to further verify the accuracy of the numerical simulation technology, the corresponding data are collected and analyzed in the industrial field. In the production process of nonoriented silicon steel, the slag inclusion rate and crack rate of casting billets under different technological conditions of the same steel grade, same production process, same furnace times and same pouring times are shown in the figure below. The optimized process is the process after limiting the depth (130 mm) and pulling speed (1.2 m/min) of the submerged water outlet. After the optimization of the nozzle, the production situation is optimized where the drawing speed is increased to 1.4 m/min after using the new water outlet (the angle of the water outlet is 25°).

From the Figure 12, it can be seen that after optimizing the process, the slag inclusion rate and crack rate of the continuous casting process of nonoriented silicon steel have decreased from the original 0.45% and 0.43% to 0.18% and 0.24%. After applying the new nozzle, the casting speed was further increased to 1.4 m/min. At this time, under the same conditions, the slag inclusion rate and crack rate of the billet were reduced to below 0.1% compared to before optimizing the nozzle. This indicates that the slag inclusion rate and crack rate of the billet have been further controlled. The industrial practice further confirms the accuracy of the numerical simulation results. The application of numerical simulation technology in industrial practice further provides theoretical guidance and technical support for practical production.

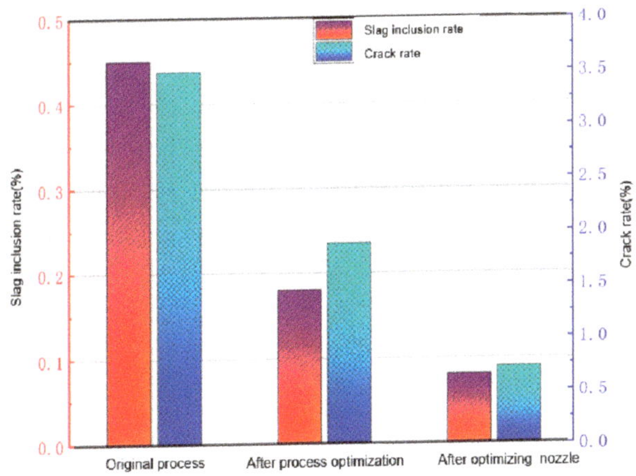


**Figure 12.** Analysis of casting billet quality under different process conditions.

## 6. Conclusions

(1) When the nozzle insertion depth remains unchanged at 120 mm, the maximum surface flow rate and wave height of the steel slag interface increase from 1.1 m/min to 1.2 m/min, and the proportion of slag entrapment liquid rapidly increases to 0.175%. When the pulling speed exceeds 1.2 m/min, more slag droplets in the mold begin to be sucked into the liquid steel. When the pulling speed remains unchanged at 1.2 m/min, the insertion depth of the water nozzle is continuously increased until it is increased to 130 mm, and the slag entrapment ratio is reduced to 0%.

(2) Under the condition of the same pulling speed (1.2 m/min), when the water nozzle insertion depth is operated within the range of 110–120 mm, there is an obvious slag entrapment phenomenon at the steel slag interface. It can also be seen that due to the distribution characteristics of the upper circulation in the mold and the fluctuation characteristics of the steel slag interface, the protective slag layer on the steel slag interface near the water outlet is thicker, resulting in the formation of a relatively obvious vortex slag entrapment behavior at 1/4 of the mold.

(3) Under the current conditions, when the pulling speed is 1.2 m/min, the insertion depth of the water nozzle is controlled at about 130 mm, which can not only reduce the occurrence of the slag entrapment phenomenon but also ensure that there is a certain activity at the steel slag interface.

(4) When the critical slag entrapment speed increased from 1.0 m/min to 1.2 m/min, the nozzle insertion depth increased linearly with the critical slag entrapment speed, but when the nozzle insertion depth exceeded 130 mm, the increasing effect of further increasing the nozzle insertion depth on the critical slag entrapment speed began to decrease. Therefore, if you want to further increase the critical slag entrapment speed to 1.3 m/min, it is necessary to increase the nozzle insertion depth to 150 mm to reduce the occurrence of slag entrapment.

(5) When the pulling speed is kept constant at 1.4 m/min, the abnormal fluctuation height of the slag interface is significantly improved when the angle of the nozzle is increased from 15° to 20°, which is reduced from the initial 15.2 mm to 7.4 mm. The proportion of slag was also reduced from 0.376% to 0.015%. When the angle of the nozzle is 25°, the slag entrapment ratio is reduced to 0%, and the steel slag interface also ensures a certain activity.

(6) The numerical simulation results were applied to the industrial site. After the on-site data statistics, the slag inclusion rate and crack rate of the billet in the continuous casting process of nonoriented silicon steel were significantly improved after the optimization process. Especially after the application of the new nozzle, the drawing

speed is further increased to 1.4 m/min, and the production efficiency is further improved under the same conditions, but the slag inclusion rate of the casting billet is reduced to less than 0.1%.

**Author Contributions:** Conceptualization, W.H. and Y.Z.; methodology, W.H.; software, W.H and C.Z.; validation, C.Z.; formal analysis, W.H.; data curation, W.H.; writing—original draft preparation, W.H. and C.Z.; writing—review and editing, W.H.; project administration, X.L.; funding acquisition, W.H. and C.Z. All authors have read and agreed to the published version of the manuscript.

**Funding:** National Natural Science Foundation of China (China); Funded person (Caijun Zhang); Funding number (57274313).

**Data Availability Statement:** Data are contained within the article.

**Conflicts of Interest:** The authors declare no conflicts of interest.

## References

1. Zhang, L.F.; Li, Y.L.; Ren, Y. Basic research on non-metallic inclusions in steel (II). Inclusion detection methods and thermodynamic basis of deoxidation. *Iron Steel* **2013**, *48*, 1–8.
2. Barna, M.; Javurek, M.; Reiter, J. Simulation of mold level velocities during continuous casting of round bloom strands with mold-electromagnetic stirring. *J. Iron Steel Res.* **2012**, *19*, 884–887.
3. Zhang, K.T.; Liu, J.H.; Cui, H. Effect of flow field on surface slag entrainment and inclusion adsorption in a continuous casting mold. *Steel Res. Int.* **2019**, *91*, 1900437. [CrossRef]
4. Zhu, L.G.; Zhang, L.M.; Xiao, P.C.; Zheng, Y.H.; Jiang, Z.Y.; Zhao, J.P. Analysis and Control of Slag Inclusion Defects in Low Carbon Steel Plate Billets. *Contin. Cast.* **2020**, *45*, 36–40.
5. Liu, J.W. Analysis and Prevention of Slag Inclusion on on the Surface and Subcutaneous Surface of Billet. *Contin. Cast.* **2013**, *90*, 32–34.
6. Wang, Z.G.; Li, M.Y.; Ma, Y.H. Analysis and control of slag inclusion defects on the surface of low-carbon and low silicon aluminum killed steel hot rolled coils. *Hebei Metall.* 21 May 2015.
7. Hou, G.T. Characteristics and Preventive Measures of Inclusion and Scarring Defects Caused by Immersion Water Nozzle Nodulation. *Met. World* **2015**, *21*, 39–45.
8. Chen, W.J. Research on Surface Slag Inclusion Defects of Continuous Casting Billets. *Baosteel Technol.* **2017**, 45–48.
9. Lu, Q.T.; Yang, R.G.; Wang, X.H. A study on slag entrapment in slab continuous casting mold and its influencing factors. *Iron Steel* **2006**, *16*, 29–32.
10. Li, B.K.; Li, D.H. Water model observation and numerical simulation of eddy current phenomenon in continuous casting mold. *J. Met.* **2002**, *38*, 315–320.
11. Liu, Z.; Sun, Z.; Li, B. Modeling of quasi-four-phase flow in continuous casting mold using hybrid Eulerian and Lagrangian approach. *Metall. Mater. Trans. B* **2017**, *48*, 1248–1267. [CrossRef]
12. Qianz, D.; Wuy, L. Large eddy simulation and control of eddy current phenomenon in continuous casting mold. *J. Met.* **2004**, *40*, 88–93.
13. Gan, M.; Pan, W.; Wang, Q.; Zhang, X.; He, S. Effect of exit shape of submerged entry nozzle on flow field and slag entrainment in continuous casting mold. *Metall. Mater. Trans. B* **2020**, *51*, 2862–2870. [CrossRef]
14. Seshardri, V.; Arruda, J.D.; Arruda, A.A.F.; da Silva De Souza, S.; Da Silva, C.A.; Da Silva, I.A. Analysis of turbulence at the metal/slag interface in the meniscus region of a continuous casting mold through physical and mathematical modelling. In Proceedings of the 7th International Symposium on High-Temperature Metallurgical Processing, Nashville, TN, USA, 14–18 February 2016; Springer: Cham, Switzerland, 2016; pp. 519–526.
15. Maurya, A.; Kumar, R.; Jha, P.K. Simulation of electromagnetic field and its effect during electromagnetic stirring in continuous casting mold. *J. Manuf. Process* **2020**, *60*, 596–607. [CrossRef]
16. Jin, X.; Meng, Z.Y.; Ren, T.Z.; Liu, Z.-W. Liquid level fluctuation in continuous casting mold based on turbulent jet theory. *Steel* **2017**, *52*, 42–48.
17. Zhang, Z.N.; Wang, Z.Y.; Liu, H.; Gao, Y.; Li, H.; Sun, B. Experimental study on entrained droplets in vertical two-phase churn and annular flows. *Int. J. Heat Mass Transf.* **2019**, *138*, 1346–1358. [CrossRef]
18. Sun, Y.H.; Wei, Y.H.; Cai, K.K. Experimental study on slag entrapment phenomenon in wide slab continuous casting mold. *Steel* **2007**, *31*, 33–52.
19. Bai, H.; Thomas, B.G. Effects of clogging, argon injection, and continuous casting conditions on flow and air aspiration in submerged entry nozzles. *Metall. Mater. Trans. B* **2001**, *32*, 707–722. [CrossRef]
20. Zhou, Q.Y.; Zhu, T.H.; Zhang, L.F.l. Large eddy simulation study on the quantitative effect of non-stationary casting on mold slag entrapment during continuous casting. *Iron Steel* **2022**, *29*, 1–13. [CrossRef]
21. Deng, Y.K.; Zhang, Y.B.; Wang, Q.Q.; Wang, Q. Study of mold oscillation parameters and modes on slag lubrication in slab continuous casting. *J. Miner. Met. Mater. Soc.* **2018**, *70*, 2909–2916. [CrossRef]

22. Cai, D.X.; Wang, W.L.; Zhang, L.; Zhang, C. A comparison study on viscosity and heat transfer property of fluorine-bearing and fluorine-free mold flux for casting silicon steel. *J. Iron Steel Res. Int.* **2019**, *26*, 385–392. [CrossRef]
23. Qi, L.I.; Zhang, L.F.; Wei, C.; Wang, Y.; Zhao, Z.; Zhang, J. Large eddy simulation on the multiphase fluid flow and slag entrainment in a continuous casting mold with electromagnetic stirring. *J. Eng. Sci.* **2022**, *44*, 690–702.
24. Huixiang, Y.; Jiongming, Z.; Wanjun, W.; Xinhua, W.; Zhenya, Y.; Lixin, Y. Numerical simulation of the effect of outlet velocity of submerged nozzle on mold flow field in slab continuous casting. *J. Beijing Univ. Sci. Technol.* **2002**, *24*, 492–496.
25. Ren, L.; Zang, L.F.; Wang, Q.Q.; Zhao, X. Research on the flow behavior of molten steel in slab continuous casting mold based on PIV technology. *J. Eng. Sci.* **2016**, *38*, 1393–1403.
26. Zhao, P.; Li, Q.; Kuang, S.B.; Zou, Z. Mathematical modeling of liquid slag layer fluctuation and slag droplets entrainment in a continuous casting mold based on VOF-LES method. *High Temp. Mater. Process* **2017**, *36*, 551–565. [CrossRef]
27. Jowsa, J.; Bielnicki, M.; Cwudzinski, A. Physical and numerical investigations of mold flux entrainment into liquid steel. *Arch. Metall. Mater.* **2016**, *61*, 2043–2049. [CrossRef]

processes

Article

# Numerical Simulation of Heat Transfer of Roller Slag in Centrifugal Preparation of Inorganic Fiber

**Chunyu Liu [1], Weixing Wang [2], Xiwei Qi [1,*], Baoxiang Wang [3,*], Wei Chen [1], Kai Zhao [1], Jie Zhen [1] and Qiaorong Zhang [1]**

1 College of Metallurgy and Energy, North China University of Science and Technology, Tangshan 063210, China; leochunyu@foxmail.com (C.L.); hblgdxzzb@163.com (W.C.); herozk81@163.com (K.Z.); herozqr@163.com (Q.Z.)
2 Department of Computer Science, The University of Sheffield, Sheffield S10 2TN, UK; 13230565596@163.com
3 College of Materials Science and Engineering, North China University of Science and Technology, Tangshan 063210, China
* Correspondence: qixiwei@ncst.edu.cn (X.Q.); wbxchw@163.com (B.W.)

**Abstract:** The synergistic preparation of aluminum silicate ceramic fibers from dust removal ash and fly ash is a newly developed process that achieves green and high added value treatment of solid waste. In this process, the centrifugal fiber forming method is used to treat molten slag to obtain aluminum silicate ceramic fibers. During the production process, the centrifugal roller, as a key component in fiber forming, is in long-term contact with high-temperature slag. The heat transfer between the two causes a huge temperature gradient inside the roller material, causing significant thermal stress inside the material, which has a significant impact on the stability of the centrifugal roller structure and its working condition. This article mainly conducts numerical simulation research on the heat transfer between the roller and the slag during the centrifugal fiber forming process, providing theoretical support for ensuring the structural stability of the roller and improving its service life. The research was carried out using the FLUENT module of the ANSYS software (V2021R1), and the heat transfer model of the slag and roller was established. The effects of the internal circulating water, different slag temperatures, different slag film widths, and different boundary layer thicknesses on the heat transfer of the roller were analyzed. The results show that the water temperature at the outlet is about 6 °C higher than that at the inlet on average; when the temperature of the slag increases by 1 °C, the temperature of the roller surface in contact with the slag increases by 0.91 °C; when the width of the slag on the roll surface is 11–17 mm and the slag thickness (boundary layer thickness) is less than 1 mm, it is beneficial for fiber production.

**Keywords:** centrifugal fiber; roller wheel; heat transfer; numerical simulation

**Citation:** Liu, C.; Wang, W.; Qi, X.; Wang, B.; Chen, W.; Zhao, K.; Zhen, J.; Zhang, Q. Numerical Simulation of Heat Transfer of Roller Slag in Centrifugal Preparation of Inorganic Fiber. *Processes* **2023**, *11*, 3225. https://doi.org/10.3390/pr11113225

Academic Editor: Blaž Likozar

Received: 16 October 2023
Revised: 11 November 2023
Accepted: 13 November 2023
Published: 14 November 2023



## 1. Introduction

Inorganic fibers are widely used as insulation materials in fields, such as in architecture, furnaces, aerospace [1–4], rock wool (using basalt and coke as the raw materials), mineral wool (using blast furnace slag, ferroalloy slag, coke, etc. as the raw materials) [5], and ceramic cotton (using fly ash, dust removal ash, coke, etc. as raw the materials). At present, the main method for the industrial production of inorganic fibers is centrifugation, with the core equipment being a multi-roll centrifuge (available in two rollers and four rollers [6]). The process of synergistic preparation of aluminum silicate ceramic fibers from dust removal ash and fly ash uses a two-roller centrifuge. After the high-temperature slag (up to 1650 °C) contacts the high-speed rotating roller, under the action of viscous and friction force, the melt spreads on the roller surface and forms a liquid film with a certain thickness. In the process of the liquid film, accompanied by the rotation of the roller, the fluid instability (Kelvin–Helmholtz instability [7] and Rayleigh–Taylor instability [8]) causes the disturbance of the liquid film surface. Subsequently, the disturbance gradually intensifies with the rotation of the liquid film and leads to the decomposition of the liquid film on the roller surface. The decomposed liquid film will stretch and pull to form liquid filaments under the action of viscosity and surface tension, and the liquid filaments will form fiber filaments of 3–10 microns after cooling and curing. However, the contact between the outer surface of the roller and the high-temperature slag during the centrifugal fiber forming process can lead to an increase in the surface temperature, adhesion of high-temperature slag, and deformation or even melting of the roller surface, which not only affects the quality and production of the fiber, but also significantly reduces the service life of the roller, increases the probability of safety accidents, and has adverse effects on production. In order to ensure the working state and structural stability of the roller, it is necessary to perform forced cooling on the roller used for centrifugal fiber forming. The usual method is to introduce circulating cooling water internally to cool the roller and maintain the normal operation of the entire centrifugal fiber forming process. The centrifugal fiber forming process and the structure of the roller water cooling system are shown in Figure 1, which results in a huge temperature gradient from the outer surface to the inner surface of the roller material, forming significant thermal stress, and ultimately causing high-temperature failure of the material as shown in Figure 2.

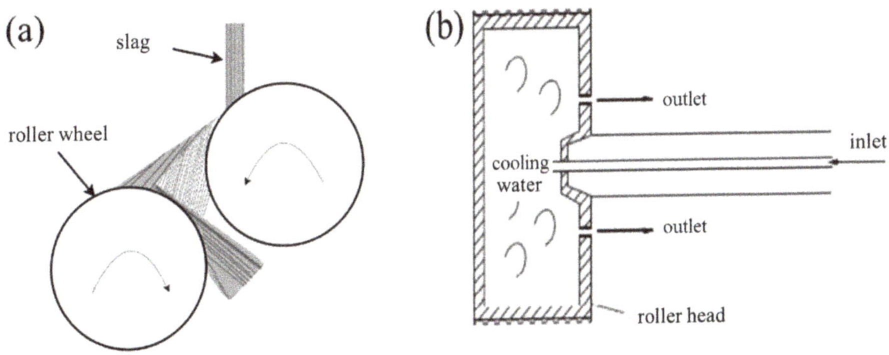


**Figure 1.** (**a**) Structure diagram of centrifugal fiber forming process; (**b**) roller water cooling system.

**Figure 2.** Roller wheel after high temperature failure (SUS304).

Although centrifugal fiber forming machines have a wide range of applications, there are few public reports on the research and modeling of these special types of rotating devices (rollers). Širok et al. [9] conducted high-speed camera visualization of molten films on industrial spinning machine rollers, analyzing the speed, and structural dynamics of early fibrosis stages. Lin Yihan et al. [10] established control differential equations and studied the rigid clamping model of four roller wheels. Jia Cili et al. [11] studied the heat treatment process of roller wheels using 2Cr13 as a raw material. Chen Jiu et al. [12] analyzed the structural design of the roller part and carried out statics analyses and strength checks on the main support arm system of the roller. Ma Wenxin et al. [13] applied ANSYS FLUENT software(V2019R1) to numerically simulate the steady-state heat transfer process of the film on a cooling drum, solving the problem of the heat transfer simulation at the moving boundary of the film. Chang Lingling et al. [14] used numerical calculation methods to calculate the temperature field of aluminum foil cooling and analyzed the variation patterns of aluminum foil cooling. Li Fushen et al. [15] conducted numerical simulations of the cooling process of the plastic film on the casting roller and obtained the temperature variation pattern of the casting roller surface. Li Yongkang et al. [16] simulated the heat flow distribution of the cooling roller and obtained the temperature field distribution of the inner and outer walls of the cooling roller. Based on FLUENT and ANSYS, Wu Shixin [17] established relevant mathematical models for new-type cooling rolls in the secondary cold and dry cooling process of slab continuous casting. The temperature field of the bearing seat of the continuous casting roller was analyzed by Dai Wei [18] using ANSYS. Based on the research and analysis of the service life of the cast roll, Li Shande [19] carried out coupling modeling and solving analysis of the thermal structure of the roll sleeve in the continuous casting and rolling process based on ANSYS.

At present, the actual production process control of the inorganic fibers mainly relies upon experience. There is no theoretical support for the determination and optimization of the various parameters in the centrifugal fiber forming process, such as the slag temperature, slag width, and thickness. Therefore, the FLUENT module is applied using the ANSYS software (V2021R1) to numerically simulate the heat transfer behavior of the centrifugal spinning roller and the slag in a collaborative preparation of the ceramic fibers from dust removal ash and fly ash, determining the impact of the heat transfer behavior on the roller, and revealing the temperature field changes of the roller and high-temperature slag during the spinning process. Combining theory with practice, along with considering the actual production and numerical simulation situation of the enterprise, can provide technical support for the maintenance of production process rollers and optimization of the cooling process parameters.

## 2. Building the Model

### 2.1. Roller Device and Geometric Parameters

The roller device studied in this article includes one water inlet, six water outlets, and an intermediate rotating bearing. The middle rotating-bearing is installed on the driving motor, which drives the roller to rotate through the rotating bearing, achieving the process of slag centrifugation and fiber formation on the roller. The actual diagram is shown in Figure 3, and the structural parameters of the roller are shown in Table 1. Among them, the surface of the roller in direct contact with the slag in the circumferential direction is called the roller surface, the circular surface where the inlet and outlet are located is called the inner end face, and the symmetrical circular surface is called the outer end face. From the figure, it can be seen that the middle rotating bearing is a hollow bearing, that is, the bearing simultaneously serves as the water inlet, and the six water outlets are evenly distributed around the periphery of the bearing.

**Figure 3.** Roller physical diagram (a) water inlet; (b) water outlet; (c) rotating-bearing; (d) roller.

Table 1. Roller structure parameters.

| Parameter | Value |
|---|---|
| Wall thickness/mm | 30 |
| Cavity diameter/mm | 240 |
| Cavity height/mm | 80 |
| Overall diameter/mm | 300 |
| Overall height/mm | 140 |
| Inlet diameter/mm | 12 |
| Outlet diameter/mm | 7 |

Based on the internal structure and structural parameters of the roller, a three-dimensional model is created using the DesignModeler module in the ANSYS software (V2021R1). The model is divided into four parts from inside to outside: internal circulating water, roller, slag, and air. The calculation domain of the circulating water is located in the internal cavity of the roller; the outer part is a roller entity; there is a layer of slag outside the roller body, and the slag thickness is calculated according to the boundary layer equation; the outside of the slag is set as the air calculation domain, considering the heat exchange between the slag and the air. The internal circulating water, slag, and air are used as fluids, and the rollers are used as solids. The three-dimensional model of the geometric structure is shown in Figure 4.

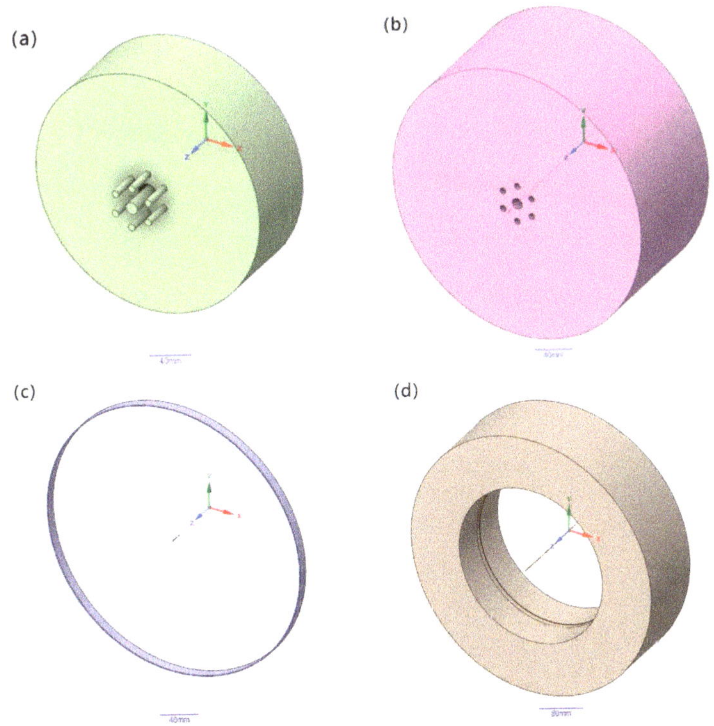


**Figure 4.** Three dimensional model of geometric structure (**a**) internal circulating water; (**b**) roller; (**c**) slag; (**d**) air.

## 2.2. Mathematical Model of Heat Transfer

### 2.2.1. Basic Assumptions

In practical situations, the slag undergoes heat exchange with the circulating cooling water through the rollers. At the beginning, the slag, rollers, and circulating cooling water are in an unstable state, and the temperature changes over time. After the equipment has been running for a period of time, the heat exchange between the slag, roller, and circulating cooling water reaches a dynamic balance, that is, the temperature of the roller basically does not change, and the heat released by the slag is equal to the heat absorbed by the circulating cooling water, achieving a stable state.

This article studies the heat transfer changes when the three reach a stable state, adopting steady-state calculations, namely $\partial/\partial\tau = 0$. In order to simplify the calculation model and obtain the key influencing factors, the following reasonable assumptions are made for the centrifugal fiber forming process:

(1) The heat of the slag is only transmitted to the rollers and air;
(2) The slag has good contact with the roller, and the gap distance approaches 0 infinitely;
(3) The thermophysical parameters of the roller material are isotropic and do not change with temperature changes;
(4) The difference in heat dissipation at various positions on the circumference is negligible, assuming that the roller has no circumferential heat transfer;
(5) During simulation, the wall is non sliding and no inertial force is calculated;
(6) Do not consider the thermal effects caused by viscous dissipation during fluid flow;
(7) The slag can form a film on the centrifugal roller surface and the film does not break.

### 2.2.2. Governing Equation

The flow of slag and cooling water in the roller follows the three most basic conservation laws, namely the conservation of mass, the conservation of momentum, and the conservation of energy. This paper treats the slag and circulating cooling water as a three-dimensional steady-state incompressible fluid and adopts the continuity equation to represent the conservation of mass, the Navier–Stokes equation to represent the conservation of momentum, and the energy conservation equation to represent the conservation of energy; for the heat transfer between the slag and the roller contact surface, temperature boundary conditions are applied, on the other hand, for the heat transfer between the slag and the air-contact surface, convection boundary conditions are used.

Continuity equation:

$$\frac{1}{\rho}\frac{d\rho}{d\tau} + \frac{\partial v_x}{\partial x} + \frac{\partial v_y}{\partial y} + \frac{\partial v_z}{\partial z} = 0 \tag{1}$$

$\rho$ is the fluid density, $kg/m^3$; $\tau$ is time, s; $v_x$, $v_y$, and $v_z$ is the fluid flow rate in the $x$, $y$, and $z$ directions, respectively, m/s.

Navier–Stokes equation is as follows:

$$\begin{aligned} \frac{dv_x}{d\tau} &= X - \frac{1}{\rho}\frac{\partial p}{\partial x} + \eta\nabla^2 v_x \\ \frac{dv_y}{d\tau} &= Y - \frac{1}{\rho}\frac{\partial p}{\partial y} + \eta\nabla^2 v_y \\ \frac{dv_z}{d\tau} &= Z - \frac{1}{\rho}\frac{\partial p}{\partial z} + \eta\nabla^2 v_z \end{aligned} \tag{2}$$

X, Y, and Z are forces per unit mass acting on the fluid, N; $\frac{\partial p}{\partial x}$, $\frac{\partial p}{\partial y}$, and $\frac{\partial p}{\partial z}$ are pressure gradients, Pa; $\eta$ is the kinematic viscosity coefficient, $m^2/s$; $\nabla^2$ is the Laplace operator.

The energy conservation equation is as follows:

$$\frac{\partial(\rho E)}{\partial \tau} + div(\rho \mu T) = div\left(\frac{\lambda}{\rho c_p} gradT\right) + S_h \quad (3)$$

$E$ is the sum of kinetic energy, internal energy, and potential energy of the fluid micro-clusters, J/kg; $T$ is the temperature, K; $\lambda$ is the thermal conductivity, W/(m·K); $S_h$ is the volumetric heat source term.

2.2.3. Selection of Internal Circulating Water Model

The Reynolds number is usually a measure of the typical size of the inertial force and viscous force. A high Reynolds number indicates that the inertial forces in the fluid are greater than the viscous forces, making it more prone to turbulent flow; conversely, a low Reynolds number signifies that the inertial forces are smaller than the viscous forces, making laminar flow more likely to occur.

Known data: ρ is the fluid density: 997 kg·m$^{-3}$; $\eta$ is the kinematic viscosity: 0.897 × 10$^{-6}$ m$^2$·s$^{-1}$; $d_s$ is the inlet diameter: 12 mm; and the flow rate Q, at the internal circulating water inlet is 60 L/h, so the flow rate at the internal circulating water inlet is 0.147 m/s, and the Reynolds number is 1967.

The critical Reynolds number varies with the shape of the fluid region. Generally, according to the size of the Reynolds number, the flow state can be divided into three types [20,21]:

(1) When Re < 2300, the flow exhibits laminar flow;
(2) When 2300 < Re < 4000, the flow pattern is unstable, which may be laminar or turbulent, known as the transition zone;
(3) When Re > 4000, the flow exhibits turbulence.

Based on the above analysis, the laminar flow model is selected for the internal circulating water flow model, and the laminal model is selected for the FLUENT solution. This model can set the flow state in the calculation domain to the laminar flow. The enhanced wall treatment method from FLUENT is selected for the wall treatment method, which is suitable for low Reynolds number models.

## 2.3. *Mesh Generation*

Mesh generation refers to dividing a spatially continuous computational domain into sufficiently small computational regions, fluid control equations are then applied to each of these regions to solve and calculate the fluid equations for all the regions, ultimately obtaining the distribution of physical quantities across all computational regions. In order to ensure the quality of the mesh generation and to improve the accuracy and speed of the numerical calculations, the geometric structure needs to be divided into four parts during the mesh generation, namely the internal circulating water, rollers, slag, and air, which need to be grid divided in sequence. The mesh module is used to divide the mesh, and the geometric size adjustment and boundary layer meshing are added to improve the mesh quality. The mesh type is a tetrahedral/hexahedral hybrid mesh.

In order to ensure the reliability of the model grid, we conducted experiments on grid convergence; the average mesh size of the model is divided into 5 mm, 6 mm, 7 mm, 8 mm, and 9 mm. The temperature of the line on the roller surface, which is parallel to the roller axis is taken as the entry point to verify the grid independence. The results are shown in Figure 5, which shows that the temperature has a weak dependence on the mesh size. The global average size of 9 mm is selected for the grid division of the model to strike a balance between computational accuracy and efficiency. The final number of the grid nodes is 89,137, and the number of grid units is 435,446. The grid cell quality of this model ranges from 0.7 to 1, and the average cell quality of the grid is 0.79224, which indicates that the quality of the grid delineation is good enough for a numerical simulation. The detailed meshing is shown in Figure 6.

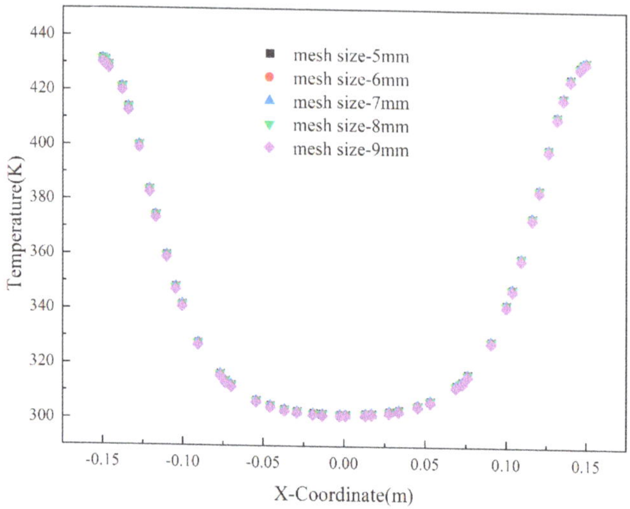


**Figure 5.** The effect of different mesh sizes on temperature.

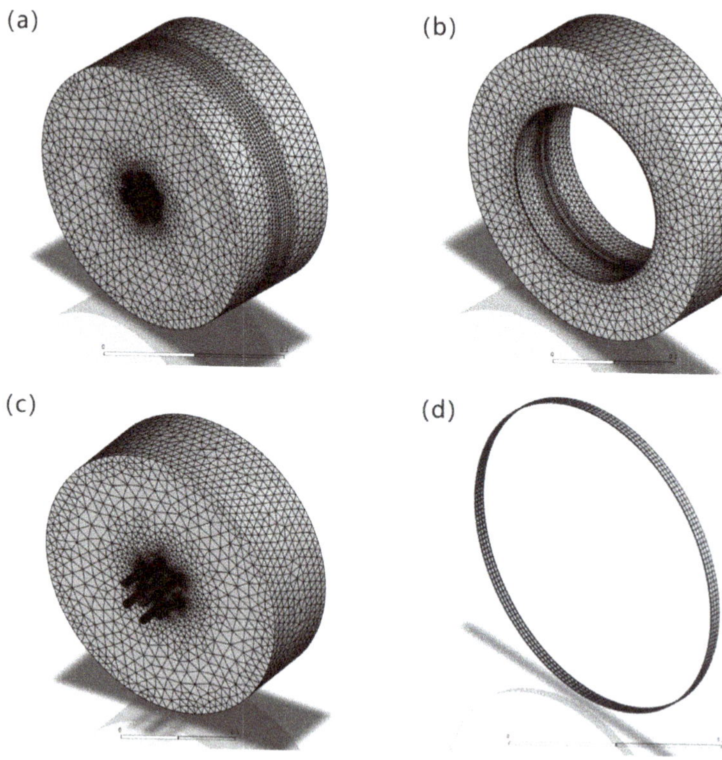


**Figure 6.** Mesh generation (**a**) roller; (**b**) air; (**c**) internal circulating water; (**d**) slag.

## 2.4. Material Property Settings

The slag, rollers, and material properties for the internal circulating cooling water in the FLUENT material database are defined and shown in Table 2.

**Table 2.** Slag and roller material properties.

| Material | Density ($kg \cdot m^{-3}$) | Specific Heat Capacity ($J \cdot kg^{-1} \cdot K^{-1}$) | Thermal Conductivity ($W \cdot m^{-1} \cdot K^{-1}$) | Dynamic Viscosity ($kg \cdot m^{-1} \cdot s^{-1}$) |
|---|---|---|---|---|
| slag | 2500 | 913.11 | 1.5 | 11.421 |
| roller wheel | 7750 | 484.96 | 25.1 | |
| circulating cooling water | 998.2 | 4182 | 0.6 | 0.001003 |

### 2.5. Boundary Condition Settings

(1) The inlet and outlet of the internal circulating cooling water adopt pressure boundary conditions with the inlet pressure set to 0.25 MPa, the outlet pressure set to 0.2 MPa, and the inlet temperature set to 25 °C;
(2) The slag temperature is set to 1500 °C, the slag width is 11 mm, and the boundary layer thickness is 1.04 mm;
(3) The slag and the roller undergo conduction heat transfer, while the slag and air undergo convective heat transfer;
(4) The initial air temperature is set to 25 °C.

### 2.6. Solving Control

The SIMPLEC algorithm is selected, which is suitable for most conventional incompressible flow problems and can quickly obtain convergence solutions in the laminar flow. The steady-state solution is adopted in the solution process, and the second-order upwind scheme is used to calculate the spatial discrete scheme. The relaxation factor and residual control remain at the default settings. In order to make efficient use of the arithmetic power and to ensure the accuracy of the calculation, the residuals of the continuous and velocity terms in the calculation are set to $10^{-2}$, while the residuals of the key energy term are set to $10^{-6}$, which is a setup that can maintain the computational speed of the numerical simulation process, as well as achieve the higher accuracy of the temperature field data on the basis of the convergence of the flow field.

## 3. Result Analysis and Discussion

### 3.1. Internal Circulation Water Cloud Diagram

The internal circulating water has a cooling effect on the roller during the entire centrifugal fiber forming process, which is of great significance in improving the roller life and ensuring the normal operation of the centrifugal fiber forming process. The internal circulating water temperature cloud diagram is shown in Figure 7.

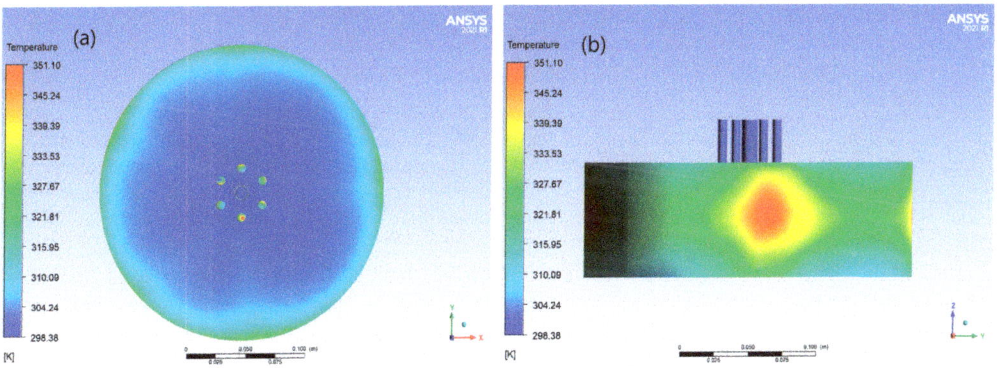


**Figure 7.** Temperature nephogram of internal circulating water (**a**) inlet and outlet; (**b**) side.

From Figure 7a, it can be seen that the inlet temperature remains at the set temperature of 25 °C. During the entire centrifugal fiber forming process, water absorbs the heat on the roller, so the outlet temperature increases and the average temperature increases by 6 °C. As shown in Figure 7b, the side that directly contacts the roller wall absorbs more heat and has a higher temperature, while the part that does not directly contact the roller wall has a lower temperature. Water is constantly circulating, so it can continuously carry away heat to maintain the normal progress of the entire process.

### 3.2. Roller Temperature Cloud Map

The roller is the main piece of equipment in the centrifugal fiber forming process and plays a decisive role in whether the process can proceed normally. The temperature cloud map of the roller is shown in Figure 8.

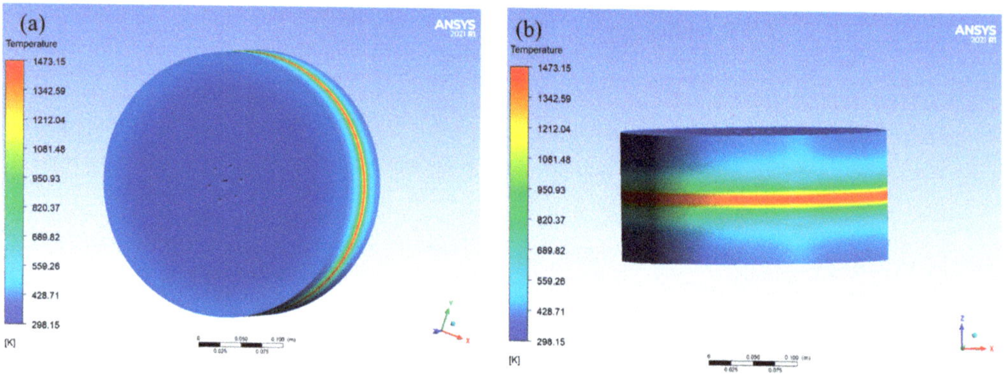


**Figure 8.** Roller temperature nephogram (**a**) inner end face; (**b**) roll surface.

From Figure 8a, it can be seen that the inner surface is far from the slag and has a small temperature influence, but the temperature also slightly increases. From Figure 8b, it can be seen that the temperature at the direct contact area between the roller and the slag is high, gradually diffusing towards the upper and lower walls, with a diminishing temperature influence. In production, the part where the slag contacts the roller is most affected by the temperature as shown in Figure 8b. Although internal circulating water can carry away a large amount of heat, this part still has the highest temperature compared to the other parts. If the slag temperature is too high, it will cause changes in the structural strength of this part, affecting the normal production process. Therefore, the slag temperature should be controlled within a reasonable range.

### 3.3. Effect of Slag Temperature on Heat Transfer of Roller

The heat transfer to the roller was studied by changing the slag temperature to 1500 °C, 1525 °C, 1550 °C, 1575 °C, and 1600 °C. The temperature cloud map corresponding to the roller under the five different working conditions is shown in Figure 9.

As shown in Figure 9, as the temperature of the slag increases, the temperature of the roller surface in contact with the slag also increases, resulting in a more pronounced impact on the temperatures at both ends of the roller. The average temperature of the roller surface in contact with the slag and the average temperature of the entire roller surface are shown in Table 3.

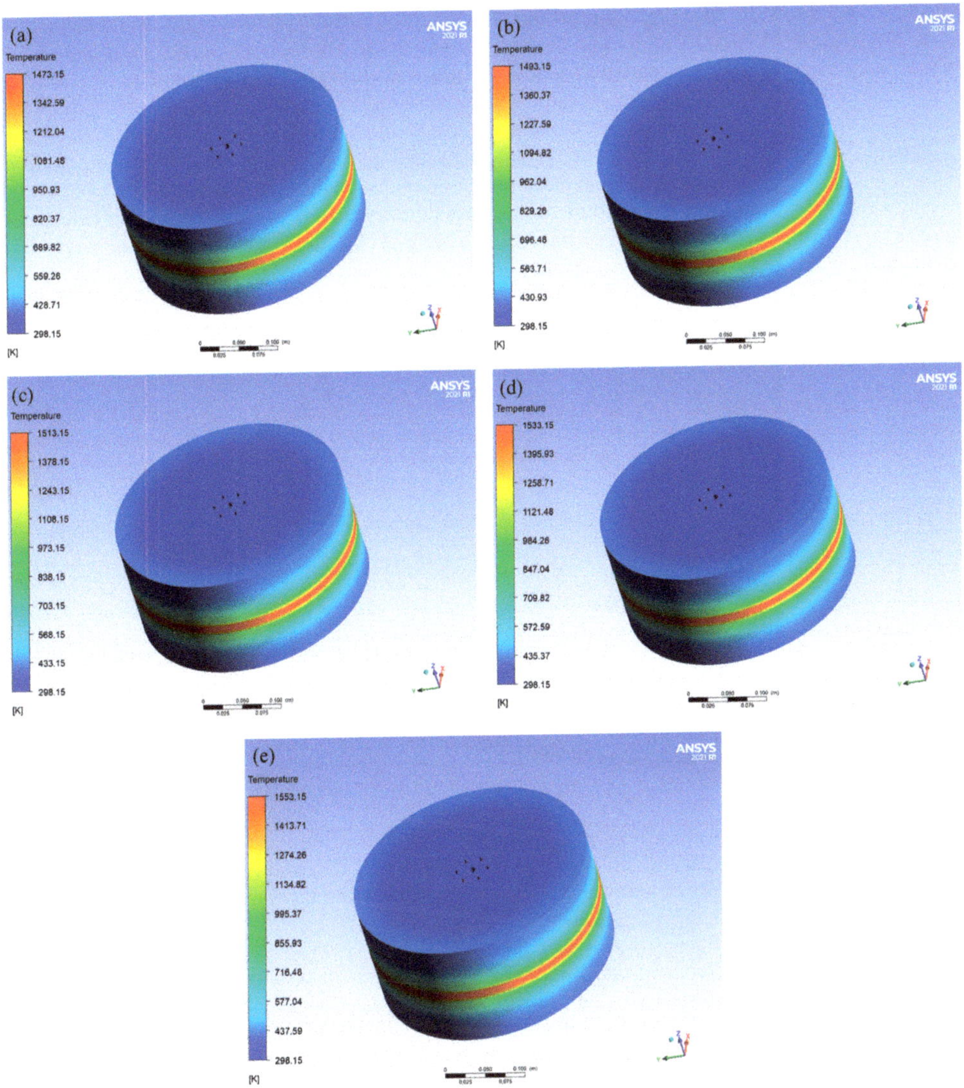


**Figure 9.** Cloud diagram of roller with different slag temperature (**a**) 1500 °C; (**b**) 1525 °C; (**c**) 1550 °C; (**d**) 1575 °C; (**e**) 1600 °C.

**Table 3.** Roller surface temperature under different slag temperature.

| Slag Temperature/°C | Average Temperature of the Part in Contact with Slag/°C | Average Temperature of the Entire Roller Surface/°C |
|---|---|---|
| 1500 | 1074.26 | 600.14 |
| 1525 | 1093.34 | 609.98 |
| 1550 | 1111.23 | 619.56 |
| 1575 | 1129.12 | 629.65 |
| 1600 | 1147.01 | 639.48 |

It can be seen that the maximum and average temperatures of the roller surface are different under the five working conditions. The variations in the average temperature of the roller surface with the slag temperature is shown in Figure 8.

It can be seen from Figure 10 that the average temperature change in the part in contact with the slag is approximately a straight line. By fitting it with a linear function, the growth rate in the figure is 0.9064, that is, when the temperature of the slag increases by 1 °C, the temperature of the roll surface in contact with the slag increases by 0.91 °C; similarly, the average temperature change growth rate of the roller surface is 0.49, which means that when the slag temperature increases by 1 °C. The average temperature of the entire roller surface increases by 0.49 °C.

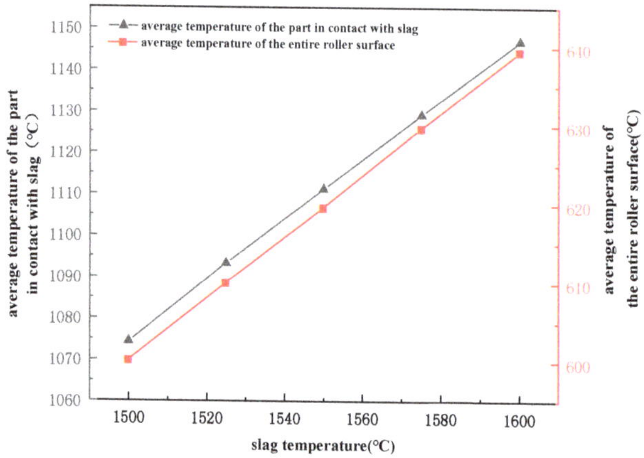


**Figure 10.** Changes in the average temperature of the contact part with the slag and the average temperature of the entire roller surface under different slag temperatures.

### 3.4. *Effect of Slag Roller Surface width on Heat Transfer of Roller Wheels*

The heat transfer of the slag to the roller was studied by keeping the slag temperature at 1500 °C and changing the slag widths to 8 mm, 11 mm, 14 mm, 17 mm, and 20 mm. The temperature cloud map corresponding to the roller under the five different working conditions is shown in Figure 11.

From Figure 11, it can be seen that the highest temperature on the roller surface does not change with the width of the slag on the roller surface, but as the width of the slag on the roller surface increases, the impact on the upper and lower wall temperatures at the direct contact area between the roll and the slag increases. With a larger slag width, the range of the high-temperature zone in the middle is larger, and the heat is transferred to both sides over a wider range, resulting in a greater impact on the temperatures on both sides. The average temperature of the roller surface at the contact area of the slag and the average temperature of the entire roller surface under these five different slag widths are shown in Table 4. The changes are shown in Figure 12.

**Table 4.** Roller surface temperature under different slag film widths.

| Width of Slag Film/mm | Average Temperature of the Part in Contact with Slag/°C | Average Temperature of the Entire Roller Surface/°C |
|---|---|---|
| 8 | 1023.32 | 544.68 |
| 11 | 1073.89 | 600.14 |
| 14 | 1074.24 | 604.11 |
| 17 | 1074.26 | 622.98 |
| 20 | 1103.89 | 672.34 |

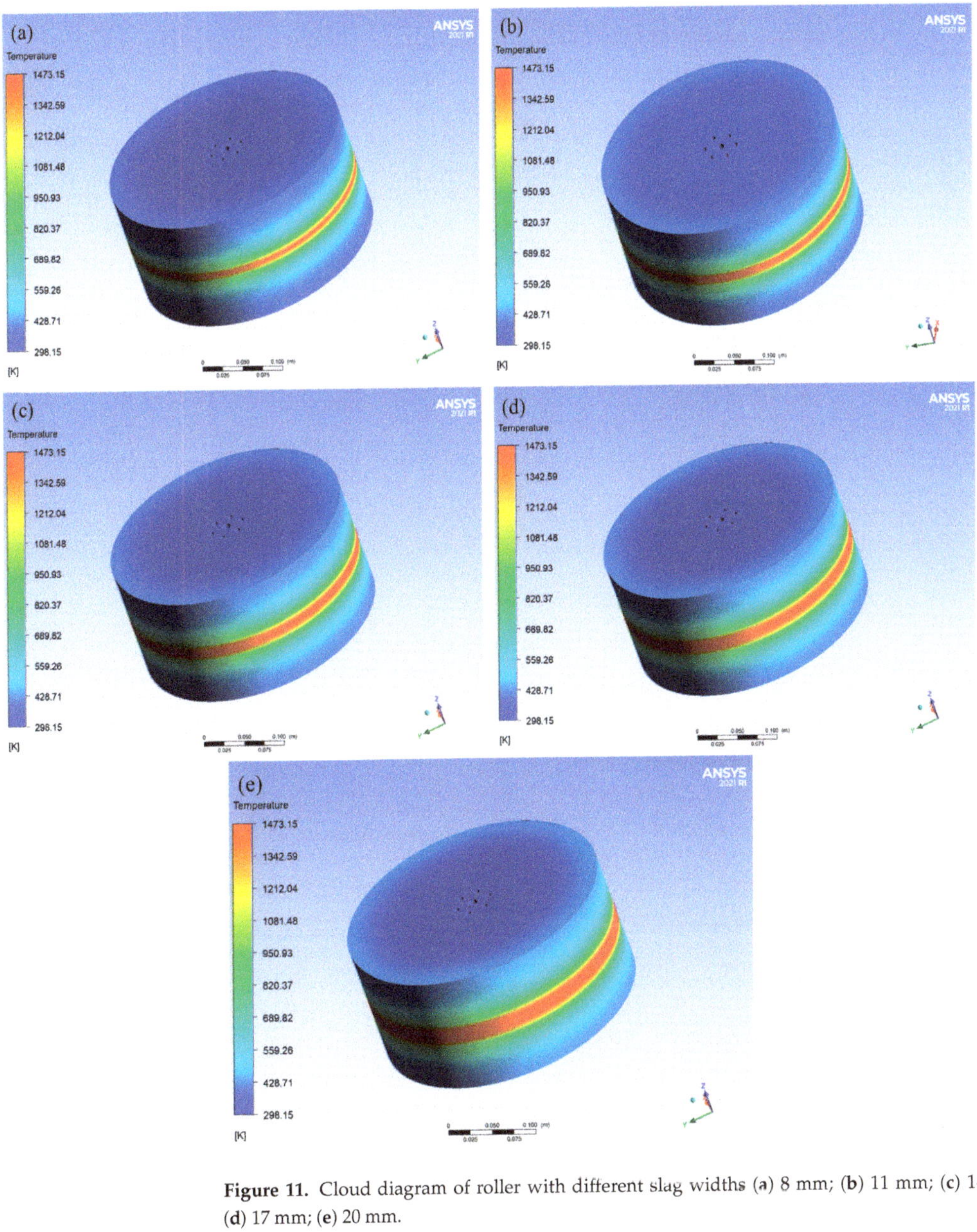


**Figure 11.** Cloud diagram of roller with different slag widths (**a**) 8 mm; (**b**) 11 mm; (**c**) 14 mm; (**d**) 17 mm; (**e**) 20 mm.

As shown in Figure 12, as the width of the slag increases, the average temperature of the roller surface in contact with the slag and the average temperature of the entire roller surface show an upward trend. However, when the width of the slag is 11 mm, 14 mm, or 17 mm, the average temperature of the part in contact with the slag is not significantly different. Therefore, during production, the width of the slag on the roller surface should be controlled to be between 11 and 17 mm to reduce the impact of temperature fluctuations on the roller and extend its service life.

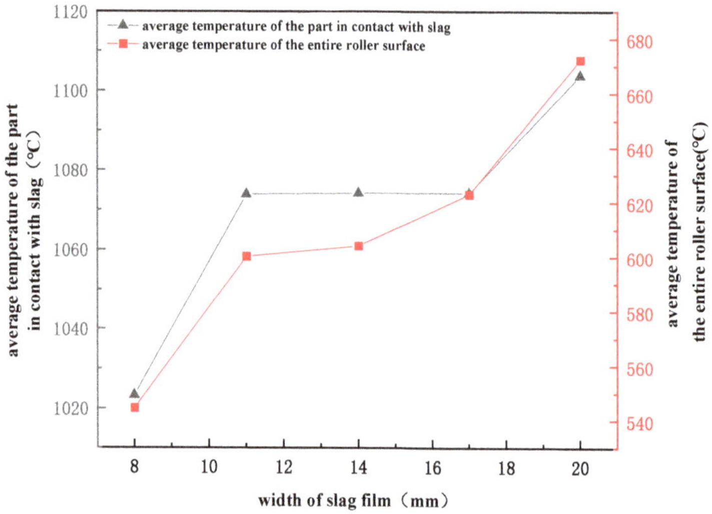


**Figure 12.** Changes in the average temperature of the roller surface in contact with the slag and the average temperature of the entire roller surface under different slag film widths.

### 3.5. *Effect of Slag Thickness on Heat Transfer of Roller*

The thickness of the boundary layer was changed to 0.91 mm, 0.94 mm, 0.97 mm, 1 mm, and 1.06 mm, and the influence on the heat transfer of the roller was studied. The impact of the different boundary layer thicknesses on the average roller surface temperature is recorded in Table 5.

**Table 5.** Roller surface temperature under different slag thicknesses.

| Slag Thickness/mm | Average Temperature of the Part in Contact with Slag/°C | Average Temperature of the Entire Roller Surface/°C |
|---|---|---|
| 0.91 | 1074.57 | 600.70 |
| 0.94 | 1074.31 | 600.24 |
| 0.97 | 1074.29 | 600.17 |
| 1.00 | 1074.20 | 600.02 |
| 1.03 | 1074.26 | 600.14 |
| 1.06 | 1074.36 | 600.38 |

From Table 5, it can be seen that the average temperature of the roller surface is different for the six different slag thicknesses. As the slag thickness increases, the average temperature of the part in contact with the slag and the average temperature of the entire roller surface first decrease and then increase. When the slag thickness is greater than 1 mm, the average temperature of the part in contact with the slag and the average temperature of the entire roller surface slightly increases. The variation in the average temperature of the roller surface with the slag thickness is shown in Figure 13.

As shown in Figure 13, a turning point appears at a slag thickness of 1 mm, so in actual production, the slag thickness should be less than 1 mm. The thickness of the slag is related to the roller speed and the ratio of the raw material. When the raw material composition is determined, the thickness of the boundary layer can be changed by adjusting the roller speed so that the temperature of the roller in contact with the slag is at a lower level, thus extending the service life of the roller.

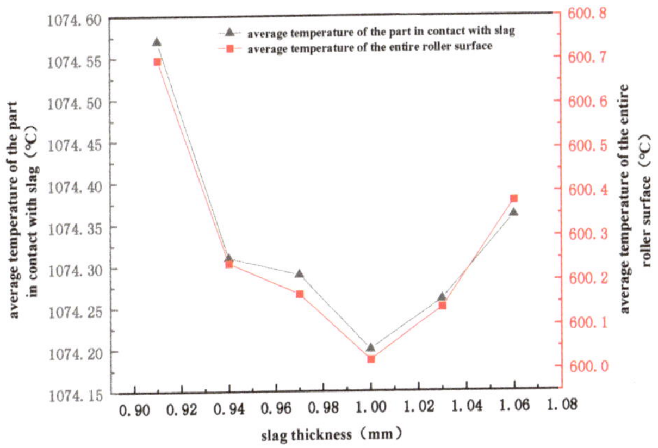


**Figure 13.** Variation in the average temperature of the contact part with the slag and the average temperature of the roller surface with the thickness of the slag.

## 4. Conclusions

This article conducts numerical simulation research on the heat transfer between the roller and slag during the centrifugal fiber forming process of the synergistic preparation of ceramic fibers from dust removal ash and fly ash. It explores the influence of internal circulating water on the heat transfer between the roller and slag, and analyzes the heat transfer changes of different slag temperatures, slag widths, and slag thicknesses during the centrifugal fiber forming process. The main conclusions are as follows:

- Internal circulating water has a significant impact on the normal operation of the centrifugal fiber forming process, and cooling with circulating water can ensure that the roller surface temperature is within a reasonable range. The simulation results show that there is a strong heat transfer between the rollers and the slag and circulating water, resulting in an increase in the outlet water temperature by about 6 °C compared to the inlet temperature, which ensures the normal production;
- The heat transfer process between the roller and the slag is uneven, with the highest temperature in the contact area between the roller and the slag, and then gradually spreading towards the upper and lower walls. The part that is most affected by temperature is the contact between the slag and the roller surface. If the slag temperature is too high, it will cause changes in the structural strength of the roller, thereby affecting the progress of production. Therefore, controlling the temperature of the slag falling onto the roller surface within a reasonable range is the key to ensuring normal production. At the same time, according to the simulation data, the manufacturer upgraded the production equipment by moving the rollers back and forth at regular intervals to change the contact surface between the slag and the rollers, which significantly improved the service life of the rollers.
- When the temperature of the slag increases by 1 °C, the temperature of the roller surface in contact with the slag increases by 0.91 °C. The simulation data show that the slag temperature, slag roller width, and the boundary layer thickness will have an impact on the roller temperature. In actual production, the slag outlet size can be adjusted to control the slag roller width to be between 11 and 17 mm to keep the temperature of the roller in contact with the slag part of the roller stable; by controlling the roller speed, the thickness of the slag can be increased or decreased, keeping the temperature of the contact part between the slag and the roller at the lowest level, reducing the impact of high temperatures on the strength of the roller, and extending the service life of the roller.

**Author Contributions:** Conceptualization, C.L. and W.C.; methodology, K.Z.; software, W.W. and J.Z.; validation, B.W., Q.Z. and X.Q.; formal analysis, J.Z.; data curation, C.L.; writing—original draft preparation, C.L. and J.Z.; writing—review and editing, C.L. and W.W.; project administration, C.L., B.W. and Q.Z.; funding acquisition, X.Q., W.C. and K.Z. All authors have read and agreed to the published version of the manuscript.

**Funding:** This study was supported by the Innovation Capability Enhancement Program of Hebei Province (V1617005355710); Central-Guide-Local Project of Hebei Provincial Science and Technology Department (236Z3806G); Science and Technology Project of Tangshan (23150218A); Tangshan Talent Funding Project (A202010006); Key Research and Development Project of Tangshan (100317).

**Data Availability Statement:** Data are contained within the article.

**Conflicts of Interest:** The authors declare no conflict of interest.

## References

1. Ran, X.; Zou, H.H.; Sun, H.W.; Ye, Y.F. The effect of heat treatment on the interface bonding characteristics and carbon fiber structure of coated carbon fibers. *J. Mater. Heat Treat.* **2015**, *36*, 136–140. [CrossRef]
2. Wang, Q.; Zhao, B.; Han, X.G. Properties and Preparation of Ceramic Fibers. *China Ceram.* **2019**, *55*, 10–18+26. [CrossRef]
3. Cooke, T.F. Inorganic Fibers—A Literature Review. *J. Am. Ceram. Soc.* **1991**, *74*, 2959–2978. [CrossRef]
4. Ishikawa, T. Advances in Inorganic Fibers. *Adv. Polym. Sci.* **2005**, *178*, 109–144. [CrossRef]
5. Li, J.; Zhang, L.L.; Zhao, G.Z.; Cang, D.Q. Pilot practice of directly modifying blast furnace slag to prepare mineral wool fibers. *Iron Steel* **2017**, *52*, 99–103. [CrossRef]
6. Zhang, Z.Q.; Zhang, Y.Z.; Xing, H.W. Analysis of slag centrifugal fiber formation test and its influencing factors on fibrosis. *Steel* **2017**, *52*, 69–74. [CrossRef]
7. Guo, Z.Y.; Yu, P.X.; Ouyang, H. The instability of flow around cylinder shear layer based on large eddy simulation. *J. Shanghai Jiaotong Univ.* **2021**, *55*, 924–933. [CrossRef]
8. Qiu, R.D.; Wang, J.Z.; Huang, R.F. Application of improved physical fusion neural networks to Rayleigh-Taylor instability problems. *Chin. J. Theor. Appl. Mech.* **2022**, *54*, 2224–2234.
9. Bizjan, B.; Širok, B.; Govekar, E. Nonlinear Analysis of Mineral Wool Fiberization Process. ASME. *J. Comput. Nonlinear Dynam.* **2015**, *10*, 21005. [CrossRef]
10. Lin, Y.H.; Li, F.; Zhang, D. A rigid clamping model for the multiple bending forming process of four roller wheels. *J. Fudan Univ.* **2002**, *5*, 495–500. [CrossRef]
11. Jia, C.L.; Shen, Y.R.; Qian, S.Q.; Li, Y.P. 2Cr13 Heat Treatment Process for Finishing Cold Roll Wheels. *Hot Work. Process* **2014**, *43*, 158–160. [CrossRef]
12. Chen, J.; Yan, W.J. Design and finite element analysis of roller components for powder mills. *Mech. Electr. Eng. Technol.* **2021**, *50*, 212–215.
13. Ma, W.X.; Yu, H.J.; Liu, Y.Y.; Qian, C.F. Numerical simulation of heat transfer process between rolling film and cooling drum. *J. Beijing Univ. Chem. Technol.* **2020**, *47*, 84–92. [CrossRef]
14. Chang, L.L.; Liu, Y.; Li, H.R.; Guan, X.R. Numerical simulation study on temperature field of hydrophilic coated aluminum foil water-cooled rolls. *J. Nanjing Univ. Technol.* **2020**, *44*, 733–738. [CrossRef]
15. Li, F.S.; Wang, S.H.; Li, X.X.; Lin, Y.H.; Sun, Y. Numerical simulation of the cooling process of plastic film on a casting roller. *Plast. Technol.* **2011**, *39*, 35–38.
16. Li, Y.K.; Yang, Y.; Song, Y.M. Numerical Simulation of Heat Flow and Temperature Field of Amorphous Belt Cooling Rollers. *Rare Met. Mater. Eng.* **2017**, *46*, 917–922.
17. Wu, S.X. Research on Temperature Field and Stress Field of New Type of Cooling Rolls for Continuous Casting of Slabs. Master's Thesis, Chongqing University, Chongqin, China, 2022. [CrossRef]
18. Dai, W. Temperature field analysis of continuous casting roll bearing housing based on ANSYS. *J. Chang. Univ.* **2014**, *11*, 59–61+4. [CrossRef]
19. Li, S.D. Thermal Structure Coupling Analysis of Casting Roll Sleeve and Its Fatigue Life Study. Master's Thesis, Zhongnan University, Changsha, China, 2007.
20. Yang, B.L. Navier-Stokes equations and generalized laminar flow—The first class dynamic equations of fluid mechanics and their application conditions. In Proceedings of the 16th National Hydrodynamics Seminar, Zhuhai and Macau, China, 17–22 November 2002; pp. 181–196.
21. Zhao, C.L. How to distinguish laminar flow and turbulence. *Coll. Phys.* **1993**, *7*, 16–17+20. [CrossRef]

processes MDPI

Article

# Diffusion Behavior of Carbon and Silicon in the Process of Preparing Silicon Steel Using Solid-State Decarburization

**Li Wen [1], Liqun Ai [1], Lukuo Hong [1,*], Yuqing Zhou [2], Guangpeng Zhu [1] and Caijiao Sun [1,*]**

1 Metallurgy and Energy College, North China University of Science and Technology, Tangshan 063210, China; wenli608824@163.com (L.W.); ailiqun@ncst.edu.cn (L.A.); zgp1boy@163.com (G.Z.)
2 HBIS Materials Technology Research Institute, Shijiazhuang 052165, China; zyq11272021@163.com
* Correspondence: honglk@ncst.edu.cn (L.H.); 15032900376@163.com (C.S.)

**Abstract:** In this study, we investigated the relationship between the decarburization effect of the solid-state decarburization method for preparing silicon steel and the atomic diffusion behavior in the matrix, focusing on 1 mm thick Fe-0.18 wt% C-Si (1.5, 3.5 wt%) alloy strips as the research object. Solid-state decarburization experiments were carried out in an Ar-$H_2O$-$H_2$ atmosphere, and the self-diffusion behavior of C and Si in Fe-C-Si alloy system was simulated by molecular dynamics. The results show that the molecular dynamics simulation results are consistent with the decarburization experimental results. When the temperature is lower than 800 °C, the atoms maintain a compact bcc structure, so the migration rate of carbon is low. When the temperature rises, the atoms move violently, resulting in the destruction of the crystal structure. Because the atomic arrangement tends towards a disordered structure, the migration rate of C is high and the diffusion coefficient increases. The experimental results showed that the decarburization rate was accelerated. At the same temperature, the diffusion activation energy Q = 48.7 kJ·mol$^{-1}$ of carbon in an Fe-3.5 wt% Si-C alloy matrix can be calculated. The diffusion activation energy of carbon Q = 47.3 kJ·mol$^{-1}$ was calculated using a molecular dynamics simulation. When the content of Si is 3.5 wt% and 1.5 wt%, the diffusion series of Si can be expressed as $D_{3.5Si,\,Si} = 8.54 \times 10^{-10} \exp(-33{,}089.7/RT)$ m$^2$/s and $D_{1.5Si,\,Si} = 2.06 \times 10^{-9} \exp(-46{,}641.5/RT)$ m$^2$/s, respectively. When the Si content is 3.5 wt%, the diffusion coefficient of Si is larger. After diffusion to the near surface, it combines with the remaining O to form a continuous strip of $SiO_2$. When the Si content is 1.5 wt%, the diffusion coefficient of Si is small. The remaining O diffuses in the matrix and will oxidize when encountering Si; it cannot aggregate in a highly-dispersed distribution. The lattice transition from face centered cubic lattice austenite to body centered cubic lattice ferrite occurred in the matrix of the thin strip. The diffusion coefficient of C in ferrite is much larger than that in austenite. Therefore, the decarburization rate suddenly increases before decarburization stagnation. With the increase in Si content, the diffusion activation energy of C decreases, which promotes the decarburization effect.

**Keywords:** Fe-C-Si alloy; solid decarburization; molecular dynamics; diffusion coefficient

**Citation:** Wen, L.; Ai, L.; Hong, L.; Zhou, Y.; Zhu, G.; Sun, C. Diffusion Behavior of Carbon and Silicon in the Process of Preparing Silicon Steel Using Solid-State Decarburization. *Processes* **2023**, *11*, 3176. https://doi.org/10.3390/pr11113176

Academic Editors: Andrea Petrella and Prashant K. Sarswat

Received: 30 September 2023
Revised: 28 October 2023
Accepted: 3 November 2023
Published: 7 November 2023



## 1. Introduction

Electrical steel, also known as silicon steel, is a type of iron–silicon binary alloy soft magnetic material widely used in various electronic and electrical power applications. Its products are widely used in various motor rotors, transformer cores, and other electrical components. With the continuous innovation of smelting technology, the production process of silicon steel is also constantly developing, and the performance of the products is gradually improving [1–3]. Traditional silicon steel production adopts converter long process, including blast furnace-converter-continuous casting and rolling-finishing rolling-cold rolling-heat treatment and other main process nodes; this process is complex, the carbon emission is high, and the environmental pollution is large. Hong et al. [4] and others proposed a new process, namely "solid-state steelmaking", in which hot blast furnace molten iron is directly solidified by the twin roll continuous casting and then decarburized after gas–solid reaction at a high temperature to produce the steel strip. This process does not need converter steelmaking and refining processes, thus greatly reducing industrial gas emissions and energy consumption. Without oxygen decarburization in converter, the oxygen potential in steel is low, effectively avoiding the generation of inclusions and bubbles, and greatly improving the cleanliness of steel.

In recent years, there has been extensive exploration and research on the new process of solid-state decarburization steelmaking [5–12]. Among them, numerous decarburization studies [13–21] have shown that the surface decarburization reaction of Fe-C alloy is the rate-limiting step in the early stage of decarburization, while the diffusion of carbon atoms from the interior to the interface is the limiting step in the middle stage of decarburization. Both decarburization rate and carbon diffusion rate are linearly related to temperature. Agren [22] summarized the relationship between the carbon diffusion coefficients and temperature in ferrite and austenite. To explore whether the solid-state decarburization effect is the same after adding silicon alloy based on the solid-state decarburization of Fe-C alloy, predecessors also have a large number of studies [23,24]. Taking 1 mm thick Fe-4.0% C-0.84% Si thin strip as the research object, the activation energy of decarburization reaction was calculated to be 111.475 kJ/mol. This is lower than the diffusion activation energy of carbon in austenite. Silicon can promote the decarburization reaction to a certain extent, mainly because silicon can increase the diffusion coefficient of carbon atoms.

However, there is no research and exploration at the micro atomic scale based on molecular dynamics. The diffusion behavior of carbon and silicon in silicon steel and the diffusion activation energy have not been reported. Scholars have calculated the atomic self-diffusion coefficients in different systems [25–27]. Ye et al. [28] used molecular dynamics to study the diffusion behavior of amorphous Si-C and Si-B-C systems at different temperatures. Sun [29] studied the diffusion behavior of oxygen isotopes in melts and calculated the changes in silicon and oxygen self-diffusion coefficients with pressure at different temperatures. Shi et al. [30] used molecular dynamics to study the diffusion behavior of oxygen in molten $SiO_2$ at different temperatures and analyzed the influence of temperature on gas diffusion. Wang et al. [31] applied molecular dynamics simulations to study interfacial atomic diffusion and calculated the mean-square displacement (MSD) of C and Ti atoms. The results showed that in the atomic diffusion at the interface between diamond and Ti, the diffusion ability and speed of C atoms were greater than those of Ti atoms.

The feasibility of preparing silicon steel by gas–solid reaction decarburization method was experimentally studied by the research group. When the initial carbon content is 0.18 wt%, the carbon content can be reduced to 0.016 wt%. Next, we need to study the diffusion behavior of carbon and silicon in the process of decarburization and its factors. The influence of atmosphere and temperature on the growth of oxide film, etc. This study focuses on Fe-0.18 wt% C—(1.5 wt%, 3.5 wt%) Si alloys, combined with experimental results, to conduct molecular dynamics simulations, FactSage thermodynamic calculations, and decarburization kinetics research. It lays a foundation for improving the decarburization conditions of silicon steel prepared by carbon fixation and decarburization method.

## 2. Experimental Program

### 2.1. Modelling

Molecular dynamics (MD) were used to simulate the diffusion behavior of carbon and silicon atoms in the Fe-C-Si alloy system. The simulations were carried out using LAMMPS software (Lammps-patch_27Oct2021) to create an FCC-structured $13a_0 \times 13 \times 13$ pure Fe supercell with 8788 atoms and $a_0 = 0.3591$ nm as the lattice parameter. To eliminate interfacial effects, three-dimensional periodic boundary conditions were adopted for the model. The Fe-C-Si alloy model, obtained by randomly replacing an equal number of Fe atoms with Si atoms and randomly inserting 0.18 wt% C atoms, is shown in Figure 1. Tobias A et al. [32] mentioned that increasing the box size has little effect on the diffusion coefficient, and changing the box size will inevitably change the insertion position of atoms. Therefore, in this simulation, the influence of random insertion position and box size on diffusion behavior is not considered. The potential function used is the MEAM potential function [33], which can well describe the interactions between various types of atoms in the Fe-C-Si alloy system. In the isothermal and isobaric (NPT) system, the initial temperature is set to be 300 K, and the temperature is increased at a rate of $1 \times 10^{11}$ K/s. To obtain an equilibrium structure, the Fe-C-Si alloy system was relaxed by the NPT system for 10,000 steps with a time step of 0.001 ps after reaching the preset temperature, and the mean-square displacements (MSD) of the C and Si atoms were calculated by running the NVT system for 200,000 steps to characterize the diffusion behavior of the atoms. Radial distribution function (RDF) [34] and common neighbor analysis (CNA) [35] in the visualization software OVITO (Open Visualization Tool, version 2.9.0) were used to characterize the phase transition process.

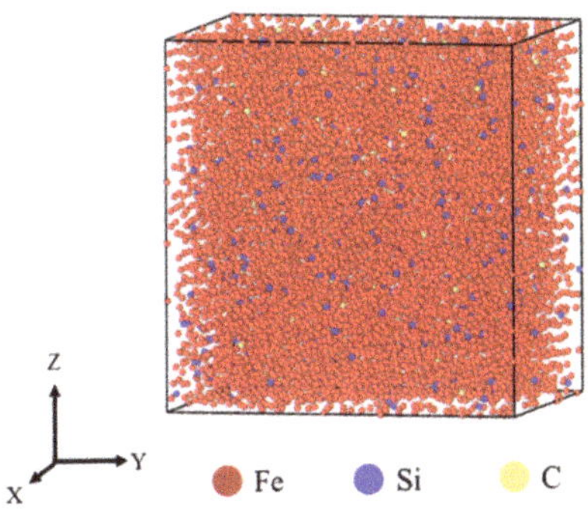


**Figure 1.** Fe-C Si alloy model.

The mean-square displacement (MSD) of the atom is obtained by molecular dynamics calculation, and the expressions of the mean-square displacement and diffusion coefficient are defined as (1) and (2), respectively:

$$\mathrm{MSD}(t) = \frac{1}{N}\left(\sum_{t=1}^{N}[r_i(t) - r_i(0)]^2\right) \quad (1)$$

$$D_{self} = \lim_{t\to\infty}\frac{1}{6N\cdot t}\left(\sum_{t=1}^{N}[r_i(t) - r_i(0)]^2\right) \quad (2)$$

where $D_{self}$ is the self-diffusion coefficient of particles, N is the total number of particles, t is time, $r_i(t)$ and $r_i(0)$ are the positions of particles at t and $t_0$, respectively.

Combining Equations (1) and (2), the self-diffusion coefficient and mean azimuthal shift relationship can be obtained as (3). According to Equation (3), the mean-square displacement is linearly correlated with the diffusion coefficient, and the self-diffusion coefficient can be obtained according to its slope [36].

$$D_{self} = \lim_{t\to\infty}\frac{1}{6t}\mathrm{MSD} \quad (3)$$

*2.2. Experimental Materials and Methods*

The solid-state decarburization experiments were carried out on thin strips with dimensions of 55 mm × 65 mm × 1 mm, with the compositions shown in Table 1.

**Table 1.** Silicon steel composition (wt%).

| Si | C | Fe | S | P | O |
|---|---|---|---|---|---|
| 3.5 | 0.18 | 96.32 | <0.02 | <0.027 | <0.066 |
| 1.5 | 0.18 | 98.32 | <0.02 | <0.027 | <0.066 |

To determine the phase composition of the sample, the Fe-C-Si thin strip sample before decarburization experiment was detected by XRD. The diffraction results are shown in Figure 2. The composition of the sample was detected under the electron microscope and the microstructure of the initial sample was observed, as shown in Figure 3.

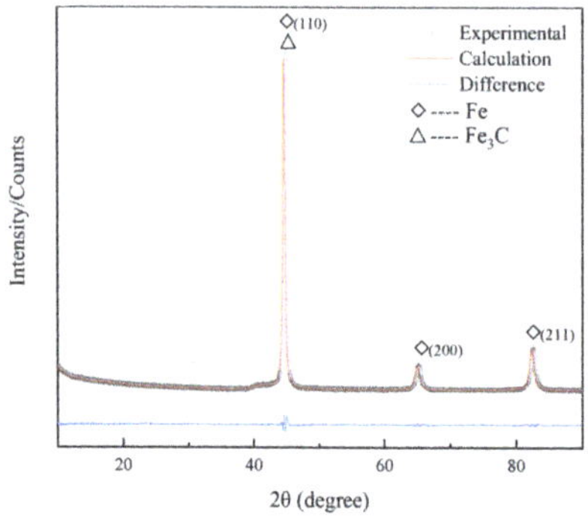


**Figure 2.** XRD phase analysis of initial sample.

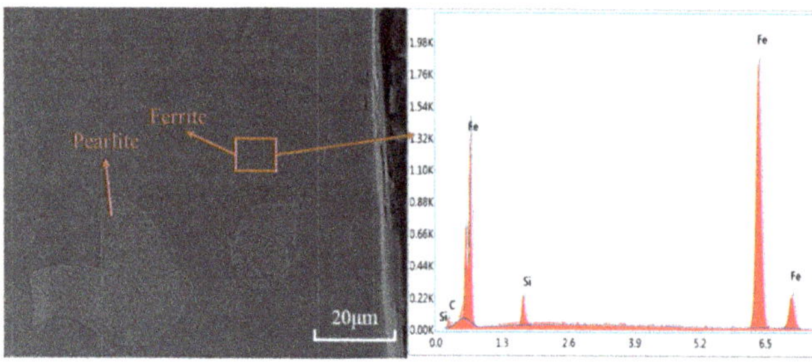


**Figure 3.** SEM backscatter micrograph and EDS map of the initial sample.

Before decarburization, the sample is composed of ferrite phase and $Fe_3C$ phase. The C in the matrix mainly exists in the form of solid solution carbon and cementite ($Fe_3C$), while the Si in the matrix exists in the form of solid solution. The composition of the sample is Fe, Si, C. The microstructure of the sample is mainly ferrite and pearlite (a layered composite of thin layers of ferrite and cementite alternately laminated).

The solid-state decarburization experiment was carried out in Ar-$H_2O$-$H_2$ atmosphere. The solid-state decarburization experiment was completed in a controllable atmosphere decarburization furnace. The furnace structure is shown in Figure 4. Before the decarburization experiment, the furnace needs to be washed by repeatedly vacuuming and charging Ar. During the heating process of the decarburization furnace, the flow rate of Ar gas into the furnace is 180 mL·min$^{-1}$. When the temperature rises to the predetermined decarburization temperature, adjust the gas flow in the furnace to the flow required for the experiment (20% $H_2$ + $H_2O$ + Ar mixed gas, with the total flow set at 1500 mL·min$^{-1}$). By adjusting the $P_{H2O}/P_{H2}$ of the gas mixture, the water vapor in the atmosphere is used to react with the carbon on the surface of the Fe-C-Si alloy in a redox reaction to decarburize it.

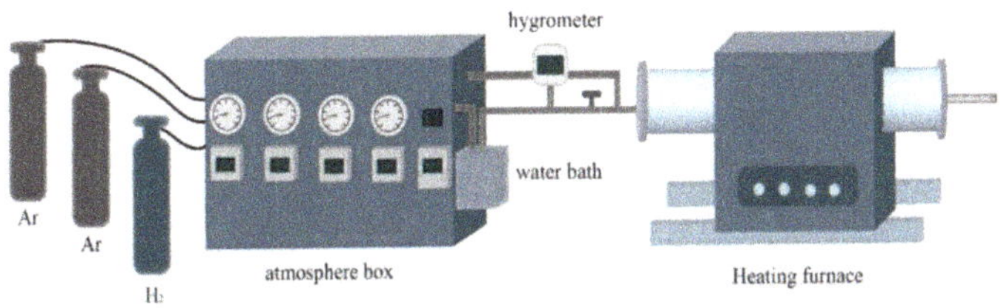


**Figure 4.** Solid-state decarburization furnace.

## 3. Results and Analysis

### 3.1. Effect of Temperature on Diffusion Behavior

According to the expression of diffusion coefficient, the diffusion coefficient has an exponential relationship with temperature. The diffusion coefficient increases with the increase in temperature, which accelerates the diffusion of carbon from the substrate to the surface. On the other hand, increasing the temperature is also beneficial to improve the reaction rate of surface decarburization. To explore the effect of temperature on decarburization of Fe-C-Si thin strips, intermittent experiments were carried out at 1000 °C, 1090 °C, 1150 °C and 1300 °C, respectively. Taking 1 mm thick Fe-0.18% C-3.5%Si thin alloy strip as the object of study, the atmosphere condition $P_{H2O}/P_{H2} = 0.31$, the content of mixed gas $H_2$ is 20%, the flow rate of mixed gas is 1500 mL·min$^{-1}$, and the decarburization time is 5–70 min. The experimental results of the decarburization were obtained as shown in Figure 5.

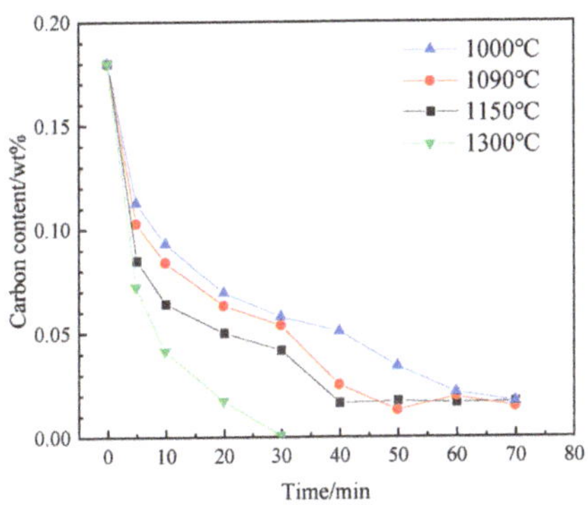


**Figure 5.** Decarburization process at different temperatures of $P_{H2O}/P_{H2} = 0.31$.

It can be seen from Figure 5 that the surface oxidation is relatively serious at 1300 °C to form a dense oxide layer and peel off at room temperature after decarburization, so the experimental results under this condition will not be discussed. At 1150 °C, decarburization stagnated after 40 min of decarburization, and the carbon content did not change. At 1090 °C, the decarburization process tended to be gentle until 50 min, while it tended to be gentle at 1000 °C for 60 min. In the range of 1000 to 1150 °C, the carbon content at the decarburization stagnation is basically the same. It shows that under the same decarburization time, the higher the temperature, the faster the decarburization rate and the better the decarburization effect.

To verify the experimental results, molecular dynamics was used to simulate the diffusion behavior of C at different temperatures. Figure 6 shows the variation in the crystal structure with temperature and Figure 7 shows the radial distribution function of the Fe-0.18% C-3.5% Si system at different temperatures during the heating process. The radial distribution function is the variation in the particle density with distance and can be used to describe the atomic structure as ordered or disordered [37]. When the crystal structure of the system changes, its radial distribution function curve will change. Figure 8 shows the mean-square displacement of C at different temperatures. The diffusion coefficients of C at 1150 °C, 1090 °C, and 1000 °C were calculated from Equation (3) to be 6.87 × $10^{-11}$ $m^2/s$, 3.92 × $10^{-11}$ $m^2/s$, and 3.15 × $10^{-11}$ $m^2/s$, respectively. From Figures 6–8, it can be seen that below 800 °C, the matrix is mostly a tightly arranged BCC structure, and the migration rate of C is low. When the temperature is higher than 800 °C, with the increase in temperature, the atomic movement intensifies, and the crystal structure changes from ferrite to austenite + ferrite. The violent motion of the atoms causes them to leave their original positions and most of the crystal structure is destroyed, forming a disordered atomic structure, which promotes the diffusion of C, i.e., the diffusion coefficient increases.

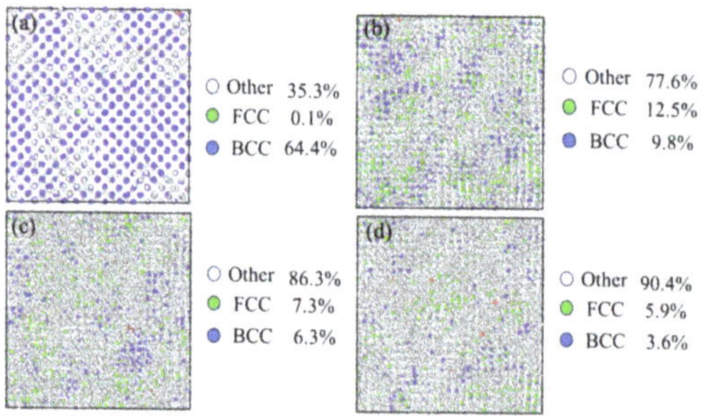


**Figure 6.** Changes in crystal structure with temperature: (**a**) 500 °C; (**b**) 800 °C; (**c**) 1150 °C; (**d**) 1300 °C.

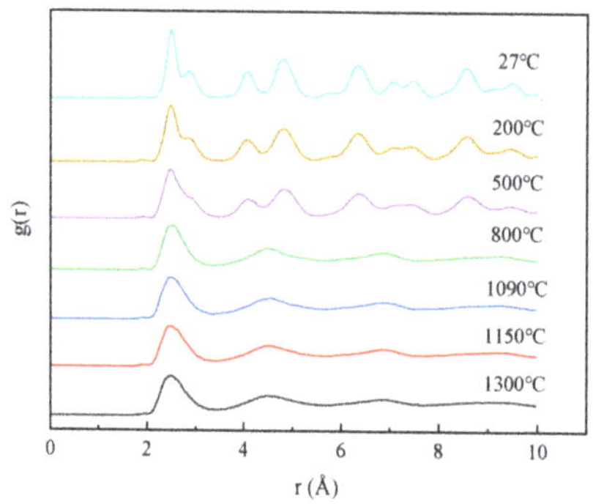


**Figure 7.** Radial distribution functions for different temperatures.

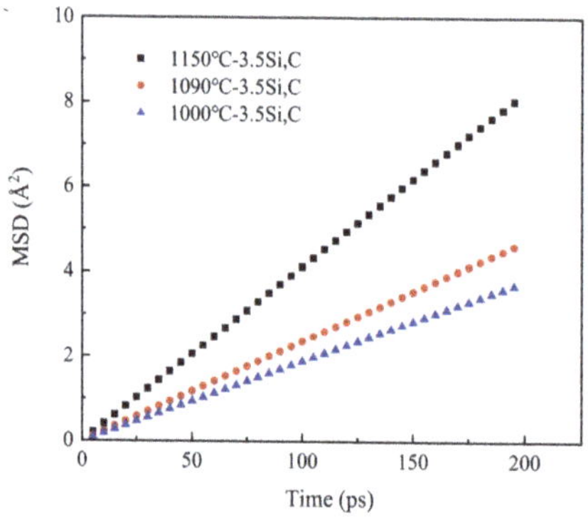


**Figure 8.** The variation in the average azimuth shift of C with temperature.

### 3.2. Verification of Simulation Results Combined with Decarburization Kinetics

In the decarburization process, surface decarburization, internal diffusion of carbon in the matrix, and oxide layer growth are the three key aspects of the decarburization process in thin strips of Fe-C-Si alloys; any of them may become the main limiting aspect. In their study of the mechanism of decarburization of Fe-C alloys, Hou et al. [38] derived mathematical expressions for the internal diffusion-controlled rate of carbon and expressions for the diffusion-controlled rate mechanism under certain conditions of surface carbon content, respectively. When the carbon content on the surface of the strip is lower than the concentration within the matrix, it is assumed that the intra-matrix diffusion of carbon within the strip is the only rate limiting aspect of the entire decarburization process. Fick's first law and mass balance can be used to simplify the analysis and obtain the following expression:

$$[\%C]_t = [\%C]_0 - kt^{1/2} + k't \tag{4}$$

where:

$[\%C]_0$—initial carbon content, wt%;

k, k′ are the coefficients and their expressions are, respectively:

$$k = \frac{2(a+b)}{ab}k_1k_2 = \frac{2(a+b)}{ab}\sqrt{2\frac{D}{\rho}(C_2 - C_3)\left[C_1 - \frac{1}{2}(C_2 + C_3)\right]} \tag{5}$$

$$k' = 4\frac{k_1{}^2k_2}{ab}k_1k_2 = \frac{8D}{ab\rho}(C_2 - C_3) \tag{6}$$

When the diffusion of carbon within the thin strip is the rate-controlled part of the whole decarburization process, the variation in carbon content can be described approximately by Equation (4). After decarburization for 30 min, the limiting link is mainly the growth of oxide layer, which is not suitable for fitting the carbon diffusion mechanism. Therefore, only the decarburization results at different temperatures 30 min before decarburization in Figure 3 are fitted according to Formula (4), and the fitting results are shown in Figure 9. The fitting expressions are shown in Table 2.

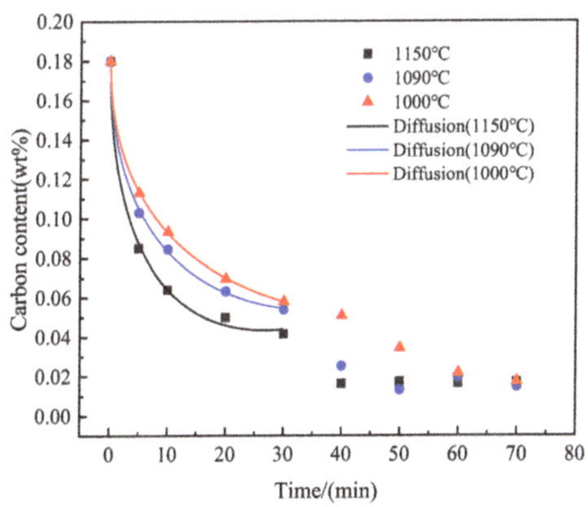


**Figure 9.** Fitting decarburization data at different temperatures.

**Table 2.** Fitting Expressions.

| Temperature (°C) | Expression | Adj. R-Square |
|---|---|---|
| 1000 | $[\%C]_t = 0.18 - 0.03499t^{1/2} + 000232t$ | 0.99989 |
| 1090 | $[\%C]_t = 0.18 - 0.0414t^{1/2} + 000333t$ | 0.99929 |
| 1150 | $[\%C]_t = 0.18 - 0.05279t^{1/2} + 00051t$ | 0.99705 |

According to Formulas (5) and (6), k is proportional to $D^{1/2}$ and k′ is proportional to D. It is known that the longer the decarburization time k′ is the more decisive, then k′ can be used to represent D.

From the expression for the diffusion coefficient of solute atoms, it can be seen that the diffusion coefficient D is exponentially related to temperature, and the diffusion coefficient increases rapidly as the temperature increases:

$$D = D_0 \exp(\frac{-Q}{RT}) \tag{7}$$

Taking logarithms on both sides of the equal sign of Equation (7) simultaneously yields the following linear equation:

$$\ln D = \ln D_0 - \frac{Q}{R}\frac{1}{T} \tag{8}$$

As shown in Figure 10, lnD is linearly related to the inverse of temperature 1/T in semi-logarithmic coordinates, and the fitting parameters are obtained as follows:

$$D_{3.5Si,\,C} = 7.84 \times 10^{-9} \exp\left(\frac{-48686.7}{RT}\right) m^2{\cdot}s^{-1},\ Q = 48.7\ kJ/mol,$$

$$D_{3.5Si,\,C,\,MD} = 8.69 \times 10^{-9} \exp\left(\frac{-47306.7}{RT}\right) m^2{\cdot}s^{-1},\ Q = 47.3\ kJ/mol.$$

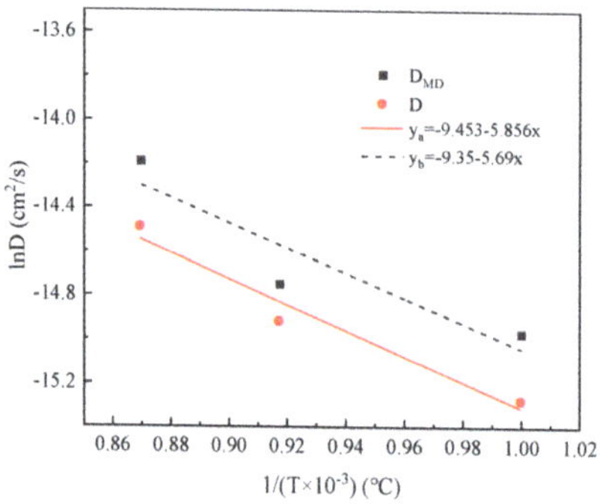


**Figure 10.** Comparison of diffusion coefficient MD value of C with experimental value.

D is calculated from decarburization experiments and $D_{MD}$ is calculated from molecular dynamics simulations. It can be seen that the simulation results are in agreement with the experimental results, the diffusion activation energy Q is basically the same, and the molecular dynamics simulation results are credible.

### *3.3. Effect of Carbon and Silicon Diffusion on Oxide Growth*

As can be seen from Figure 5, the carbon content no longer decreases after 40 min of decarburization at 1150 °C. It was experimentally deduced that Fe-C-Si alloys are related to the growth of oxide layer during the solid-state decarburization [39]. The higher Si content in the matrix can improve the activity of C, and the diffusion coefficient of C in the matrix is much higher than that of Si. Therefore, there is a competitive oxidation relationship between C and Si elements during the decarburization process. The oxide layer is too thick in the late stage of decarburization, which hinders the diffusion of C to the surface of the thin strip, thus leading to the stagnation of decarburization.

To verify the experimental inference, molecular dynamics simulations were used to calculate the diffusion coefficients of Si at different temperatures, with Si contents of 1.5 wt% and 3.5 wt%, respectively. The mean-square shifts of Si at different temperatures with Si content of 1.5 wt% and 3.5 wt% are shown in Figure 11. The diffusion coefficients of Si at 1150 °C, 1090 °C, and 1000 °C for Si content of 3.5 wt% were calculated from

Equation (3) to be $2.83 \times 10^{-11}$ m$^2$/s, $2 \times 10^{-7}$ cm$^2$/s, $1.67 \times 10^{-11}$ m$^2$/s, and the diffusion coefficients of Si for Si content of 1.5 wt% were $1.73 \times 10^{-11}$ m$^2$/s, $1.02 \times 10^{-11}$ m$^2$/s, and $8.02 \times 10^{-12}$ m$^2$/s, respectively. From Equation (8), the relationship between lnD and 1/T is shown in Figure 12. After fitting, the expression of Si diffusion coefficient can be obtained when the Si content is 1.5 wt% and 3.5 wt%, respectively.

$$D_{3.5Si,\,Si} = 8.54 \times 10^{-10} \exp\left(\frac{-33089.7}{RT}\right) m^2 \cdot s^{-1},\ Q = 33.1\ kJ/mol,$$

$$D_{1.5Si,\,Si} = 2.06 \times 10^{-9} \exp\left(\frac{-46641.5}{RT}\right) m^2 \cdot s^{-1},\ Q = 46.6\ kJ/mol.$$

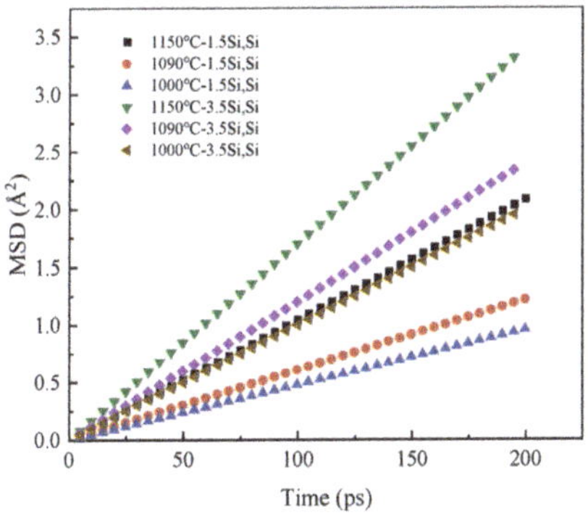


**Figure 11.** The change in average azimuth shift of Si with temperature.

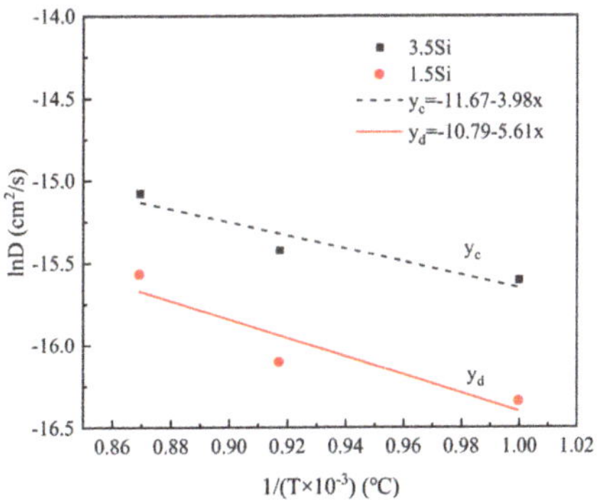


**Figure 12.** The relationship between lnD and the reciprocal of temperature (1/T).

It can be seen that the diffusion coefficient of C is greater than the diffusion coefficient of Si at different temperatures. C diffuses to the surface before Si, and C consumes a large amount of O. A small amount of O combines with the Si that later diffuses to the surface to form $SiO_2$, which hinders the diffusion of C and leads to stagnation in the later stages of decarburization. In the previous study on the effect of Si content on the growth of the oxide layer, it was found that, other conditions being equal, different Si contents resulted in large differences in the morphology of the oxide layer.

When the silicon content is 1.5 wt%, the SEM backscatter micrographs and EDS maps of oxide layer evolution under different decarburization times are shown in Figure 13. The SEM micrographs show that the structure and morphology of the oxide layer are the same under different decarburization times. Taking the EDS maps of oxide layer after decarburization for 40 min as an example, the structure and element distribution of oxide layer were explored (The same below). The oxide layer of the sample section is composed of small granular circular $SiO_2$ and is highly dispersed. With the increase in decarburization time, the oxide layer thickens obviously. When the Si content is 1.5 wt%, the diffusion coefficient of Si is very small, the migration rate is too slow, and the remaining o will oxidize when encountering Si in the matrix; it cannot aggregate to form banded $SiO_2$, which can only be highly dispersed.

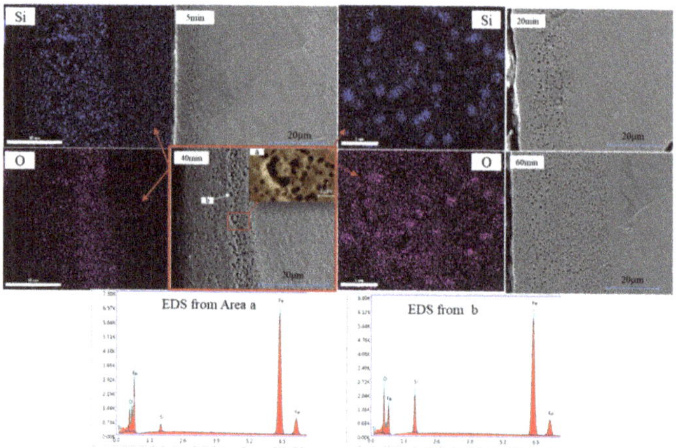


**Figure 13.** SEM micrographs and EDS maps of oxidation layer evolution under different decarburization time when silicon content is 1.5 wt%: a stands for EDS area; b stands for EDS point scan.

When the silicon content is 3.5 wt%, the SEM backscatter micrographs and EDS maps of oxide layer evolution under different decarburization time is shown in Figure 14. After decarburization, an obvious continuous strip of $SiO_2$ was formed at the junction of the substrate and the oxide layer, and spherical oxides were distributed in the middle of the oxide layer, with large scale and small quantity. It can be seen that when the content of Si is 3.5 wt%, the diffusion coefficient of Si is large, and the migration speed to the surface is faster. After diffusion to the near surface, it combines with the remaining o to form $SiO_2$. With the increase in decarburization time, $SiO_2$ gradually aggregates to form an obvious continuous strip of $SiO_2$.

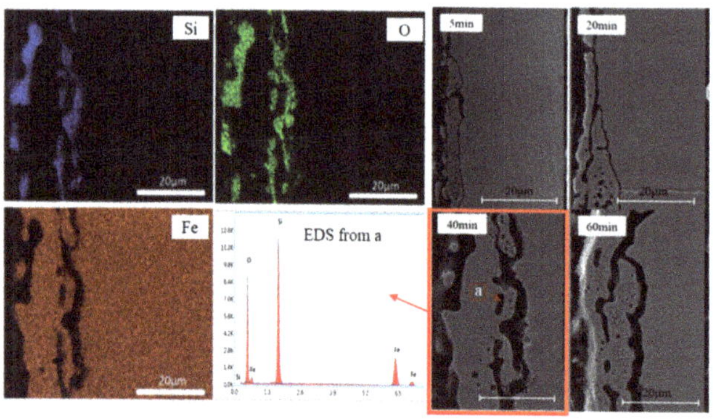


**Figure 14.** SEM micrographs and EDS maps of oxidation layer evolution under different decarburization time when silicon content is 3.5 wt%: a stands for EDS point scan.

### 3.4. *Effect of the Phase Transition on Diffusion Behavior*

In the middle and late decarburization period (30–40 min) before decarburization reached the stagnation state, there was a process of rate increase at different temperatures

(Figure 5). To analyze the reasons for the increase in the rate, FactSage software (FactSage 8.1.0) was used to draw the Fe-C-Si ternary phase diagram, and the decarburization process was analyzed, as shown in Figure 15.

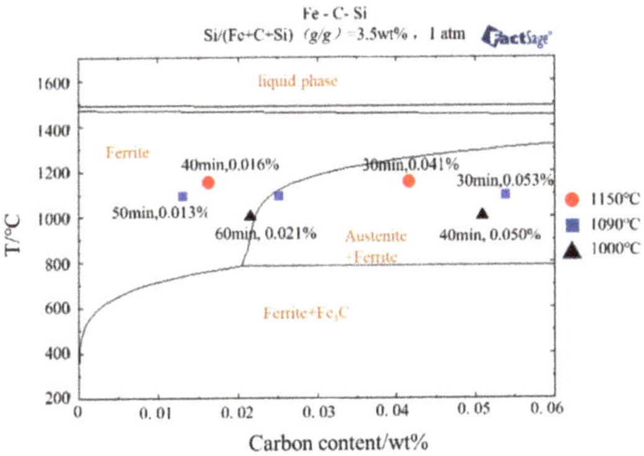


**Figure 15.** Phase diagram of Fe-C-3.5Si alloy.

At 30 min of decarburization, the carbon content was 0.0416 wt% at 1150 °C. At this time, the organization in the strip at high temperature was mainly ferrite and austenite. When decarburized for 40 min, the carbon content was 0.0164 wt%. At this time the organization is ferrite at high temperatures. The process of ferrite + austenite biphasic transformation to pure ferrite phase exists between 30 and 40 min, as shown in Figure 16. Similarly, the phase transformation exists between 30 and 50 min at 1090 °C, as shown in Figure 17. At 1000 °C, the phase transformation process is completed at 60 min of decarburization. Therefore, it can be deduced that at the middle stage of decarburization (corresponding to the 30th or 40th min at different temperatures), the thin strip matrix undergoes a crystallographic transition from the face-centered cubic lattice austenite to body-centered cubic lattice ferrite. The diffusion coefficient of C in ferrite is much larger than that in austenite, which leads to an increase in the diffusion rate by the phase change and thus a window of high decarburization rate before decarburization stagnation.

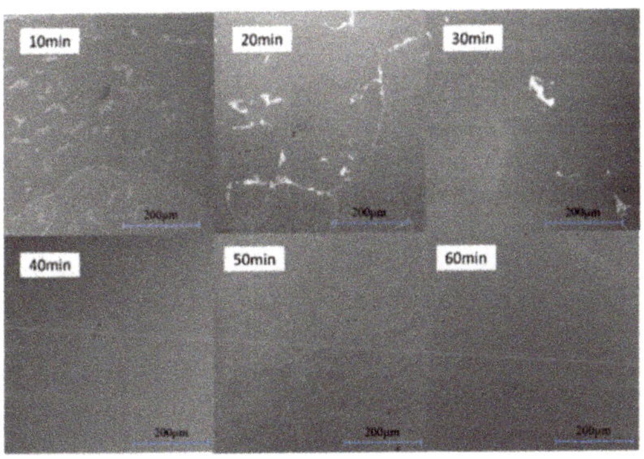


**Figure 16.** SEM micrographs of microstructure evolution with time at 1150 °C.

To verify this inference, molecular dynamics simulations were used to calculate the diffusion coefficients of C in the ferrite and austenite lattices, respectively. The mean-square displacement of C in ferrite and austenite lattices is shown in Figure 18. After calculating the diffusion coefficients from Equation (3), it can be seen that the diffusion coefficients of C in ferrite lattice are $9.8 \times 10^{-11}$ $m^2/s$ and in austenite lattice are $1.7 \times 10^{-11}$ $m^2/s$. The diffusion coefficient of C in ferrite is greater than in austenite at the same temperature. It can be explained that the phase transformation of the thin strip matrix from the face centered cubic lattice austenite to body centered cubic lattice ferrite occurred in the experiment, and the diffusion rate increased due to the phase transformation, and then the decarburization rate increased before decarburization stagnation.

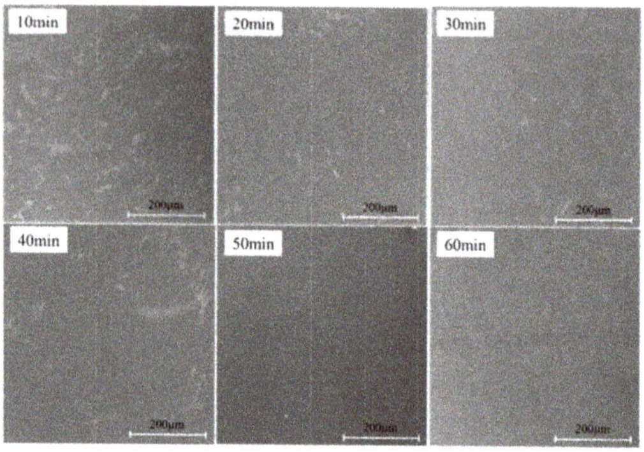


**Figure 17.** SEM micrographs of microstructure evolution with time at 1090 °C.

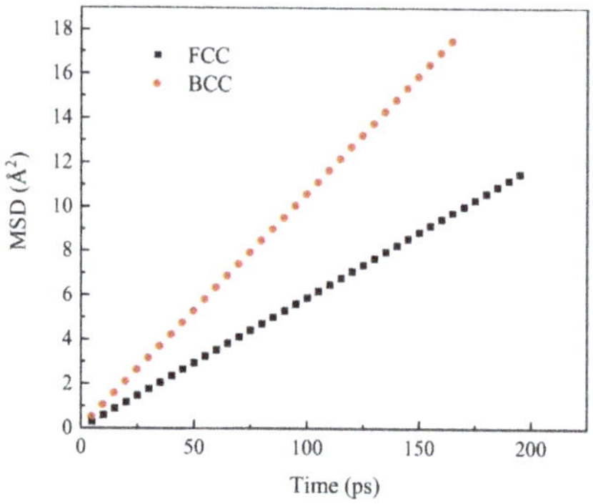


**Figure 18.** Mean-square displacement of different phase regions C.

*3.5. Effect of Silicon Content on Diffusion Behavior*

To study the effect of silicon content on decarburization of the thin alloy strip, experiments were carried out on 1 mm thick Fe-0.18C-Si thin strip specimens with silicon content of 1.5 wt% and 3.5 wt%, respectively. The atmosphere condition was $P_{H2O}/P_{H2} = 0.31$, the gas flow rate was set at 1500 mL·min$^{-1}$, and the decarburization temperature was 1150 °C. The experimental results after decarburization are shown in Figure 19.

For the same decarburization time, the amount of decarburization was greater in the first 40 min for Si = 3.5 wt% compared to Si = 1.5 wt%. The mechanism behind this lies in the fact that silicon elements can enhance the activity of dissolved carbon in the alloy. As the silicon content increases, the diffusion activation energy for carbon decreases, leading to an increase in the rate of carbon diffusion, thereby accelerating the decarburization process.

To validate this hypothesis, molecular dynamics simulations were employed to calculate the diffusion coefficient of carbon at various temperatures with a silicon content of 1.5 wt%. The mean-square displacement of carbon within the matrix at different temperatures with a silicon content of 1.5 wt% is depicted in Figure 20. Using Equation (3), the calculated diffusion coefficients for C with a silicon content of 1.5 wt% at temperatures of 1150 °C, 1090 °C, and 1000 °C are $2.64 \times 10^{-11}$ m$^2$/s, $1.79 \times 10^{-11}$ m$^2$/s, and $1.25 \times 10^{-11}$ m$^2$/s, respectively. The relationship between lnD and 1/T is shown in Figure 21, and fitting the data yielded the expression for the diffusion coefficient of C at a silicon content of 1.5 wt% as follows:

$$D_{1.5Si,\,C} = 8.69 \times 10^{-10} \exp\left(\frac{-55204.9}{RT}\right) m^2 \cdot s^{-1},\ Q = 55.2\ kJ/mol.$$

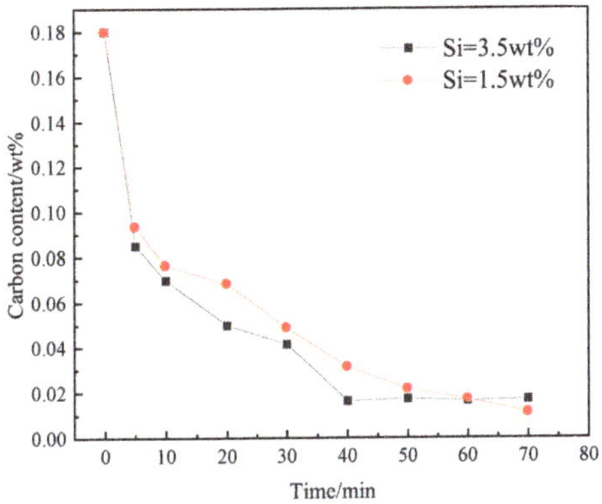


**Figure 19.** Decarburization effect of different silicon contents.

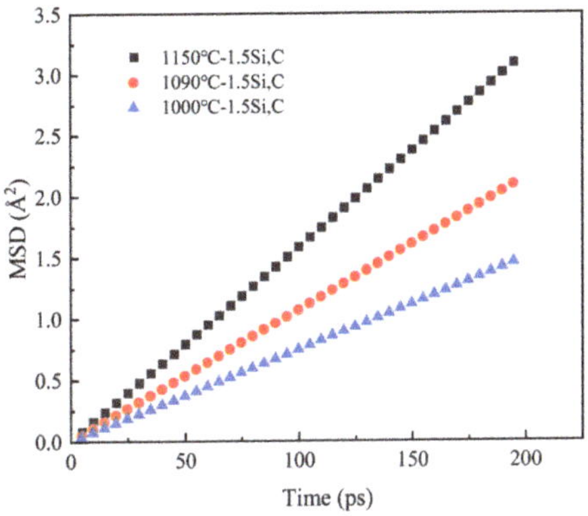


**Figure 20.** Mean-square displacement of C at a Si content of 1.5 wt%.

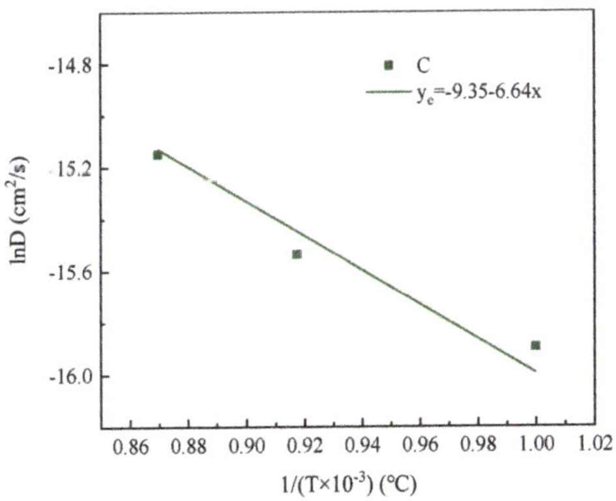


**Figure 21.** The relationship between lnD and 1/T at a Si content of 1.5 wt%.

Luo et al. [40] and Yang et al. [41], in their investigations of decarburization in silicon steel, solely considered the diffusion coefficient of C in the iron matrix, without exploring the impact of varying amounts of Si on the C diffusion. It is known that the diffusion activation energy of C in $\gamma$-Fe is 140 $kJ \cdot mol^{-1}$. Combined with the above molecular dynamics simulation of the diffusion coefficient of C when the Si content is 3.5 wt%, it can be seen that the diffusion activation energy of C decreases with the increase in matrix silicon content. The diffusion coefficient of C increases, which is beneficial for the promotion of decarburization.

## 4. Conclusions

The diffusion behavior of carbon in the decarburization process of 1 mm thick 0.18 wt%C—(3.5 wt%, 1.5 wt%) Si thin strip was studied by decarburization experiment, FactSage thermodynamic calculation, and molecular dynamics simulation. The research conclusions are as follows:

(1) The experimental results show that with the same decarburization time, the higher the temperature (1000–1300 °C), the faster the decarburization rate; Results from molecular dynamics simulations indicate that at temperatures below 800 °C, atoms maintain a tightly packed bcc structure, resulting in a low migration rate of carbon. When the temperature is higher than 800 °C, the atoms move violently and the crystal structure is destroyed. Some atoms tend to be disordered, the migration rate of C is high, and the diffusion coefficient increases. The thin strip substrate undergoes a lattice transformation from face-centered cubic lattice austenite to body-centered cubic lattice ferrite. Notably, the diffusion coefficient of C in ferrite significantly exceeds that in austenite, giving rise to an observed enhancement in decarburization rate just before decarburization stagnation (30 min to 50 min);

(2) Based on a decarburization model that considers carbon diffusion within the matrix as the rate-limiting step in the early stages of decarburization, the diffusion coefficient for carbon was determined. Subsequently, the diffusion activation energy for C in the Fe-3.5% Si-C alloy matrix was calculated as $Q = 48.7\ kJ \cdot mol^{-1}$. Molecular dynamics simulations yielded a diffusion activation energy of 47.3 $kJ \cdot mol^{-1}$ for C in the Fe-3.5% Si-C alloy matrix, which closely aligns with experimental findings;

(3) The diffusion coefficients of carbon were found to be $D_{3.5Si,\,C} = 7.84 \times 10^{-9}$ $\exp(-48{,}686.7/RT)\ m^2/s$ and $D_{1.5Si,\,C} = 8.69 \times 10^{-10} \exp(-55{,}204.9/RT)\ m^2/s$ when the silicon content was 3.5 wt% and 1.5 wt%, respectively. In comparison to Si = 1.5 wt%, at the same decarburization time, Si = 3.5 wt% exhibited a higher decarburization rate during the initial 40 min. At a decarburization temperature of 1150 °C, the diffusion coefficients of C were determined to be $2.64 \times 10^{-11}\ m^2/s$ and $6.87 \times 10^{-11}\ m^2/s$ for Si contents of 1.5 wt% and 3.5 wt%, respectively. This suggests that an increase in silicon content leads to a reduction of the activation energy for the C diffusion and an increase in the carbon diffusion coefficient, thereby promoting the decarburization process;

(4) Molecular dynamics simulations reveal that the diffusion coefficients of silicon were determined to be $D_{3.5Si,\,Si} = 8.54 \times 10^{-10} \exp(-33{,}089.7/RT)\ m^2/s$ and $D_{1.5Si,\,Si} = 2.06 \times 10^{-9} \exp(-46{,}641.5/RT)\ m^2/s$ when the Si content was 3.5 wt% and 1.5 wt%, respectively. At different temperatures, the diffusion coefficient of carbon was consistently higher than that of Si. Carbon diffuses to the surface before Si, consuming a significant amount of oxygen. When the Si content is 3.5 wt%, the diffusion coefficient of Si is relatively high, allowing it to diffuse to the near-surface region, where it combines with the remaining O to form a continuous banded $SiO_2$ structure. In contrast, when the Si content is 1.5 wt%, the Si diffusion coefficient is significantly lower, causing the remaining O to diffuse within the matrix and oxidize upon encountering Si. In this case, $SiO_2$ does not aggregate into banded structures but instead exhibits a highly-dispersed distribution.

**Author Contributions:** Methodology, L.A.; software, L.W. and G.Z.; validation, L.W., L.A. and L.H.; formal analysis, Y.Z. and C.S.; data curation, Y.Z. and G.Z.; writing—original draft preparation, L.W.; and writing—review and editing, L.A. and L.H. All authors have read and agreed to the published version of the manuscript.

**Funding:** This study was supported by the Natural Science Foundation of Hebei Province (E2021209101; E2022209112); Central Government Guided Local Science and Technology Development Fund Project (236Z1006G); Science and Technology Research Projects of Higher Education Institutions in Hebei Province (ZD2022125); Tangshan Talent Funding Project (A20220212).

**Data Availability Statement:** Data are contained within the article.

**Conflicts of Interest:** The authors declare no conflict of interest.

## References

1. Wu, K.M. Production process and development of non-oriented electrical steel. *China Metall.* **2012**, *22*, 2430–2435.
2. Zhang, Y.; Zhu, G.H.; Wu, Q. Development prospect of CSP production process for non-oriented electrical steel. *Mod. Metall.* **2013**, *41*, 1–4.
3. Xiang, L.; Chen, S.L.; Pei, Y.H. Present situation and research progress of electrical steel produced by thin slab continuous casting and rolling in China. In Proceedings of the 11th China Iron and Steel Annual Conference, Beijing, China, 27 May 2017.
4. Hong, L.K.; Ai, L.Q.; Cheng, R. Experimental study on Decarburization of Fe-C alloy strip by gas-solid reaction. *Steel* **2016**, *51*, 27–31.
5. Mcdonald, C. Solid state steelmaking: Process technical and economic viability. *Ironmak. Steelmak.* **2012**, *39*, 487–489. [CrossRef]
6. Park, J.O.; Long, T.V.; Sasaki, Y. Feasibility of solid-state steelmaking from cast iron-decarburization of rapidly solidified cast iron. *Tetsu Hagane* **2012**, *98*, 151–160. [CrossRef]
7. Lee, W.H.; Park, J.O.; Lee, J.S. Solid state steelmaking by decarburization of rapidly solidified high carbon iron sheet. *Ironmak. Steelmak.* **2012**, *39*, 530–534. [CrossRef]
8. Liang, M.; Cheng, J.; Lin, W.M. Thermodynamic Study on Solid Phase Decarburization of High Carbon Ferromanganese Powder. *Ferroalloy* **2009**, *40*, 14–17.
9. He, H.Q.; Wei, X.J.; Kang, G.Z. Study on solid state decarburization of high carbon ferromanganese in carbon dioxide atmosphere. *Ferroalloy* **2010**, *41*, 5–8.
10. Li, Q. *Gas Solid Decarburization of Fe—3.2% C Alloy Strip*; North China University of Technology: Beijing, China, 2017.
11. Li, Y.Q. *Effect of Atmosphere Conditions on Gas Solid Reaction Decarburization of Iron Carbon Alloy Strip*; North China University of Technology: Beijing, China, 2018.
12. Cheng, R. *Kinetic Analysis of Decarburization of Fe-C Alloy Strip by Gas-Solid Reaction*; North China University of Technology: Beijing, China, 2018.
13. Ai, L.Q.; Hou, Y.B.; Hong, L.K. Kinetics of gas-solid decarburization of Fe-C alloy strip in $H_2/H_2O$ atmosphere. *J. Eng. Sci.* **2021**, *43*, 816–824.
14. Zhou, M.J. *Decarburization Kinetics of Fe-C Alloy Strip by Gas-Solid Reaction in Ar-CO-$CO_2$ Atmosphere*; North China University of Technology: Beijing, China, 2022.
15. Guo, Y.; Dai, F.Q.; Gao, Y. Study on decarburization process model of grain-oriented silicon steel. In Proceedings of the 2016 Seminar on Production Technology and Application of Domestic High Performance Electrical Steel, Dalian, China, 25 August 2016.
16. Yang, S.Z.; Dai, F.Q.; Guo, Y. Numerical simulation analysis on decarburization process of grain-oriented silicon steel. *J. Wuhan Univ. Sci. Technol.* **2016**, *39*, 253–258.
17. Hao, J.J.; Chen, J.; Han, P.D. Solid Phase Decarburization Kinetics of High-Carbon Ferrochrome Powders in the Microwave Field. *Steel Res. Int.* **2014**, *85*, 461–465. [CrossRef]
18. Larssen, T.A.; Senk, D.; Tangstad, M. Reaction Rate Analysis of Manganese Ore Prereduction in CO-$CO_2$ Atmosphere. *Metall. Mater. Trans. B* **2021**, *52*, 2087–2100. [CrossRef]
19. Li, H.; Liang, J.L.; Li, Y.G. Study on the diffusion behavior of Si during annealing. *Ind. Heat.* **2014**, *43*, 30–32+41.
20. Zhu, Y.B. Mathematical model of decarburization of steel during heating. *Spec. Steel* **1989**, *3*, 1–4.
21. Zhang, C.; Xiang, L.; Dong, Y.L. Experimental study on decarburization annealing of grain-oriented silicon steel. *J. Iron Steel Res.* **2009**, *21*, 55–58.
22. Agren, J. Computer simulations of the austenite/ferrite diffusional transformations in low alloyed steels. *Acta Metall.* **1982**, *30*, 841–851. [CrossRef]
23. Zhou, Y.Q.; Ai, L.Q.; Hong, L.K. Experimental study on Decarburization of Fe-C-Si alloy in solid state. *J. Mater. Metall.* **2023**, *22*, 45–51+66.
24. Liu, Y.B.; Zhang, W.; Tong, Q. Effects of Si and Cr on complete decarburization behavior of high carbon steels in atmosphere of 2 vol.% $O_2$. *J. Iron Steel Res.* **2016**, *23*, 1316–1322. [CrossRef]

25. Li, N. *Molecular Dynamics Simulation of Tensile Behavior of $Ti_2Cu$ Inter Metallics*; Changchun University of Technology: Changchun, China, 2022.
26. Wang, W.T.; Wang, H.; Lu, X.G. Molecular dynamics simulation of tracer diffusion coefficient of liquid Ni Al Binary Alloy. *Shanghai Met.* **2021**, *43*, 108–111.
27. Tapasa, K.; Barashev, A.V.; Bacon, D.J. Computer simulation of carbon diffusion and vacancy–carbon interaction in α-iron. *Acta Mater.* **2007**, *55*, 1–11. [CrossRef]
28. Ye, Y.J.; Zhang, L.T.; Cheng, L.F.; Xu, Y.D. Molecular dynamics simulation of diffusion behavior of amorphous Si-B-C system. *J. Inorg. Mater.* **2006**, *21*, 843–847.
29. Sun, Y.C. *Molecular Dynamics Simulation of Oxygen Isotope Diffusion in Molten $SiO_2$*; Nanjing University: Nanjing, China, 2012.
30. Shi, X.T.; Gao, X.G.; Zhong, Y. Molecular dynamics simulation of oxygen diffusion in molten $SiO_2$. *J. Nanjing Univ. Aeronaut. Astronaut.* **2021**, *53*, 821–826.
31. Wang, G.L.; Yu, A.B.; Yuan, J.D. Molecular dynamics simulation of diamond/Ti hot press diffusion process. *Surf. Technol.* **2022**, *51*, 435–442.
32. Timmerscheidt, A.T.; Appen, V.J.; Dronskowski, R. A molecular-dynamics study on carbon diffusion in face-centered cubic iron. *Comput. Mater. Sci.* **2014**, *91*, 235–239. [CrossRef]
33. Aslam, I.; Baskes, M.I.; Dickel, D.E. Thermodynamic and kinetic behavior of low-alloy steels: An atomic level study using an Fe-Mn-Si-C modified embedded atom method (MEAM) potential. *Materialia* **2019**, *8*, 100473. [CrossRef]
34. Allen, M.P.; Tildsley, D.J. *Computer Simulation of Liquid*; Oxford University Press: New York, NY, USA, 1987; Volume 28.
35. Honeycutt, D.J.; Andersen, H.C. Molecular dynamics study of melting and freezing of small Lennard–Jones clusters. *J. Chem. Phys.* **1987**, *91*, 4950–4963. [CrossRef]
36. Wang, S.G.; Liu, C.X.; Jian, Z.Y. Molecular dynamics simulation of diffusion coefficient of Al Cu alloy. *J. Xi'an Technol. Univ.* **2018**, *38*, 559–564.
37. Ye, Q.; Zhao, S.J. Molecular dynamics simulation of continuous cooling transformation in Fe based alloys. *Shanghai Met.* **2012**, *34*, 9–12.
38. Hou, Y.B. *Decarburization of Fe-C Alloy by Staged Heating in CO-$CO_2$ Atmosphere*; North China University of Technology: Beijing, China, 2021.
39. Zhou, Y.Q.; Ai, L.Q.; Hong, L.K. Evolution of oxide layer in solid decarburization process of Fe-C-Si alloy[J/OL]. *Steel* **2023**.
40. Luo, H.W.; Xiang, R.; Chen, L.F. Modeling decarburization kinetics of grain-oriented silicon steel. *Chin. Sci. Bull.* **2014**, *59*, 1778–1783. [CrossRef]
41. Yang, S.Z. *Study on Decarburization of Grain-Oriented Silicon Steel*; Wuhan University of Science and Technology: Wuhan, China, 2016.

processes

MDPI

*Article*

# Study on Mushy Zone Coefficient in Solidification Heat Transfer Mathematical Model of Thin Slab with High Casting Speed

**Zhijun Ding [1], Yuekai Xue [1,*], Limin Zhang [2,*], Chenxiao Li [1,*], Shuhuan Wang [1] and Guolong Ni [1]**

1 College of Metallurgy and Energy, North China University of Science and Technology, Tangshan 063009, China; wshh88@ncst.edu.cn (S.W.)
2 College of Mechanical Engineering, Tangshan Polytechnic College, Tangshan 063299, China
* Correspondence: xueyuekai965@163.com (Y.X.); zhanglimindota1@163.com (L.Z.); lichenxiao34@163.com (C.L.)

**Abstract:** When the casting speed of the thin slab continuous caster is increased, the ratios of the solid and liquid phases in the solidification front of the molten steel in the mold change, which affects the thickness of the solidified shell. In order to accurately calculate the thickness of the solidified shell and determine the value range of the mushy zone coefficient suitable for the mathematical model of solidification heat transfer at high casting speed, this paper established the solidification heat transfer mathematical model in thin slab funnel mold, and the influence of different mushy zone coefficients on the accuracy of solidification heat transfer mathematical model was analyzed and compared with the actual solidified shell thickness. The results showed that, when the casting speed was increased to 4~6 m/min and the coefficient of the mush zone coefficient was $3 \times 10^8$~$9 \times 10^8$ kg/(m$^3$·s), the thickness of solidified shell calculated by the solidification heat transfer model was in good agreement with that measured in practice. The research in this paper provides an important reference for the establishment of the solidification heat transfer mathematical model at high casting speed in the future.

**Keywords:** high casting speed; solidification heat transfer mathematical model; mushy zone coefficient; solidified shell; thickness

**Citation:** Ding, Z.; Xue, Y.; Zhang, L.; Li, C.; Wang, S.; Ni, G. Study on Mushy Zone Coefficient in Solidification Heat Transfer Mathematical Model of Thin Slab with High Casting Speed. *Processes* **2023**, *11*, 3108. https://doi.org/10.3390/pr11113108

Academic Editor: Carlos Sierra Fernández

Received: 4 October 2023
Revised: 25 October 2023
Accepted: 27 October 2023
Published: 30 October 2023



## 1. Introduction

In the process of continuous casting, the flow condition of molten steel in the mold, the temperature distribution and the growth of the initial solidification shell all have great influence on the quality of the slab. With the improvement of the computer computing performance and the development of commercial software, numerical simulation has become an important means to study the transmission behavior of the continuous casting process, in which finite volume software Fluent is widely used in the numerical simulation of physical phenomena related to molten steel flow in the continuous casting process.

The solidification process of molten steel is carried out in a temperature range (liquidus temperature to solidus temperature), which is macroscopically manifested as a solid–liquid two-phase zone—that is, a mushy zone [1,2]. The forming slab generally has a surface chill layer, a central equiaxed crystal region and a middle columnar crystal region. It can be seen that, in the solidification process, the mushy region consists of columnar crystals or (and) equiaxed crystals and a liquid phase group. Among them, some columnar crystals are interrupted by the flowing molten steel and move with the molten steel; the rest of the columnar crystals move with the solidified shell and grow forward, and the equiaxed crystals are deposited with the flow of the liquid steel or downward. Columnar crystals and equiaxed crystals can hinder the flow of molten steel in the mushy zone and then affect the heat transfer and solidification process of molten steel. Two methods are generally used to deal with the flow of molten steel in the mushy zone, namely the variable viscosity method and the Darcy source term method. Because the viscosity data of the mushy zone are not easy to obtain when using the variable viscosity method, the Darcy source term method is widely used. The dendrites in the mushy zone are regarded as porous media that move along the solidification front and are filled with the liquid phase and a fraction of the solid phase changes continuously by the Darcy source term method [3–7]. Among them, permeability is an important parameter to reflect the transmission performance of porous media, which can reflect the flow characteristics of the liquid phase with a certain momentum in porous media, mainly affected by the structural characteristics and physical properties of porous media [8,9]. Pfeiler et al. [10,11] calculated the permeability using the Blake–Kozeny law and considered that the permeability was a function of the primary dendrite spacing. The expression given by Beckermann et al. [12,13] also believed that the permeability is related to the primary dendrite spacing. However, most scholars [14–16] use the Carman–Kozeny equation to deal with permeability and regard permeability as a function of secondary dendrite spacing. The mushy zone coefficient ($A_{mush}$) is a physical quantity related to permeability. For general fluid software such as Fluent, Open FOAM, COMSOL, etc., in the process of establishing the solidification heat transfer model, the $A_{mush}$ introduces a damping term into the porous medium model. The $A_{mush}$ has a great influence on the numerical calculation of the flow and solidification behavior of molten steel, but the correct expression of the $A_{mush}$ is rarely given in the literature. Most scholars ignore the effect of the mushy zone coefficient on the solidification heat transfer model and usually use the default value of $1 \times 10^5$ kg/($m^3$·s) as the $A_{mush}$, resulting in the calculation results being inconsistent with reality [17,18]. Some scholars, Refs. [19–21], use a smaller $A_{mush}$, resulting in the calculation results of overheating dissipation too fast, the liquid surface is too cold and so on. Trindade et al. [22,23] adopted the default value of Fluent $1 \times 10^5$ kg/($m^3$·s) as the $A_{mush}$, resulting in the liquid phase ratio of the steel at the center of the mold outlet being lower than 1. Hietanen et al. [24,25] studied the effect of $A_{mush}$ at $1 \times 10^3$~$1 \times 10^7$ kg/($m^3$·s) on the solidification of molten steel, and the results showed that, when the $A_{mush}$ value was low, the temperature of molten steel dropped too fast, which was inconsistent with reality. When the $A_{mush}$ was $1 \times 10^7$ kg/($m^3$·s), the calculated results were more realistic. However, when the $A_{mush}$ is increased to $1.5 \times 10^7$ kg/($m^3$·s), the calculation is difficult to converge. Wang Qiangqiang [26] pointed out that, when the $A_{mush}$ was $1 \times 10^5$ kg/($m^3$·s), the molten steel temperature in the mold was mostly lower than the liquidus temperature, and there was almost no formed solidified shell near the impact zone of the narrow surface flow. It can be seen that it is very important to select the appropriate $A_{mush}$ for the understanding and grasp of the flow and solidification phenomenon of molten steel in the mold.

At present, with the increasing of the casting speed of the thin slab continuous casting caster, the range of $A_{mush}$ values satisfying the mathematical model of solidification heat transfer at high casting speed is worth discussing. Therefore, this study only considers the scope of application of $A_{mush}$ in the solidification process of thin slab continuous casting with high casting speed and focuses on studying the temperature field of molten steel in the mold and the law of solidified shell growth. Moreover, the thickness of the solidified shell is compared with that calculated by numerical simulation through actual measurement. The value range of $A_{mush}$ in the heat transfer solidification model with high casting speed is determined, which provides a theoretical basis and reference for industrial production.

## 2. Experiment and Simulation

### *2.1. Basic Assumptions*

Based on the complex metallurgical physical phenomenon in the thin slab continuous casting mold, combined with the characteristics of the thin slab continuous casting process, the following basic assumptions are made about the flow of molten steel and solidification heat transfer in the funnel mold to simplify the calculation process [27,28]:

(1) The flow process of molten steel is a viscous incompressible fluid, and the effects of the phase change, vibration, protective slag and solidification shrinkage are not considered.

(2) The effects of the fluctuation of the mold meniscus, taper, phase transformation of Fe-C alloy, solidified shell shrinkage and vibration on solidification heat transfer are ignored.
(3) The continuous casting process is regarded as a steady state, and the turbulent effect of molten steel is simulated by the low Reynolds number standard k-ε model.
(4) The mushy zone is regarded as a porous medium, and the flow in the mushy zone obeys Darcy's law.
(5) The movement speed of the solidified shell is consistent with the casting speed.

*2.2. Governing Equation*

In this part, the enthalpy–porous medium method is used to deal with the flow, heat transfer and solidification of molten steel in a funnel mold. The differential equation for heat transfer control is shown in (1).

$$\rho C_P \frac{dT}{dt} + \nabla \cdot \left( \rho \vec{v} H \right) = \nabla \cdot (\lambda \nabla T) + S_h \tag{1}$$

Among them,

$$H = h_{ref} + \int_{T_{ref}}^{T} C_p dT + \beta L_N \tag{2}$$

Porosity is equal to the liquid phase volume fraction $\beta$, as shown in (3).

$$\beta = \begin{cases} 0 \ (T < T_s) \\ \frac{T - T_s}{T_l - T_s} \ (T < T_s) \\ 1 \ (T > T_l) \end{cases} \tag{3}$$

Formula: $T_l$ is the liquidus temperature, K; $T_s$ is the solidus temperature, K; $S_h$ is the heat source term, J/(m$^3$·s); $C_p$ is the specific heat capacity, J/(kg·K); $L_N$ is the latent heat of the solidification, J/kg; the reference enthalpy is $h_{ref}$ with respect to the $T_{ref}$; $H$ is the total enthalpy of the system, J; $\beta$ is the liquid phase volume fraction.

The liquidus temperature $T_L$ and solidus temperature $T_S$ of molten steel are calculated by the empirical Formulas (4) and (5).

$$T_L = 1539 - 70w[\%C] + 8w[\%Si] + 5w[\%Mn] + 30w[\%P] - 25w[\%S] + 4w[\%Ni] - 1.5w[\%Gr] \tag{4}$$

$$T_s = 1536 - 415.3w[\%C] - 12.3w[\%Si] - 6.8w[\%Mn] - 124.5w[\%P] - 183.9w[\%S] - 4.3w[\%Ni] - 1.4w[\%Gr] - 4.1w[\%Al] \tag{5}$$

Formula: $w$ is the mass fraction, %.

The basic principle of the enthalpy–porous medium method is to regard the metal solid–liquid two-phase region as a porous medium and to characterize the proportion of the liquid phase in the unit by calculating the temperature of each unit. The sum of the proportion of the liquid phase in the two phases is between 0 and 1, and the porosity is different in each cell, so the fluid resistance settings are different [29].

The momentum and turbulent kinetic energy lost during the solidification of the fluid will be added to the momentum and turbulence equations in the form of source terms.

The source term to be added to the momentum equation is shown in Equation (6).

$$S_m = -\frac{(1-\beta)^2}{(\beta^3 + \varepsilon)} A_{\text{mush}} \left( \vec{u} - \vec{u_p} \right) \tag{6}$$

The source term to be added in the turbulence equation is shown in Equation (7).

$$S_q = -\frac{(1-\beta)^2}{(\beta^3+\varepsilon)} A_{mush}\left(\vec{u} - \vec{u_p}\right) \quad (7)$$

Formula: $\vec{u_p}$ is the casting speed, m/s; $A_{mush}$ is the mushy zone coefficient, kg/(m$^3$·s).

The size of the $A_{mush}$ will affect the flow in the two-phase region. To prevent the divisor from being 0, set the ε to 0.001.

*2.3. Boundary Conditions and Physical Parameters*

This section studies the solidification heat transfer process of molten steel in the mold. The main process parameters are as follows: grade of the steel is SS400, its main component is $0.18\% \leq C \leq 0.20\%$, $S \leq 0.005\%$ and $P \leq 0.018\%$, with a section (width × thickness) of $1520 \times 90$ mm$^2$. Other assumptions are as follows:

(1) The superheat of molten steel is 20 K, and the inlet temperature of molten steel is 1823 K, and it is kept constant.
(2) Due to the good heat preservation effect of the liquid surface mold slag, the free liquid surface is set as the adiabatic condition, and the submerged entry nozzle (SEN) wall is also set as the adiabatic wall.
(3) The temperature gradient perpendicular to the symmetric plane is zero.
(4) The moving speed of a narrow face and wide face is consistent with the casting speed, which is given by Patch. The boundary heat flux of the mold area is given according to the real-time monitoring data of the actual production computer; that is, the boundary condition of the mold is defined by the second type of boundary condition of heat transfer.

The heat transfer and solidification of molten steel are simulated by the commercial software Fluent. The initial inlet velocity of the model was calculated according to the inner diameter of the SEN (80 mm), the casting speed and the size of the funnel mold. The numerical calculation method is used with the standard K-ε turbulence model, opening the energy equation and calculating in combination with the solidification and melting model. The superheat and specific heat of molten steel are added through the operation interface settings, and the heat flow is applied through the user-defined function. The hydrothermal physical properties of molten steel used in the simulation are shown in Table 1 [30]. It should be noted that, since the two-phase region is approximately treated as a porous medium, the heat transfer in the two-phase region can be approximately treated as the heat transfer in a porous medium.

**Table 1.** Related parameters for the solidification heat transfer model.

| Thermophysical Property Parameters | Value |
|---|---|
| Molten steel density | 7020 (kg/m$^3$) |
| Molten steel viscosity | 0.0064 (Pa·s) |
| Casting speed | (4, 5, 6) m/min |
| Initial value of velocity (casting speed: (4, 5, 6) m/min) | (1.82, 2.27, 2.72) m/s |
| Turbulent kinetic energy (casting speed: (4, 5, 6) m/min) | (6.32, 9.31, 1.28) m$^2$/s$^2$ |
| Turbulent dissipation rate (casting speed: (4, 5, 6) m/min) | 1.48, 2.63, 4.24 |
| Superheat of molten steel | 20 K |
| Solidus temperature ($T_S$) | 1763 K |
| Liquidus temperature ($T_L$) | 1803 K |
| Specific Heat Capacity ($Cp$) | 680 J/(kg·K) |
| Latent heat of solidification ($L_N$) | 270,000 J/kg |
| Wide surface heat flux ($\bar{q}_w$ : 4 m/min, 5 m/min, 6 m/min) | (2.0, 2.1, 2.2) × 10$^6$ W/m$^2$ |
| Narrow surface heat flux ($\bar{q}_n$: 4 m/min, 5 m/min, 6 m/min) | (1.8, 1.9, 2.0) × 10$^6$ W/m$^2$ |
| Specific heat capacity ($C_W$) | 4.2 × 10$^3$ J/(kg·K) |

### 2.4. Grid and Computing Domain

The three-dimensional (3D) geometric model of two-phase flow 1/2 combined with a funnel mold and the SEN is shown in Figure 1. The fluid domain is molten steel.

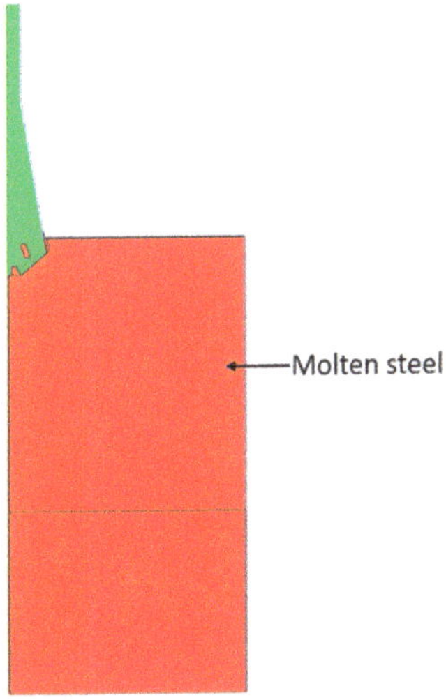


**Figure 1.** Three-dimensional geometric model of mold.

The MESH module in ANSYS software is used to mesh the 3D geometric model of SEN and the funnel mold. Due to the complexity of the structure of the SEN and the funnel mold, the model adopts hexahedral nonstructural mesh division. In order to consider the accuracy of the calculation results and the stability of the calculation process, the mesh encryption process was carried out at the molten steel surface of the fluid domain in the mold, and the total number of the mesh was about 1.4 million. The reflux at the outlet of the mold seriously affects the calculation accuracy and convergence. After repeated trial and error, the length of the calculation domain of the mold was determined to be 2200 mm. The local grid division is shown in Figure 2.

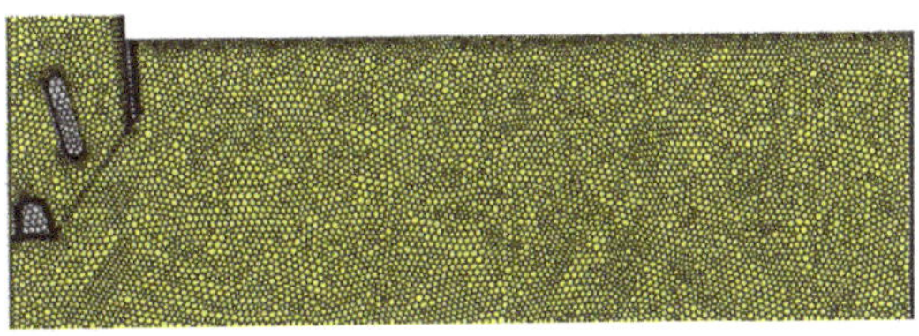

**Figure 2.** Local mesh division of the funnel mold geometric model.

### 2.5. Numerical Methods

Considering that there are many coupling models, the residual curve is prone to oscillations and difficult to converge, so the Couple algorithm is used for transient calculation, where the time step gradually increases from 0.001 s to 0.005 s. The momentum equation adopts the second-order upwind discrete scheme. Convergence is considered when the convergence standard is set as $\leq 1.0 \times 10^{-5}$, and the ratio between the difference of inlet and outlet flow and the value of inlet flow is $\leq 1\%$.

### 2.6. Test Scheme

The expression of $A_{mush}$ is shown in Equation (8).

$$A_{mush} = \frac{u_l}{K} \quad (8)$$

Formula: $K$ is the permeability; $u_l$ is the laminar viscosity, Pa·s.

In the solidification process of molten steel, many scholars and experts have obtained different permeability expressions through experiments. The expression given by Beckermann shows that the permeability is related to the primary dendrite spacing, as shown in Equation (9).

$$K = 6 \times 10^{-4} \lambda_1^2 \frac{\beta^3}{(1-\beta)^2} \quad (9)$$

Formula: $\lambda_1$ is the primary dendrite spacing size, m.

Figure 3 shows the measured relationship between the primary dendrite spacing of a continuous casting slab and the thickness of the slab. When $\beta = 0.3$, the variation rule of $A_{mush}$ with the thickness of the slab can be obtained. From the surface of the slab to the thickness center, $A_{mush}$ gradually decreases from $3.1 \times 10^8$ kg/(m$^3$·s) to $1.7 \times 10^8$ kg/(m$^3$·s).

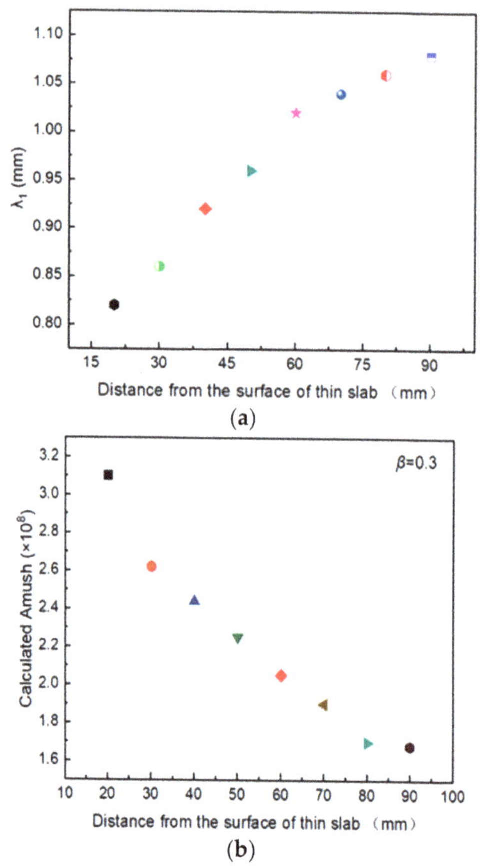


**Figure 3.** The change of primary dendrite spacing and $A_{mush}$ thickness in the casting slab: (**a**) primary dendrite spacing; (**b**) $A_{mush}$ value.

Previous studies have found that permeability is closely related to the secondary dendrite spacing and solidification state of a continuous casting slab, as shown in Equation (10).

$$K = \frac{\lambda_2^2 u_l \beta^3}{180(1-\beta)^2} \tag{10}$$

Formula: $\lambda_2$ is the secondary dendrite spacing, m.

Due to the selective crystallization in the solidification process, some components are enriched in the liquid phase, which affects the liquidus temperature. Therefore, temperature and solute redistribution need to be considered in order to accurately predict the solidification process, and permeability is an important parameter for accurately predicting tissue properties [31]. In order to make the calculation results of the solidification heat transfer model of thin slabs with high casting speed more accurate, based on previous studies, this paper set six groups of parameters for the $A_{mush}$ value in the solidification heat transfer model of thin slabs with high casting speed, as shown in Table 2. The influence of the $A_{mush}$ value on the solidification heat transfer model was quantitatively studied. The paper provides a reference for the coefficient $A_{mush}$ of the mushy zone to establish the solidification heat transfer model of thin slabs with high casting speed in the future.

**Table 2.** $A_{mush}$ value ranges.

| Number | Amush (kg/(m$^3$·s)) |
|---|---|
| A | $1 \times 10^5$ |
| B | $1 \times 10^7$ |
| C | $1 \times 10^8$ |
| D | $3 \times 10^8$ |
| E | $6 \times 10^8$ |
| F | $9 \times 10^8$ |

## 3. Results and Discussion

### *3.1. Description of the Effect of $A_{mush}$ on Heat Transfer and Solidification of Molten Steel*

As can be seen from Table 3, when the $A_{mush}$ is set to a smaller value of $1 \times 10^5$ kg/(m$^3$·s), the liquid-phase fraction of the molten steel near the liquid level at the SEN outlet is about 0.8, and between the solidification front of the narrow surface of the mold and the molten steel, blue bars are obviously visible from the mold liquid level to the SEN outlet, indicating that the mushy zone is large and there is almost no formed solidified shell on the narrow face.

When the $A_{mush}$ is $1 \times 10^7$ kg/(m$^3$·s), a large part of the region with a low liquid fraction near the liquid surface of the mold and SEN outlet still exists, but the solidified shell thickness gradually increases from below the meniscus, and the solidified shell also begins to solidify on the impact path of the flow stream, but the mushy zone is still obvious. When the $A_{mush}$ is $1 \times 10^8$ kg/(m$^3$·s), the simulation results are similar to those when $A_{mush}$ is $1 \times 10^7$ kg/(m$^3$·s), but the mushy zone is still clearly visible in the lower part. When the $A_{mush}$ is $\geq 3 \times 10^8$ kg/(m$^3$·s), the lower liquid fraction region disappears, but when the $A_{mush}$ is $3 \times 10^8$ kg/(m$^3$·s), the solidified shell thickness is not obvious in the upper part of the mold, and the solidified shell formation is thin, but the mushy zone is significantly reduced. When the $A_{mush}$ is $9 \times 10^8$ kg/(m$^3$·s), the thickness of the solidified shell starts from the liquid side down, and the thickness increases obviously. In order to further determine the numerical range of the $A_{mush}$, the numerical simulation results were compared with the actual measurement of solidified shell thickness, and the value of the $A_{mush}$ was determined.

**Table 3.** Picture of the solid- and liquid-phase distribution in the middle surface of the funnel mold thickness. Unit: kg/(m$^3$·s).

| $A_{mush}$ | A: $1 \times 10^5$ | B: $1 \times 10^7$ | C: $1 \times 10^8$ |
| --- | --- | --- | --- |
| Picture of solid- and liquid-phase distribution |  |  |  |
| $A_{mush}$ | D: $3 \times 10^8$ | E: $6 \times 10^8$ | F: $9 \times 10^8$ |
| Picture of solid and liquid phase distribution |  |  |  |

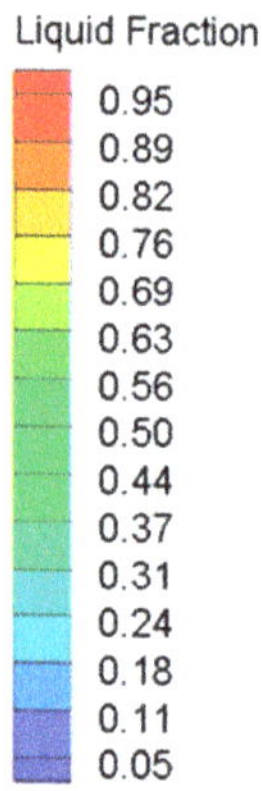


### 3.2. *$A_{mush}$ Parameter Determination and Heat Transfer Solidification Model*

#### 3.2.1. Election of $A_{mush}$ Parameters

At present, there are three methods to determine the thickness of a solidified shell: namely, the empirical method, the heat balance method and the test method. Due to the restriction conditions of the empirical method and heat balance method, such as open casting, protective casting, molten steel composition, liquidus temperature and solid-phase temperature, they have great influence on the thickness of a solidified shell. At present, the test determination method should be widely used, such as "punching and draining liquid method", "nail method", "trace method" and other methods, but these methods are difficult to measure, and the measurement accuracy is relatively low. In recent years, the method of measuring solidified shell thickness using the test method has developed rapidly, and the method is simple and has high precision. Therefore, this part uses the test method to measure the thickness of the solidified shell—that is, using the solidified shell that is leaked in the production process, cutting the solidified shell along different heights and then measuring the average thickness of each position.

In this part, a continuous casting slab with a cross-section (width × thickness) of 1520 × 90 mm$^2$, SEN immerse depth of 160 mm and casting speed of 5 m/min was cut and measured. The measuring positions were at one-fourth of the width of the solidified shell and at the middle of the narrow surface along the thickness, respectively. The distance from the liquid surface of the mold was 0.3 m, 0.7 m and 0.9 m. Then, the thickness of the solidified shell was measured at the symmetric position of the outer arc side and inner arc side, each position was measured three times and the average value of the thickness data was taken. The measurement process is shown in Figure 4.

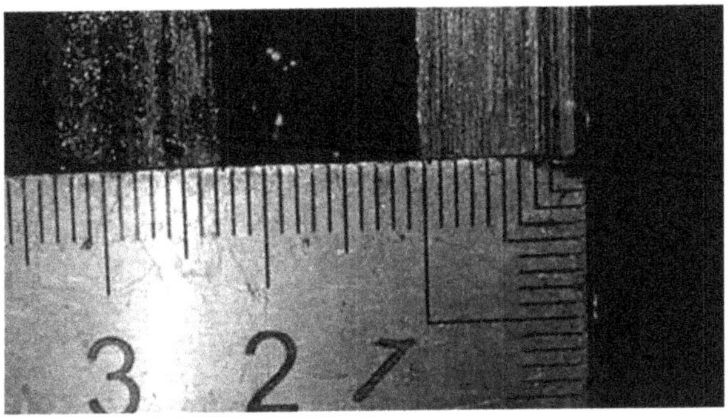


**Figure 4.** Measurement of the solidified shell of molten steel leakage.

The thickness of the solidified shell at the three measuring points on the narrow surface was about 4.2 mm, 6.9 mm and 10.2 mm from the liquid side down. The thickness of the wide solidified shell was about 4.0 mm, 8.2 mm and 10.6 mm in turn. As can be seen in Figure 4, no matter the narrow face or the wide face, the position of the measuring point from the liquid level down to the first measuring point and the third measuring point were wrapped by three solidified shell thickness curves calculated by numerical simulation with $A_{mush}$ values of D, E and F, followed by C, and A and B being the worst. The second measurement point was in poor agreement with the numerical simulation calculation of the solidified shell thickness, because the measurement point was located at the impact position of the main jet and the turbulent flow area near the main jet, and the formation of the solidified shell thickness had great instability. In summary, when the value of $A_{mush}$ ranges from $3 \times 10^8$ to $9 \times 10^8$ kg/(m$^3$·s), the solidified shell thickness calculated by the mathematical model is more accurate. Therefore, when calculating the solidification mathematical model in this section, the value of the $A_{mush}$ was $6 \times 10^8$ kg/(m$^3$·s).

### 3.2.2. Verification of the Heat Transfer Solidification Model

According to the curve of the solidified shell thickness calculated when the value of $A_{mush}$ is $6 \times 10^8$ kg/(m$^3$·s), it can be seen that the solidified shell thickness shows a gradual increasing trend along the mold height direction, which can be said to be linear growth, and the solidified shell thickness reaches about 14 mm at the outlet position. However, as can be seen from the narrow face comparison diagram in Figure 5a, the growth rate of the solidified shell thickness decreased significantly at a distance of 600~800 mm from the meniscus of the mold, and a turning point appeared. The thickness of the solidified shell was only about 5.2 mm at the thinnest point, and then, the solidified shell thickness began to increase. Both the measured values and the simulated values indicated that the high-speed jet in the mold impaction on the narrow surface and the probability of steel leakage and slag inclusion was greater there. The reason is that there is a large temperature gradient and intense heat exchange at and near the impact point of the high-speed molten steel flow, and the strong scour effect of the high-speed molten steel flow makes heat replenishment there in time and the temperature is high, resulting in a growth lag of the solidified shell. The width face comparison diagram of the mold shows that the thickness of the solidified shell fluctuates greatly in the area of about 400~900 mm, but the overall growth is relatively stable. The reason is that the molten steel flowing out of the SEN outlet at high speed is very active in the flow path area, and because of the wide and thin section of the mold, the heat replenishment of the wide surface solidified shell is timely, resulting in slow solidification of the molten steel and a thin solidified shell. The flow of the molten steel in the center of the vortex is poor, and the temperature is low, which leads to the thickness of the solidified shell. The measurement results at 0.7 m away from the meniscus are not in good agreement with the mathematical simulation results, because the molten steel flow here is disordered, and there is no obvious rule. On the whole, the simulated calculation of the solidified shell thickness in this study is basically consistent with the actual measurement of the solidified shell thickness, which indicates that the solidification mathematical model established in this study has high reliability.

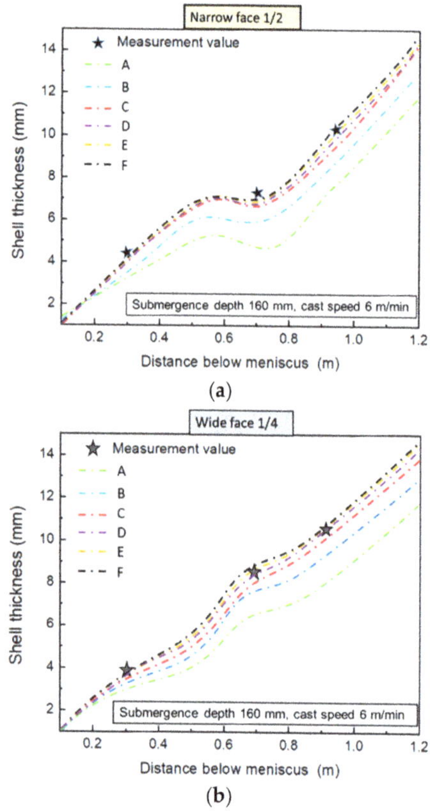


**Figure 5.** Comparison between the calculated value of the solidified shell thickness and the measured value: (**a**) thickness of the narrow face solidified shell; (**b**) thickness of the wide face solidified shell.

## 4. Conclusions

By developing a 3D solidification heat transfer mathematical model of a thin slab continuous casting mold, the effects of the mushy zone coefficient $A_{mush}$ on the flow, heat transfer and solidification process of molten steel in the two-phase zone at high casting speed were analyzed, and the thickness of the solidified shell was compared with that of an actual thin slab. We came to the following conclusions:

1. The smaller the mushy zone coefficient is, the higher the liquid–solid-phase ratio of steel near the liquid level, which is inconsistent with the actual situation. The larger the mushy zone between the solid and liquid in the front of the solidified shell, the thinner the solidified shell. The larger the mushy zone coefficient is, the lower the

solid-phase ratio near the liquid surface, and the smaller the mushy zone area is, the thicker the solidified shell.
2. The mushy zone coefficient of $3 \times 10^8$~$9 \times 10^8$ kg/(m$^3$·s) can reliably reveal the actual solidification phenomenon in the mold of thin slab continuous casting at high casting speed.

**Author Contributions:** All authors contributed to the study's conception and design. Material preparation, review and editing were performed by Z.D., Y.X., L.Z. and S.W. The first draft of the manuscript was written by Z.D. All authors commented on previous versions of the manuscript. All authors have read and agreed to the published version of the manuscript.

**Funding:** This research received no external funding.

**Institutional Review Board Statement:** Not applicable.

**Data Availability Statement:** Data can be obtained from the corresponding author upon reasonable request.

**Conflicts of Interest:** The authors declare no conflict of interest.

## References


1. Xuan, M.; Chen, M. Numerical investigation for the influence of turbulent heat transfer of mushy zone on shell growth in the slab mold: Casting and solidification. *ISIJ Int.* **2022**, *62*, 142–148. [CrossRef]
2. Ghaemifar, S.; Mirzadeh, H. Refinement of banded structure via thermal cycling and its effects on mechanical properties of dual phase steel. *Steel Res. Int.* **2018**, *89*, 1700531. [CrossRef]
3. Fujisaki, K. Magnetohydrodynamic solidification calculation in Darcy flow. *IEEE Trans. Magn.* **2003**, *39*, 3541. [CrossRef]
4. Sun, H.B.; Zhang, J.Q. Study on the macrosegregation behavior for the bloom continuous casting: Model development and validation. *Metall. Mater. Trans.* **2014**, *45*, 1133. [CrossRef]
5. Masud, A.; Hughes, T.J.R. A stabilized mixed finite element method for darcy flow. *Comput. Methods Appl. Mech. Eng.* **2002**, *191*, 4341–4370. [CrossRef]
6. To, V.T.; To, Q.D.; Monchiet, V. On the Inertia Effects on the Darcy Law: Numerical Implementation and Confrontation of Micromechanics-Based Approaches. *Transp. Porous Media* **2016**, *111*, 171–191. [CrossRef]
7. Bernardi, C.; Maarouf, S.; Yakoubi, D. Spectral discretization of darcy's equations coupled with the heat equation. *IMA J. Numer. Anal.* **2015**, *36*, 1193–1216. [CrossRef]
8. Li, H.M.; Liang, Y.F.; Chen, J.F.; Li, H.G.; Chen, X.Q. Relationship between pore structural characteristics and physical-mechanical properties of sandstone in shendong mining area. *J. Henan Polytech. Univ.* **2018**, *37*, 9–16.
9. Kumar, S.S.; Rubio, E.J.; Noor-A-Alam, M.; Martinez, G.; Manandhar, S.; Shutthanandan, V. Structure, morphology, and optical properties of amorphous and nanocrystalline gallium oxide thin films. *J. Phys. Chem. C* **2013**, *117*, 4194–4200. [CrossRef]
10. Pfeiler, C.; Thomas, B.G.; Wu, M. Solidification and particleentrapment during continuous casting of steel. *Steel Res. Int.* **2008**, *79*, 599–607. [CrossRef]
11. Pacella, H.E.; Eash, H.J.; Frankowski, B.J.; Federspiel, W.J. Darcy permeability of hollow fiber bundles used in blood oxygenation devices. *J. Membr. Sci.* **2011**, *382*, 238–242. [CrossRef] [PubMed]
12. Gu, J.P.; Beckermann, C. Simulation of convection and macroseg-regation in a large steel ingot. *Metall. Mater. Trans. A* **1999**, *30*, 1357–1366. [CrossRef]
13. Aboutalebi, M.R.; Guthrie, R.I.L.; Seyedein, S.H. Mathematical modeling of coupled turbulent flotw and solidification in a single belt caster with electromagnetic brake. *Appl. Math. Model.* **2007**, *31*, 1671–1689. [CrossRef]
14. Madhani, S.P.; D'Aloiso, B.D.; Frankowski, B.; Federspiel, W.J. Darcy permeability of hollow fiber membrane bundles made from membrana polymethylpentene fibers used in respiratory assist devices. *Asaio J.* **2016**, *62*, 329–331. [CrossRef]
15. Valdes-Parada, F.J.; Ochoa-Tapia, J.A.; Alvarez-Ramirez, J. Validity of the permeability carman–kozeny equation: A volume averaging approach. *Phys. A Stat. Mech. Appl.* **2009**, *388*, 789–798. [CrossRef]
16. Costa, A. Permeability-porosity relationship: A reexamination of the kozeny-carman equation based on a fractal pore-space geometry assumption. *Geophys. Res. Lett.* **2006**, *33*, 5134. [CrossRef]
17. Ceretani, A.N.; Tarzia, D.A. Determination of one unknown thermal coefficient through a mushy zone model with a convective overspecified boundary condition. *Math. Probl. Eng.* **2015**, *21*, 637852. [CrossRef]
18. Peng, P.; Lu, L.; Zheng, W.; Wang, J. Macrosegregation and channel segregation formation of faceted mushy zone in solidification of sn-ni hypereutectic alloy in a temperature gradient. *Mater. Chem. Phys.* **2021**, *264*, 124447. [CrossRef]
19. Sun, H.B.; Zhang, J.Q. Effect of feeding modes of molten steel on the mould metallurgical behavior for round bloom casting. *ISIJ Int.* **2011**, *51*, 1657. [CrossRef]
20. He, M.L.; Wang, N.; Chen, M.; Xuan, M. Physical and numerical simulation of the fluid flow and temperature distribution in bloom continuous casting mold. *Steel Res. Int.* **2017**, *88*, 1600447. [CrossRef]

21. Trindade, L.B.; Nadalon, J.E.A.; Contini, A.C.; Barroso, R.C. Modeling of solidification in continuous casting round billet with mold electromagnetic stirring (M-EMS). *Steel Res. Int.* **2017**, *88*, 1600319. [CrossRef]
22. Mohamed, F.; Eames, P.C. Numerical investigation of the influence of mushy zone parameter Amush on heat transfer characteristics in vertically and horizontally oriented thermal energy storage systems. *Appl. Therm. Eng.* **2019**, *151*, 90–99.
23. Won, Y.M.; Thomas, B.G. Simple model of microsegregation during solidification of steels. *Metall. Mater. Trans. A* **2001**, *32*, 1755–1767. [CrossRef]
24. Hietanen, P.T.; Louhenkilpi, S.; Yu, S. Investigation of solidification, heat transfer and fluid flow in continuous casting of steel using an advanced modeling approach. *Steel Res. Int.* **2017**, *88*, 1600355. [CrossRef]
25. Ji, Y.; Tang, H.Y.; Lan, P.; Shang, C.; Zhang, J. Effect of dendritic morphology and central segregation of billet castings on the microstructure and mechanical property of hot-rolled wire rods. *Steel Res. Int.* **2017**, *88*, 1600426. [CrossRef]
26. Wang, Q.Q. *Stundy on the Multiphase Flow, Heat Transfer and Solidification, Motion and Entrapment of Inclusions During Continuous Casting*; University of Science and Technology: Beijing, China, 2016.
27. Yue, Q.; Zhou, L.; Wang, J.J. Numerical simulation of the coupling of molten Steel flow, heat transfer and solidification in a slab mold with Double transverse Stability Electromagnetic Braking. *J. Process Eng.* **2011**, *11*, 187–192.
28. Zhang, L.M.; Xiao, P.C.; Wang, Y.; Liu, Z.X.; Zhu, L.G. Effect of EMBr on flow in slab continuous casting mold and industrial experiment of nail dipping measurement. *Metals* **2023**, *13*, 167. [CrossRef]
29. Yang, C.L. *Simulation Research on Solidification Heat Transfer and Film Formation of Protective Slag in Continuous Casting Mold*; Chongqing University: Chongqing, China, 2018.
30. Alizadeh, M. Correlation between the continuous casting parameters and secondary dendrite arm spacing in the mold region. *Mater. Lett.* **2013**, *91*, 146–149. [CrossRef]
31. Minakawa, S.; Samarasekera, I.V.; Weinberg, F. Centerline porosityin plate castings. *Metall. Trans. B* **1985**, *16*, 823–829. [CrossRef]

processes

MDPI

Article

# The Kinetic Mechanism of the Thermal Decomposition Reaction of Small Particles of Limestone at Steelmaking Temperatures

**Chenxiao Li [1], Yun Zhang [1], Yuekai Xue [1,*], Kaixuan Zhang [1], Shuhuan Wang [1], Huakang Sun [1] and Huaqing Xie [2,*]**

[1] Metallurgy and Energy College, North China University of Science and Technology, Tangshan 063210, China
[2] School of Metallurgy, Northeastern University, Shenyang 110819, China
* Correspondence: xueyuekai965@163.com (Y.X.); huaqing_2008@163.com (H.X.); Tel.: +86-150-3159-5965 (Y.X.); +86-024-8367-2216 (H.X.)

**Abstract:** Converter blowing limestone powder making slag steelmaking process has the advantages of low carbon and high efficiency, and can realize the resource utilization of $CO_2$ in the metallurgical process, which is in line with the development direction of green metallurgy. Based on a thermogravimetric-differential thermal analyzer, the kinetic mechanism of decomposition of small limestone at steelmaking temperatures was investigated by a modified double extrapolation method. The results showed that with a higher rate of heating, limestone decomposition lagged, and decomposition temperature increased. Furthermore, the smaller the limestone particle size, the lower the activation energy of decomposition. Compared with $N_2$, air, and $O_2$, small limestone powder used for converter blowing could complete more rapid decomposition, and the time required for decomposition shortened by about 1/3, although the decomposition temperature increased in the $CO_2$. The limestone decomposition rate increased and then decreased at low to high $CO_2$ partial pressures. With a limiting link, the inhibition was more significant under high $CO_2$ partial pressure, but the reaction can be fully completed by 1000 °C. The decomposition type modeled was stochastic nucleation and subsequent growth. As the partial pressure of $CO_2$ increased from 25% to 100%, the number of reaction stages, n, increased.

**Keywords:** converter steelmaking; small-particle limestone; decomposition kinetics; carbon dioxide partial pressure; low carbon

**Citation:** Li, C.; Zhang, Y.; Xue, Y.; Zhang, K.; Wang, S.; Sun, H.; Xie, H. The Kinetic Mechanism of the Thermal Decomposition Reaction of Small Particles of Limestone at Steelmaking Temperatures. *Processes* **2023**, *11*, 2712. https://doi.org/10.3390/pr11092712

Academic Editor: Carlos Sierra Fernández

Received: 28 August 2023
Revised: 8 September 2023
Accepted: 8 September 2023
Published: 11 September 2023



## 1. Introduction

The steel industry, as a highly intensive, highly industrialized, and wide-ranging industry, accounts for about five percent of the gross domestic product (GDP) of China and plays an indispensable role in stimulating economic and industrial development, making it a veritable pillar industry. At the same time, the rapid development of the iron and steel industry has brought about the environmental problems of high carbon emissions and high energy consumption. At present, the iron and steel industry is mainly based on the "blast furnace–converter" process of steelmaking. In the context of the "carbon neutrality and carbon peak" era, the development of the iron and steel industry is bound to gradually transform to low carbon and green. As an important part of the process, the converter steelmaking process is facing the serious challenge of carbon reduction and energy saving to realize deep carbon reduction [1–3]. Converter smelting links mostly use lime slag for steelmaking, but the production and transportation of lime for steelmaking produces a lot of pollution. In the process of converter application due to the low activity of lime, the dephosphorization rate is generally around 40–60% [4]. Part of the input of large pieces of lime is to a certain extent caused by slag difficulties [5–8], but the subsequent smelting also has a certain impact. The main component of limestone is $CaCO_3$, which is decomposed into CaO and $CO_2$ at high temperatures. CaO can be used as slagging material in the converter process, and the released $CO_2$ can participate in the melting pool reaction. For this reason, limestone is added into the converter as a slagging material. At the same time, the task of slag making and dephosphorization is accomplished; however, this can also be accomplished through the resource utilization of $CO_2$, which greatly reduces the cost of $CO_2$ emissions. Li et al. [9] proposed a converter limestone technique instead of lime-slagging steelmaking, capitalizing on the use of converter steelmaking conditions to achieve the production and application of high-activity lime. Compared with the traditional slagging steelmaking process, the use of limestone instead of lime-slagging steelmaking in converter steelmaking to meet the needs of smelting is based on an effective increase in the rate of dephosphorization. The dephosphorization process is also more stable and material consumption is reduced. Moreover, the $CO_2$ generated from the decomposition of limestone in the converter can react with the elements in the hot liquid iron to generate CO, which increases the amount of gas recovery and the concentration of CO [10–13]. Mao et al. confirmed through industrial tests that the slagging of limestone in the converter can lead to the volatilization of some silicon elements in the hot liquid iron, thus reducing the amount of slag [14–16]. Moreover, during the decomposition of limestone in the converter, the $CO_2$ produced can not only be used as part of the endogenous oxygen source to replace oxygen in the steelmaking reaction, but the escape of $CO_2$ can also lead to the improvement of the molten pool kinetic conditions [17–19]. However, while limestone shows certain advantages in rapid slagging and phosphorus removal, limestone particle size also affects melting in the converter; Deng et al. [20] found that the decomposition of the surface layer of massive limestone at steelmaking temperatures produces a dense structure of lime and affects the further decomposition of limestone. Lu et al. [21] found that the addition of large blocks of limestone is not conducive to the improvement of phosphorus removal efficiency. Therefore, the study of the decomposition degree and decomposition mechanism of limestone at steelmaking temperatures is one of the hot spots in the study of slagging steelmaking by replacing lime with converter limestone. Liang et al. [22], through a thermal state experimental study, found that limestone replacement of lime for slagging steelmaking is practicable. A limestone substitution ratio of 40% to 60% has a significant impact on the effect of dephosphorization, and the limestone particle size should not be too large to ensure that it can be quickly decomposed. Tang et al. [23], through the corresponding physical simulation, found that the powder penetration ratio increases with an increase in powder particle size. Furthermore, limestone particles in the spray travel through the gas–liquid interface into the melt pool when there is a critical diameter. Zhang et al. [24], through the use of a spray gun water simulation test, found that the limestone particle size is one of the main influencing factors affecting the powder blowing speed; a change in the blowing speed will thus have a chain of effects on the blowing effect. Too large a blowing particle size will lead to an increase in the blowing resistance, dispersing part of particles and making it so the particles cannot pass through the gas–liquid interface into the molten pool. Wang et al. [25] studied the decomposition mechanism of 4–25 mm limestone at high temperatures and found that the reaction of large-size limestone follows the model of random nucleation and subsequent growth, and the number of reaction stages was 3/4; Cao et al. [26] used a differential thermal analyzer to study the reaction mechanism of the high-temperature calcination of limestone under an air atmosphere of air and a mixed air–carbon dioxide atmosphere and found that different reaction atmospheres change the model of limestone decomposition, and the decomposition temperature and activation of the decomposition of limestone under the mixed atmosphere are also changed. The limestone decomposition temperature and decomposition activation energy are higher than in a pure air atmosphere. Zhang et al. [27] used a double extrapolation method to study the kinetic parameters of the limestone thermal decomposition reaction under a high concentration of carbon dioxide and found that the reaction follows the stochastic nucleation and stochastic growth model, and the number of reaction stages, n, is different under different carbon dioxide concentrations. Chen et al. [28] studied the decomposition mechanism of limestone with a particle size of 10 mm or less through thermogravimetric experiments and found that the higher the temperature and the smaller the particle size, the faster the decomposition rate of limestone, and that the particle size and the calcination temperature also have a significant effect on the thermal decomposition mechanism of limestone. Several studies have been conducted to confirm the advantages of converter limestone slagging for phosphorus removal; however, the limestone decomposition process consumes a large amount of heat, especially when large quantities are added centrally from the top through the silo. In addition, the additional limestone diameter is generally between 10 and 40 mm. A larger limestone block is not easy to calcine, and if the limestone and the initial slag contact to form a harder slag shell, the reaction surface activity is reduced, which is not conducive to the subsequent slag process, resulting in $CO_2$ that is not utilized. Furthermore, the density of limestone is about 2.7 g/cm$^3$; after directly added it through the top of the converter, even with the stirring effect of top-blowing oxygen, part of the limestone will float in the upper part of the molten pool, and the $CO_2$ generated by the decomposition of limestone cannot fully contact the molten pool, so the $CO_2$ utilization rate is not high.

Li et al. [29,30] proposed the use of converter-blown limestone powder slag steelmaking, in which carrier gas is used to blow small-particle-size limestone into the molten pool. The study found that to avoid the defects caused by block limestone, and at the same time give full play to the advantages of limestone in the slag steelmaking link, the slag dephosphorization conditions and the efficiency of the reaction need to be improved. Sun et al. [31], in their research on the rapid calcination decomposition behavior of small particles of limestone at converter steelmaking temperatures, found that the conversion rate of the decomposition of small particles of limestone is faster and more likely to produce high activity than lime. In addition, the dephosphorization effect of an average particle size of 0.44 mm of small particles of limestone using the carrier gas jet blowing into the converter melting pool is better. Although the high temperature decomposition behavior of small particles of limestone and the converter dephosphorization effect have been the subjects of a large number of preliminary studies, at present, the kinetic mechanism of the decomposition of small particles of limestone under converter steelmaking temperatures has undergone less research. By studying the thermal decomposition behavior of small-grained limestone at the temperature of converter steelmaking, the authors clarified the kinetic parameters of the rapid calcination and decomposition of small-grained limestone at high temperatures and the reaction mechanism of converter blowing of small-grained limestone for slag-making and steelmaking functions by using the improved double extrapolation method in the hope of providing a theoretical basis for further applications.

## 2. Materials and Methods

### 2.1. Materials

The raw material used in the test was limestone for steelmaking from Shijiazhuang, and the chemical composition of the sample was determined by XRF, as shown in Table 1. After the calcination of the limestone, the mass fraction of CaO in the limestone reached 54.38% and the impurity content was low, indicating the high quality of limestone.

**Table 1.** Chemical composition of limestone (% wt.).

| CaO | MgO | $SiO_2$ | $SO_3$ | $Al_2O_3$ | $Fe_2O_3$ | $K_2O$ | Others | Loss |
|---|---|---|---|---|---|---|---|---|
| 54.38 | 0.96 | 0.17 | 0.01 | 0.234 | 0.258 | 0.08 | 0.056 | 43.875 |

### 2.2. Methods

The raw materials used in the tests were limestone with particle sizes of 0.18–0.68 mm and 2–3 mm, with average particle sizes of 0.44 mm and 2.5 mm, respectively. Raw limestone was simply cleaned of some of the impurities present on the surface with anhydrous ethanol before being crushed and left to dry naturally. Then, the experimental raw materials were crushed by a jaw crusher, and the limestone particles with different particle size ranges were sieved using standard sieves of different mesh sizes. The sieved limestone particles were washed with fresh water, dried in a drying oven for 48 h, and then vacuum-sealed in a sealed bag.

The temperature range of the thermogravimetric-differential thermal analyzer used in the test can reach 25–1600 °C. An alumina crucible was selected, the sample mass was weighed with a precision balance as (20 ± 0.1) mg, and the experimental atmospheres were the standard gas of nitrogen, oxygen, and carbon dioxide, standard carbon dioxide, and nitrogen mixed with carbon dioxide at carbon dioxide partial pressures of 0%, 25%, 50%, 75%, and 100%, respectively. The gas flow rate was set to 20 ML/min, and the sample was heated from 25 °C to 1100 °C after setting the heating rate so that the limestone was completely transformed to a stable state and the complete thermal decomposition curve of limestone was obtained.

### 2.3. Calculation Methodology

The conversion rate, α, of the limestone samples was calculated from Equation (1) through the weight loss of limestone.

$$\alpha = \frac{w_0 - w_t}{w_0 - w_x} \times 100\% \tag{1}$$

where α is the conversion rate of the limestone, $w_0$ is the original weight of the limestone sample, $w_t$ is the weight of the limestone sample at a point t of the reaction, and $w_x$ is the mass of the sample at the complete decomposition.

In performing the computational solution of the kinetic parameters and the most probabilistic mechanism function, the double extrapolation method combining the Flynn–Wall–Owaza method and the Coats–Redfern method is often applied nowadays [32]. In this paper, we adopt the improved double extrapolation method, using the combination of the Starink differential and the Coats–Redfern integral with higher precision, mainly because the Starink differential equation makes a detailed categorization and comparison for the Flynn–Wall–Owaza method. The partial integral and Doyle's approximation of the temperature integral are analyzed, and the temperature integral term is rationally corrected by linear regression. Thus, it will fit more accurately and achieve a smaller relative error in $E_{\alpha\to0}$. We fix the conversion rate, α, and extrapolate α to 0 by fitting to obtain the value of $E_{\alpha\to0}$ without any interference of side reactions and assuming the system is in the pristine state, i.e., the activation energy of the nucleation of the new phases. Furthermore, we fix the rate of heating, β, and obtain the kinetic parameter $E_{\beta\to0}$ when β is extrapolated to 0 by fitting to clarify the most probable mechanism function of limestone decomposition.

The kinetic equation for the gas–solid chemical reaction is:

$$\frac{d_\alpha}{d_t} = kf(\alpha) \tag{2}$$

where $k$ is the reaction rate constant, $t$ is the reaction time, and $f(\alpha)$ is the reaction mechanism function in differential form.

The reaction rate constant $k$ versus the thermodynamic temperature $T$ was obtained according to the Arrhenius formula for a non-isothermal, non-homogeneous system:

$$k = Ae^{-E/RT} \tag{3}$$

where $E$ is the activation energy, $A$ is the finger forward factor, and $R$ is the gas reaction constant, 8.3145 J/mol·K.

The kinetic equation for the non-isothermal non-homogeneous phase is thus obtained:

$$\frac{d_\alpha}{d_T} = \frac{A}{\beta}e^{-E/RT}f(\alpha) \tag{4}$$

where $\beta$ is the heating rate.

The kinetic model function in integral form is obtained:

$$G(\alpha) = \int_{T_0}^{T} \frac{A}{\beta} e^{-E/RT} dT = \frac{A}{\beta} \int_{T_0}^{T} e^{-E/RT} dT = (AE/\beta R)P(u) \quad (5)$$

where $P(u)$ is the temperature integral, $P_{(u)} = \int_{\infty}^{u} -(e^{-u}/u^2)du$, $u = E/RT$ and $T_0$ is the initial reaction temperature.

The Starink differential [19] is:

$$\ln(\beta/T^{1.8}) = C_s - 1.0037E/RT \quad (6)$$

where $C_s$ is the univariate function of $G(\alpha)$ when the conversion rate $\alpha$ is fixed, $G(\alpha)$ is fixed, and $C_s$ is constant.

We fixed $\ln(\beta/T^{1.8})$ as the $Y$-axis and 1/T as the X-axis, and made a linear fit to a straight-line graph according to Y = AX + B. According to the slope of the fitted straight line, the apparent activation energy corresponding to a fixed conversion rate $\alpha$ was obtained. Using Equation (7), $\alpha$ was extrapolated to zero, and the value of $E_{\alpha\to0}$ was obtained, which was the value of $E_{\alpha\to0}$ without any interference of side reactions when the system was in the original state.

$$E = \alpha_1 + b_1\alpha + c_1\alpha^2 + d_1\alpha^3 \quad (7)$$

The Coats–Redfern integral is written as:

$$\ln(G(\alpha)/T^2) = \ln(AR/(\beta E)) - E/(RT) \quad (8)$$

We fixed $\ln(G(\alpha)/T^2)$ as the $Y$-axis and 1/T as the X-axis, according to Y = AX + B for a linear-fitting straight-line graph, analyzing the slope of the fitted straight line to find the apparent activation energy, E, corresponding to a fixed rate of temperature increase, $\beta$. The closer the linear correlation coefficient $R^2$ is to 1, the higher the degree of fit, and the more the functional form of $G(\alpha)$ represents the real situation of the reaction process. Then, 47 commonly used kinetic mechanism functions [32] were utilized, from which the $G(\alpha)$ equation with high linear correlation and kinetic parameters in line with the general law of thermal decomposition reaction was selected, and the corresponding kinetic parameters were calculated from it. Using Equation (9) to extrapolate $\beta$ to zero, the kinetic parameters of the sample in the thermal equilibrium state and the value of $E_{\beta\to0}$ were obtained.

$$E = \alpha_2 + b_2\beta + c_2\beta^2 + d_2\beta^3 \quad (9)$$

When comparing the two extrapolated $E_{\alpha\to0}$ values with the $E_{\beta\to0}$ values, if they are equal or similar, the corresponding $G(\alpha)$ mechanism function equation can be considered the most probable mechanism function in the thermal decomposition process of limestone.

## 3. Results

### 3.1. *Effect of Particle Size on the Thermal Decomposition Reaction of Small-Particle Limestone*

To study the effect of particle size on the decomposition kinetics of limestone under rapid heating conditions, we selected the average particle sizes of 0.44 mm limestone and 2.5 mm (the maximum particle size allowed in the test) limestone as comparison points to observe their decomposition performance. Combined with the converter-blowing limestone-powder slagging steelmaking technology, taking into account 1 mol of carrier gas carrying 1 mol of limestone powder into the molten pool inside, and 1 mol $CaCO_3$ in the thermal decomposition of 1 mol $CO_2$, we selected a decomposition furnace atmosphere of 50% $CO_2$/50% $N_2$ for the test. We selected the heating rate to simulate the temperature environment of the converter as far as possible and combined it with the actual operation of the thermogravimetric-thermal equipment. Figure 1 below shows the thermal decomposition curves of 0.44 mm limestone.

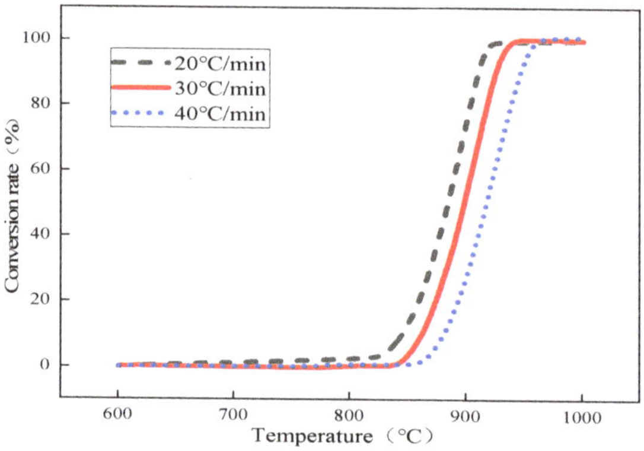


**Figure 1.** The conversion rate curve of 0.44 mm small particles of limestone with temperature changes.

The thermal decomposition of limestone with an average particle size of 0.44 mm was analyzed as an example, and the corresponding temperatures of the conversion rates of limestone at different heating rates were obtained according to the conversion rate curves of limestone, as shown in Table 2 below. The conversion rate was fixed first, and according to the Starink differential equation $\ln(\beta/T^{1.8}) = C_s - 1.0037E/RT$, the inverse 1/T of the values of small limestone particles at different heating rates and corresponding reaction temperatures were substituted, the linear fitting relationships between $\ln(\beta/T^{1.8})$ and 1/T, and the corresponding fitting coefficients $R^2$ according to the slope of the linear equations obtained by fitting the activation energy E under the corresponding conversion rate are shown in Table 3 below. Then, according to Equation (7), $\alpha$ was extrapolated to 0, and the activation energy $E_{\alpha \to 0}$ = 250.32 kJ·mol$^{-1}$ was obtained without any interference from side reactions when the system was in the original state; the fitting results are shown in Figure 2.

**Table 2.** Reaction temperatures for each conversion rate at different heating rates.

| Conversion Rate (%) | Temperature/°C (20 °C/min) | Temperature/°C (30 °C/min) | Temperature/°C (40 °C/min) |
|---|---|---|---|
| 20 | 863.37 | 878.9 | 892.51 |
| 30 | 870.91 | 886.93 | 901.13 |
| 40 | 878.8 | 895.12 | 909.64 |
| 50 | 886.5 | 902.22 | 917.12 |
| 60 | 890.66 | 908.29 | 923.82 |
| 70 | 896.87 | 915 | 931.06 |
| 80 | 902.88 | 921.46 | 937.77 |
| 90 | 909.59 | 929.98 | 946.35 |

**Table 3.** Corresponding linear fitting equations for different conversion rates in 50%$CO_2$.

| Conversion Rate (%) | Linear Fitting Equation | Fitting Factor ($R^2$) | E (kJ·mol$^{-1}$) |
|---|---|---|---|
| 20 | $\ln(\beta/T^{1.8}) = -29516\frac{1}{T} + 16.312$ | 0.9968 | 244.51 |
| 30 | $\ln(\beta/T^{1.8}) = -28801\frac{1}{T} + 15.504$ | 0.9965 | 238.58 |
| 40 | $\ln(\beta/T^{1.8}) = -28594\frac{1}{T} + 15.14$ | 0.9963 | 236.87 |
| 50 | $\ln(\beta/T^{1.8}) = -28189\frac{1}{T} + 14.636$ | 0.9962 | 233.51 |
| 60 | $\ln(\beta/T^{1.8}) = -27064\frac{1}{T} + 13.554$ | 0.9967 | 224.19 |
| 70 | $\ln(\beta/T^{1.8}) = -26496\frac{1}{T} + 12.935$ | 0.9965 | 219.49 |
| 80 | $\ln(\beta/T^{1.8}) = -26212\frac{1}{T} + 12.568$ | 0.9968 | 217.14 |
| 90 | $\ln(\beta/T^{1.8}) = -25087\frac{1}{T} + 11.475$ | 0.9992 | 207.82 |

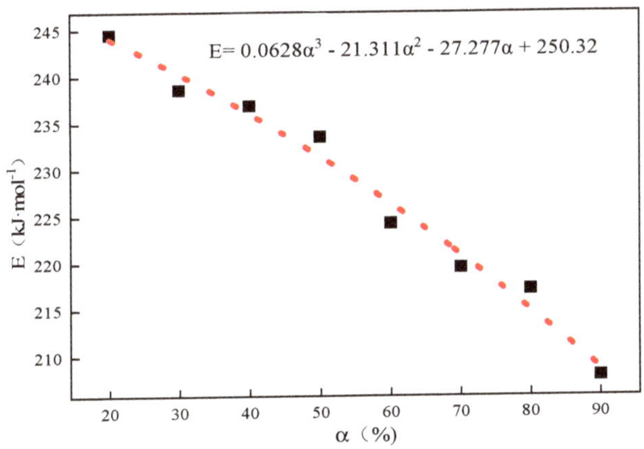


**Figure 2.** The curve of activation energy E to α for the thermal decomposition of 0.44 mm limestone.

Similarly, we analyzed the kinetic parameters of limestone with an average particle size of 2.5 mm for extrapolation. The apparent activation energy of the limestone decomposition was 328.27 kJ·mol$^{-1}$; with the increase in limestone particle size, the apparent activation energy of limestone thermal decomposition increases and the decomposition reaction difficulty increases. The particle size of limestone used in steelmaking is 10–40 mm, which is much larger than that of the limestone selected in the test; therefore, with the further increase in particle size, the activation energy of decomposition will increase more and more, and the difficulty of decomposition will be greater. The decomposition of small-grained limestone will be better under the environment of steelmaking in the converter to provide better conditions for slag dephosphorization.

### 3.2. *Effect of Atmosphere on the Thermal Decomposition Reaction of Small-Particle Limestone*

Converter-blowing limestone steelmaking technology uses a carrier gas to carry small particles of limestone to the molten iron pool, where the heat transfer process is no longer restricted and the limestone can be rapidly heated for decomposition and calcination to generate a large surface area with large porosity, small pile density, and the high activity of high-quality lime. Furthermore, the slagging effect and dephosphorization rate are improved, and the $CO_2$ released in the decomposition acts as a source of endogenous oxygen to participate in the steelmaking reaction, improving the dynamic conditions of $CO_2$ resource utilization. Therefore, this subsection investigates the thermal decomposition behavior of small-grained limestone in different decomposition furnace atmospheres. To make the heat of the small-particle limestone more uniform when studying the effect of the thermal decomposition of small-particle limestone under different decomposition atmospheres, the selected heating rate was 10 °C/min, and the average particle size of the selected limestone was 0.44 mm. The TG-DTG curves of the small-particle limestone in four reaction atmospheres, namely, $N_2$, air, $O_2$, and $CO_2$, are shown in Figure 3.

From Figure 3a, it can be seen that in the $N_2$, air, and $O_2$ atmospheres, the limestone decomposition onset and completion temperature were roughly equivalent; the initial decomposition temperature was close to 643 °C, the endpoint of the decomposition temperature, when the decomposition was completed, was at about 847 °C. The initial decomposition temperature of limestone under the pure $CO_2$ atmosphere was about 900 °C, and the terminal decomposition temperature was close to 970 °C, with a weight loss rate of about 43.5%. Small particles of limestone in all four atmospheres were fully reacted within 1000 °C. The change in the limestone decomposition temperature under different carrier gas atmospheres was mainly because limestone produces part of the $CO_2$ during the decomposition process. Therefore, compared with other atmospheres, the presence of $CO_2$

in the pure $CO_2$ atmosphere significantly inhibits the limestone decomposition behavior, so that the activation energy required for decomposition is increased, the reaction is not easy to carry out, and the decomposition temperature is increased.

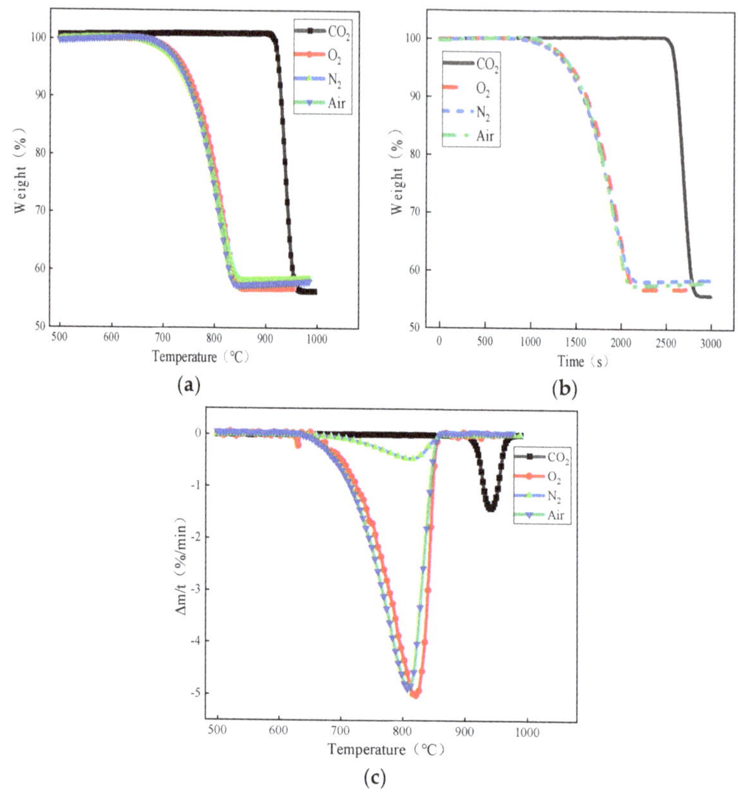


**Figure 3.** Effect of carrier gas atmosphere on limestone decomposition: (**a**) weight loss versus temperature curves under different atmospheres; (**b**) curves of weight loss with decomposition time under different atmospheres; and (**c**) DTG curves of limestone decomposition under different atmospheres.

In addition, as can be seen from Figure 3b, under the atmospheres of $N_2$, air, and $O_2$, the decomposition of limestone from the beginning to the completion of the decomposition took about 1100 s, while under the atmosphere of pure $CO_2$, the initial decomposition temperature of the small particles of limestone was increased, but the decomposition time required was instead shortened by about 1/3, and the decomposition was completed in about 390 s. Combined with the actual environment of the converter steelmaking, in which the converter temperature can reach more than 1350 °C and small particles of limestone in the carrier gas jet are blown into the converter pool, the heat transfer rate would be much greater than the experimental rate of heating, the limestone decomposition would be more intense and rapid, and the decomposition time would be much shorter, with the decomposition quickly completed to obtain the high activity of lime.

From Figure 3c, it can be seen that in the limestone decomposition process, there is an "S" shaped peak. In the initial reaction stage, the limestone decomposition rate is slower; in the middle, the limestone decomposition rate began to accelerate to a certain level; and the end of the reaction, the limestone decomposition rate gradually slows down until the reaction is complete. This is because with the increase in temperature, the limestone heat transfer efficiency increases, which makes the limestone decomposition faster. In addition, compared to the other atmospheres, the peak of the DTG curve in the $CO_2$ atmosphere lags behind the peak weight loss rate of the other three curves, peaking at about 937 °C.

### 3.3. Effect of $CO_2$ Partial Pressure on the Thermal Decomposition Behavior of Small-Particle Limestone

To further study the effect of small-particle limestone on the reaction limit and decomposition behavior under different $CO_2$ partial pressures, based on the above study, experiments were carried out using an atmosphere of carbon dioxide and nitrogen mixed with standard gas and increasing the $CO_2$ partial pressure for three groups of experiments of 25%, 50%, and 75%. The heating rate was still selected to be 10 °C/min, with an average limestone size of 0.44 mm. The pyrolysis characteristic curves of small-particle limestone in a total of five reaction atmospheres of 0% $CO_2$, 25% $CO_2$, 50% $CO_2$, 75% $CO_2$, and 100% $CO_2$ are shown in Figure 4.

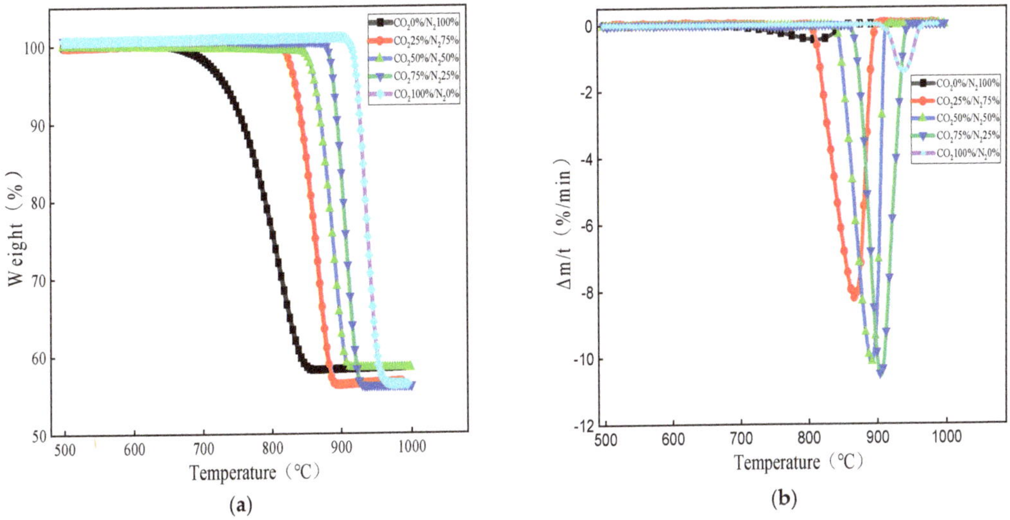


**Figure 4.** Effect of $CO_2$ partial pressures on limestone decomposition: (**a**) weight loss versus temperature curves at different partial pressures and (**b**) DTG curves of limestone decomposition under different partial pressures.

As can be seen from Figure 4a, under the five atmospheres shown, the limestone decomposition onset temperature is between 643 °C and 903 °C, and with the increase in $CO_2$ partial pressure, the limestone decomposition onset and completion temperatures produce a hysteresis. As the $CO_2$ concentration in the reaction atmosphere increases, the decomposition curve shifts to the higher temperature region. According to the Le Chatelier principle, when the partial pressure of $CO_2$ in the atmosphere is lower than the equilibrium partial pressure, calcium carbonate decomposes in the forward direction; conversely, the reaction reverses to produce calcium carbonate. Therefore, the decomposition reaction is controlled by the partial pressure of $CO_2$ and the reaction temperature. The limestone produces part of $CO_2$ during decomposition, and as the partial pressure of $CO_2$ increases, limestone decomposition is inhibited, and the process of $CO_2$ escape is greatly inhibited, resulting in a lag in decomposition temperature. In a high concentration $CO_2$ atmosphere, calcium carbonate can decompose, because in non-constant temperature and pressure conditions, the calcium carbonate decomposition temperature of $CO_2$ is much higher than the temperature of the carrier gas, so it can be desorbed into solids to escape, ensuring that the decomposition of calcium carbonate can continue. In addition, as can be seen from Figure 4b, with an increase in $CO_2$ partial pressure, the peak temperature of limestone decomposition also corresponds to the hysteresis. As the $CO_2$ partial pressure progressed from 0 to 50%, the limestone decomposition reaction rate growth is very large, the peak is steeper, and the peak reaction rate increases; as the $CO_2$ partial pressure progresses from 50% to 75%, the growth of the limestone decomposition reaction rate is greatly reduced compared to the lower concentrations and the peak reaction rate increases slightly. Combined with Figure 3a,b, this indicates that the $CO_2$ partial pressure does produce a certain inhibition on the decomposition of limestone, although the inhibition is not obvious when the $CO_2$ partial pressure is lower. There is a limited link between $CO_2$ partial pressure and the decomposition rate of limestone, and the inhibition is more significant at high $CO_2$ partial pressure, but the reactions are all completed by 1000 °C.

### 3.4. *Effect of $CO_2$ Partial Pressure on the Thermal Decomposition Kinetics of Small-Particle Limestone*

Based on the study of $CO_2$ partial pressure on the thermal decomposition behavior of limestone in Section 2.3, in order to further calculate the kinetic parameters and kinetic model for the rapid thermal decomposition of small-particle limestone at the temperature of converter steelmaking, the atmospheric partial pressures of 0%, 25%, 50%, 75%, and 100% $CO_2$ and heating rates of 20 °C/min, 30 °C/min, and 40 °C/min were used. The average limestone particle size was 0.44 mm. In this subsection, the improved double extrapolation method is used to calculate the conversion rate of limestone using the TG-DTG curve obtained from the thermal decomposition of limestone to solve the activation energy of limestone thermally decomposed under different partial pressures of carbon dioxide. Furthermore, the temperature increase rate is extrapolated to zero to obtain the activation energy of limestone in thermal equilibrium. The apparent activation energy of limestone in the initial state clarifies the most generalized mechanism model of the limestone decomposition reaction, taking 25% $CO_2$ as an example. The following Figure 5 shows the curve of the thermal decomposition of limestone under different heating rates when the $CO_2$ partial pressure is 25%.

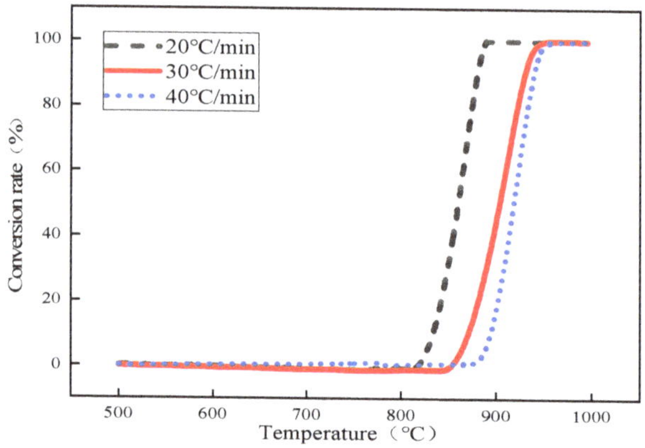


**Figure 5.** Conversion rate versus temperature curve at 25% $CO_2$ partial pressure.

According to the conversion rate curve of limestone, the temperatures corresponding to the conversion rate of limestone at different heating rates under 25% $CO_2$ partial pressure were obtained, as shown in Table 4 below.

By utilizing the same research method as in Section 2.1, the conversion rate was fixed, and the corresponding activation energies, E, at different conversion rates were found according to the Starink differential equation, $\ln(\beta/T^{1.8}) = C_s - 1.0037E/RT$, and $E = \alpha_1 + b_1\alpha + c_1\alpha^2 + d_1\alpha^3$. The corresponding activation energies at different conversion rates were obtained and fitted to obtain $E_{\alpha\to 0}$ = 231.81 kJ·mol$^{-1}$ without any interference of the side reactions when the system was in the pristine state, and the results are shown in Table 5 and Figure 6, respectively.

**Table 4.** Reaction temperatures for each conversion at different heating rates.

| Conversion Rate (%) | Temperature/°C (20 °C/min) | Temperature/°C (30 °C/min) | Temperature/°C (40 °C/min) |
|---|---|---|---|
| 20 | 832.07 | 852.91 | 865.33 |
| 30 | 841.74 | 862.12 | 877.46 |
| 40 | 849.45 | 873.19 | 887.74 |
| 50 | 857.65 | 882.62 | 897.57 |
| 60 | 867.56 | 888.85 | 910.15 |
| 70 | 873.15 | 899.02 | 917.94 |
| 80 | 878.95 | 907.32 | 924.15 |
| 90 | 886.21 | 916.87 | 934.31 |

**Table 5.** Corresponding linear fitting equations at different conversion rates in 25% $CO_2$.

| Conversion Rate (%) | Linear Fitting Equation | Fitting Factor ($R^2$) | E ($kJ \cdot mol^{-1}$) |
|---|---|---|---|
| 20 | $\ln(\beta/T^{1.8}) = -23984\frac{1}{T} + 12.074$ | 0.9965 | 198.68 |
| 30 | $\ln(\beta/T^{1.8}) = -22782\frac{1}{T} + 10.797$ | 0.9993 | 188.72 |
| 40 | $\ln(\beta/T^{1.8}) = -21369\frac{1}{T} + 9.3819$ | 0.9972 | 177.02 |
| 50 | $\ln(\beta/T^{1.8}) = -20723\frac{1}{T} + 8.6581$ | 0.9964 | 171.67 |
| 60 | $\ln(\beta/T^{1.8}) = -219897\frac{1}{T} + 7.7847$ | 0.9956 | 164.82 |
| 70 | $\ln(\beta/T^{1.8}) = -19023\frac{1}{T} + 6.9108$ | 1 | 157.58 |
| 80 | $\ln(\beta/T^{1.8}) = -18822\frac{1}{T} + 6.6348$ | 0.9947 | 155.92 |
| 90 | $\ln(\beta/T^{1.8}) = -17808\frac{1}{T} + 5.6463$ | 0.9941 | 147.52 |

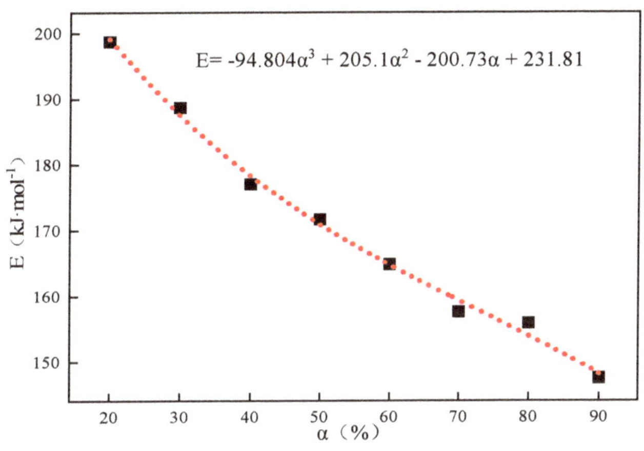


**Figure 6.** Thermal decomposition of limestone at 25% $CO_2$ partial pressure corresponding to the E-α relationship curve.

The same method was used to obtain the results of the limestone activation energy changes at partial pressures of 0%, 25%, 50%, 75%, and 100% $CO_2$, and the activation energy data obtained were fitted and extrapolated to obtain the activation energy of the limestone in its original state, as shown in Table 6.

Based on Table 5, the apparent activation energy E was obtained by fixing the heating rate β, substituting the Coats–Redfern integral equation, and then extrapolating β to zero to obtain the value of $E_{\beta \to 0}$ at thermal equilibrium under the partial pressure of 25% $CO_2$. The following Table 7 shows a few kinetic parameters with good fits in the 47 reaction mechanism functions corresponding to $E_{\beta \to 0}$ and $R^2$.

**Table 6.** Activation energies corresponding to limestone in its pristine state at different $CO_2$ partial pressures.

| Partial Pressure of $CO_2$ | $E_{\alpha\to0}$ ($kJ\cdot mol^{-1}$) |
|---|---|
| 0 | 201.9 |
| 25% | 231.81 |
| 50% | 250.32 |
| 75% | 310.42 |
| 100% | 347.11 |

**Table 7.** Apparent activation energies and correlation coefficients corresponding to mechanism functions with good linearity at different warming rates.

| $G(\alpha)$ | 0 °C/min | | 20 °C/min | |
|---|---|---|---|---|
| | $E_{\beta\to0}$ ($kJ\cdot mol^{-1}$) | $R^2$ | E ($kJ\cdot mol^{-1}$) | $R^2$ |
| $\alpha+(1-\alpha)\ln(1-\alpha)$ | 911.78 | 1 | 638.39 | 0.9888 |
| $[1-(1-\alpha)^{1/3}]^2$ | 1047.6 | 1 | 744.56 | 0.9946 |
| $(1-2\alpha/3)-(1-\alpha)^{2/3}$ | 956.45 | 0.996 | 673.22 | 0.9916 |
| $1-(1-\alpha)^{1/2}$ | 477.68 | 0.999 | 336.79 | 0.9923 |
| $1-(1-\alpha)^{1/3}$ | 515.08 | 0.999 | 362.87 | 0.9943 |
| $-\ln(1-\alpha)$ | 588.44 | 1 | 420.69 | 0.9937 |
| $[-\ln(1-\alpha)]^{2/3}$ | 386.47 | 1 | 274.19 | 0.9934 |
| $[-\ln(1-\alpha)]^{1/2}$ | 285.49 | 0.998 | 200.94 | 0.9931 |
| $[-\ln(1-\alpha)]^{1/3}$ | 184.51 | 0.997 | 127.69 | 0.9924 |
| $[-\ln(1-\alpha)]^{2/5}$ | 228.92 | 1 | 156.99 | 0.9928 |
| $[-\ln(1-\alpha)]^{1/4}$ | 134.02 | 0.999 | 91.06 | 0.9917 |
| $[-\ln(1-\alpha)]^{3/4}$ | 436.99 | 1 | 310.82 | 0.9935 |
| $G(\alpha)$ | 30 °C/min | | 40 °C/min | |
| | E ($kJ\cdot mol^{-1}$) | $R^2$ | E ($kJ\cdot mol^{-1}$) | $R^2$ |
| $\alpha+(1-\alpha)\ln(1-\alpha)$ | 565.93 | 0.9813 | 536.31 | 0.989 |
| $[1-(1-\alpha)^{1/3}]^2$ | 662.24 | 0.9926 | 626.06 | 0.9947 |
| $(1-2\alpha/3)-(1-\alpha)^{2/3}$ | 597.51 | 0.9861 | 565.75 | 0.9918 |
| $1-(1-\alpha)^{1/2}$ | 298.34 | 0.9875 | 281.23 | 0.9924 |
| $1-(1-\alpha)^{1/3}$ | 321.49 | 0.9921 | 303.28 | 0.9943 |
| $-\ln(1-\alpha)$ | 374.03 | 0.9967 | 352.16 | 0.9936 |
| $[-\ln(1-\alpha)]^{2/3}$ | 242.94 | 0.9965 | 228.27 | 0.9933 |
| $[-\ln(1-\alpha)]^{1/2}$ | 177.39 | 0.9963 | 166.33 | 0.9929 |
| $[-\ln(1-\alpha)]^{1/3}$ | 111.85 | 0.9958 | 104.39 | 0.992 |
| $[-\ln(1-\alpha)]^{2/5}$ | 138.06 | 0.9961 | 129.17 | 0.9925 |
| $[-\ln(1-\alpha)]^{1/4}$ | 79.07 | 0.9953 | 73.42 | 0.991 |
| $[-\ln(1-\alpha)]^{3/4}$ | 275.71 | 0.9965 | 259.25 | 0.9934 |

According to the values of $E_{\beta\to0}$ obtained in Table 7 compared with the previously obtained $E_{\alpha\to0}$ = 231.81 $kJ\cdot mol^{-1}$, the value of $E_{\beta\to0}$ = 228.92 $kJ\cdot mol^{-1}$ in Table 7 is the closest and the fit is high. Therefore, the corresponding model for the thermal decomposition of limestone at a partial pressure of 25% $CO_2$ is a random nucleation and subsequent growth model, and the number of reaction levels is 2/5. Similarly, the values of $E_{\beta\to0}$ of limestone extrapolated from the partial pressures of 0%, 25%, 50%, 75%, and 100% $CO_2$ are summarized as shown in Table 8.

**Table 8.** Activation energies and mechanism functions corresponding to the thermal decomposition of limestone at different $CO_2$ partial pressures.

| Partial Pressure of $CO_2$ | $E_{\beta\to0}$ ($kJ\cdot mol^{-1}$) | $R^2$ | $G(\alpha)$ | n |
|---|---|---|---|---|
| 0 | 204.99 | 0.999 | $[-\ln(1-\alpha)]^{2/3}$ | 2/3 |
| 25 | 228.92 | 1 | $[-\ln(1-\alpha)]^{2/5}$ | 2/5 |
| 50 | 249.48 | 0.998 | $[-\ln(1-\alpha)]^{2/5}$ | 2/5 |
| 75 | 309.03 | 0.999 | $[-\ln(1-\alpha)]^{1/2}$ | ½ |
| 100 | 340.2 | 1 | $[-\ln(1-\alpha)]^{1/2}$ | ½ |

Comparing the mechanism function of the thermal decomposition of limestone under different $CO_2$ partial pressures in high temperature and rapid calcination, the reaction models are stochastic nucleation and subsequent growth models. Limestone decomposition initially occurs in some localized areas, and then as the reaction proceeds, these adjacent decomposition products will be aggregated to form a new phase, CaO; then, the surrounding $CaCO_3$ molecules continue to undergo the interfacial reaction until the entire solid phase is completely decomposed. The functional equation of the mechanism is $G(\alpha) = [-\ln(1-\alpha)]^n$, and the $CO_2$ partial pressure is different when the number of levels of the reaction changes; As the $CO_2$ partial pressure of 25% increased to 100%, the activation energy of thermal decomposition gradually increased, and the number of reaction levels, n, increased.

## 4. Conclusions

(1) With the increase in heating rate, the decomposition temperature of small particles of limestone increases. The apparent activation energies of the decomposition of limestone with average particle sizes of 0.44 mm and 2.5 mm is 250.32 $kJ\cdot mol^{-1}$ and 328.27 $kJ\cdot mol^{-1}$, respectively, and with the increase in limestone particle size, the apparent activation energy of the thermal decomposition of limestone increases, and the difficulty of the decomposition reaction increases.

(2) Compared with $N_2$, air, and $O_2$ atmospheres, in the pure $CO_2$ atmosphere, limestone thermal decomposition reaction completion temperature increases, but the time required for decomposition is shortened. With the increase in the partial pressure of carbon dioxide, limestone decomposition is inhibited and the decomposition temperature lags. However, the decomposition reaction occurs more intensely as the limestone decomposition speed is accelerated, and time required to complete the decomposition is shortened.

(3) At low $CO_2$ partial pressures, the limestone decomposition temperature produces a hysteresis as the partial pressure increases and the reaction rate increases by a large amount. Under high $CO_2$ partial pressure, the inhibition is more significant, and the reaction rate slows down, but the reaction can be fully reacted within 1000 °C.

(4) The thermal decomposition reactions of small-grained limestone at different $CO_2$ partial pressures at steelmaking temperatures are all consistent with the model of both stochastic nucleation and subsequent growth, and the functional equation of the mechanism is $G(\alpha) = [-\ln(1-\alpha)]^n$; the activation energy of the decomposition reaction gradually increases when the $CO_2$ partial pressure is increased from 25% to 100%, the number of reaction stages n increases, and the difficulty of the reaction increases.

However, subsequent studies on the reaction mechanism of small-grained limestone at steelmaking temperatures, especially those related to the decomposition behavior of small-grained limestone under different $CO_2$ partial pressures, will continue. Also, the subject of the slagging of converter-blasted small limestone will continue to be fully investigated under the prerequisites of ensuring proper blowing size, atmosphere conditions, and optimum parameters for phosphorus removal. This will be demonstrated by thermodynamic calculations and thermal experiments in our subsequent work.

**Author Contributions:** C.L. and Y.X. conceived and designed the study; Y.Z. and H.S. performed the calculations; Y.Z. conducted the experiment; K.Z., H.X. and S.W. analyzed the experimental data; and C.L. and Y.Z. wrote the paper. All authors have read and agreed to the published version of the manuscript.

**Funding:** This research was funded by the National Natural Science Foundation of China, grant number 52274334 and 52004097, and the Science and Technology Research Program for Higher Education Institutions in Hebei Province, grant number BJK2022059.

**Data Availability Statement:** Not applicable.

**Acknowledgments:** Thanks to X. Meng and H. Shen of North China University of Science and Technology for checking and revising the manuscript.

**Conflicts of Interest:** The authors declare no conflict of interest.

## References

1. Zhang, Q.; Shen, J.-L.; XU, L.-S. Carbon peak and low-carbon transition path of China's iron and steel industry. *Iron Steel* **2021**, *56*, 152–163.
2. Cui, Z.-F.; Xu, A.-J.; Shangguan, F.-Q. Low-carbon development strategy analysis of the domestic and foreign steel industry. *Chin. J. Eng.* **2022**, *44*, 1496–1506.
3. Zhang, F.-M.; Liu, Q.-M. Development and understanding on low carbon technology based on BF-BOF steel manufacturing processes. *China Metall.* **2023**, *33*, 1–17.
4. Somnath, B.; Ashok, K.L.; Seshadri, S.; Halder, J. Change in phosphorus partition during blowing in a commercial BOF. *ISIJ Int.* **2007**, *47*, 766.
5. Ye, G.-F.; Yang, J.; Lu, X.-W.; Wang, H.; Liu, W.-S.; Zhang, R.-H.; Yang, W.-K. Effects of operation parameters on dephosphorization results of BOF at low temperature and low basicity. *Steelmaking* **2019**, *35*, 16–23.
6. Zhang, R.-H.; Yang, J.; Ye, G.-F.; Sun, H.; Yang, W.-K. Latest development of dephosphorization process in converter steelmaking. *Steelmaking* **2022**, *38*, 1–13.
7. Zeng, J.-Q.; Yang, L.-B.; Wang, J.; Liu, X.-L.; Yao, T.-L.; Dai, S.-F. Effect of bottom blowing on dephosphorization technology in combined blowing converter. *Iron Steel* **2017**, *52*, 40–44+51.
8. Zhi, J.-G.; Wu, W.; Gao, Q.; Xu, T.; Luo, H.-M.; Zhang, X.-F. Effect of improving dephosphorization of molten steel by optimizing slagging in large converters. *Iron Steel* **2020**, *55*, 72–77.
9. Chen, L.; Diao, J.; Wang, G.; Xie, B. Assessment of dephosphorization during vanadium extraction process in converter. *JOM* **2018**, *70*, 46–51. [CrossRef]
10. Li, H.; Qu, Y. Discussion on limestone addition instead of lime for energy-saving and emission reduction in BOF steelmaking. *China Metall.* **2010**, *20*, 45–48.
11. Katherine, L.; Yang, Y.D.; Mansoor, B.; McLean, A. Dephosphorization of levitated silicon-iron droplets for production of Solar-Grade Silicon. *Metall. Mater. Trans. B* **2018**, *49*, 21–25.
12. Diao, J.; Shao, L.; Liu, D.; Qiao, Y.; Tan, W.; Wu, L.; Xie, B. Removal of Phosphorus from Leach Liquor of Steel Slag: Adsorption Dephosphorization with Activated Alumina. *JOM* **2018**, *70*, 36–40. [CrossRef]
13. Davide, M.; Carlo, M.; Silvia, B.; Gruttadauria, A.; Sosio, R.; Valentino, G.; Ancona, V. Model for phosphorus removal in LD converter and design of a valuable operative practice. *Steel Res. Int.* **2018**, *89*, 46–51.
14. Mao, W.-W.; Li, C.-X.; Lu, H.; Hu, L.-F.; Li, H.; Cao, Z.-M. Silicon volatilization in the form of SiO during slagging by limestone in BOF. *Ironmak. Steelmak.* **2017**, *44*, 389–393. [CrossRef]
15. Mao, W.; Li, C.; Lu, H.; Li, H.; Xie, W. Limestone dissolution and decomposition in steelmaking slag. *Ironmak. Steelmak.* **2018**, *45*, 720–726. [CrossRef]
16. Lu, H.; Li, C.-X.; Mao, W.-W.; Li, H. Laboratory study of $CaCO_3$ decomposition, influence of BOF converter slag. *Metall. Res. Technol.* **2017**, *114*, 20–26. [CrossRef]
17. Dong, K.; Wang, X.-L. $CO_2$ utilization in the ironmaking and steelmaking process. *Metals* **2019**, *9*, 273. [CrossRef]
18. Yi, C.; Zhu, R.; Chen, B.-Y.; Yi, C.; Zhu, R.; Chen, B.-Y.; Wang, C.-R.; Ke, J.-X. Experimental research on reducing the dust of BOF in $CO_2$ and $O_2$ mixed blowing steelmaking process. *ISIJ Int.* **2009**, *49*, 1694–1699. [CrossRef]
19. Lv, M.; Zhu, R.; Wei, X.; Wang, H.; Bi, X. Research on top and bottom mixed blowing $CO_2$ in converter steelmaking process. *Steel Res. Int.* **2012**, *83*, 11–15. [CrossRef]
20. Deng, T.; Patrice, N.; Ek, M.; Du, S. Limestone dissolution in converter slag at 1873 K (1600 °C). *Metall. Mater. Trans. B* **2013**, *44*, 98. [CrossRef]
21. Lu, W.-G.; Zhu, R.; Yu, H. Research on slagging by limestone during BOF steelmaking process. *Chin. J. Eng.* **2016**, *38*, 78–82.
22. Liang, Y.-C.; Tang, H.-Y.; Li, J.-S.; Guo, S.-L.; Wang, G.-F.; Dan, J.-Q.; Yang, H.-B.; Li, H. Laboratory study on dephosphorization with limestone substituting for some lime. *J. Chongqing Univ.* **2015**, *38*, 83–88.
23. Tang, B.; Wang, X.-M.; Zou, Z.-S.; Sun, G.-Q. Physical simulation of BOF limestone particles blowing via top lance. *J. Northeast. Univ. (Nat. Sci.)* **2014**, *35*, 695–699.

24. Zhang, B.; Wu, W.; Wu, W.; Wang, T.-M.; Luo, L.-G.; Wang, Y.-H. Water model of freely swing lance injecting limestone powder to dephosphorization. *Iron Steel* **2019**, *54*, 28–32.
25. Wang, L.-Y.; Xue, Z.-L.; Chen, K.-F.; Hu, B. High-temperature decomposition kinetics of large particle-size limestone. *J. Chongqing Univ.* **2020**, *43*, 32–46.
26. Cao, J.; Qiao, X.-C.; Liu, C.-L.; Zhang, J.-S. A study on kinetics of limestone decomposition in air and $CO_2$ atmospheres. *Inorg. Chem. Ind.* **2016**, *48*, 32–36.
27. Zhang, W.-X.; Liu, L.-S.; Cao, H.-J.; Wu, B.-K.; Cheng, Z.-P. Effect on kinetics of limestone decomposition under different $CO_2$ atmospheres. *Inorganic Chem. Ind.* **2020**, *52*, 59–63.
28. Chen, H.; Zhang, S.-H.; Yang, H.-P.; Li, P.; Chen, H.-P.; Zeng, J. Study on thermal decomposition kinetics of limestone with large particle size. *Inorg. Chem. Ind.* **2013**, *45*, 11–14. [CrossRef]
29. Li, C.-X.; Li, H.; Zhou, B.; Lv, Y.-C.; Qin, D.-P.; Wei, S.-H. Experimental study on steelmaking using limestone instead of lime as the slagging material in 100t converter. *China Metall.* **2015**, *25*, 22–26+49.
30. Lu, H.; Li, C.-X.; Mao, W.-W.; Li, H. The effect of $CO_2$ oxidation on hot metal arising from $CaCO_3$ decomposition in BOF. *Steelmaking* **2017**, *33*, 49–55.
31. Sun, H.-K.; Li, C.-X.; Wang, S.-H.; Zhang, K.-X.; Tong, S.; Zhang, Y. Research on microstructure evolution behavior of small particle limestone during rapid calcination at high temperature. *Steelmaking* **2022**, *38*, 38–42+58.
32. Hu, R.-Z.; Gao, S.-L.; Zhao, F.-Q. *Thermal Analysis Kinetics*; Beijing Science Press: Beijing China, 2008; pp. 29–30, 119–120, 138–140, 151–155.

processes

Article

# Promoting Effect of Microwave Field on Gas Phase Diffusion Limited Magnetite Reduction in Carbon Monoxide

**Meijie Zhou, Liqun Ai, Lukuo Hong *, Caijiao Sun and Shuai Tong**

College of Metallurgy and Energy, North China University of Science and Technology, Tangshan 063210, China
* Correspondence: honglk@ncst.edu.cn

**Abstract:** To investigate the effect of microwave irradiation on the rate of magnetite reduction while increasing the gas phase diffusion rate limit, the microstructure and kinetics of CO reduction of magnetite powder were studied. The investigation was conducted through microwave irradiation and conventional heating at 900~1100 °C. Under the two heating methods, the iron crystal is selectively reduced and gradually expanded along the direction normal to the length of the ore powder, forming a strip of iron crystal that penetrates the powder and expands outward across the width. The microwave field can effectively improve the sintering of minerals. The changes in Avrami exponents $m$ and $k$ in the reduction process were determined by combining the Johnson–Mehl–Avrami (JMA) model with the *lnln* method. The microwave field did not change the limiting step. Microwave irradiation proves to be the most effective means to enhance both the initial reduction rate and the rate during the primary iron crystal precipitation phase. The morphology of the iron crystal takes on a dense punctate shape, influenced by the rate of diffusion control.

**Keywords:** CO reduction; magnetite; microwave irradiation; microstructure characterization; kinetics

**Citation:** Zhou, M.; Ai, L.; Hong, L.; Sun, C.; Tong, S. Promoting Effect of Microwave Field on Gas Phase Diffusion Limited Magnetite Reduction in Carbon Monoxide. *Processes* **2023**, *11*, 2709. https://doi.org/10.3390/pr11092709

Academic Editors: Xin Yao and Huaqing Xie

Received: 9 August 2023
Revised: 2 September 2023
Accepted: 4 September 2023
Published: 11 September 2023



## 1. Introduction

The blast-furnace converter process is the prevailing method in steelmaking. This is a relatively mature technology that achieves high production efficiency. The main characteristics of the process are the excessive reduction of silicon, phosphorus, and sulfur in molten iron, and the carbon metallurgy process of carbon saturation. Blast-furnace ironmaking depends on the coke's structure and its function as a reducing agent. Amidst the trend of low-carbon metallurgy and enhancements in electric furnace capacity, gas-based reduction ironmaking offers the potential to eliminate dependence on coke resources and mitigate impurity elements in electric furnace steelmaking. This more environmentally friendly approach reduces carbon emissions, making it the most popular choice for a low-carbon process.

The most representative processes for large-scale industrial production of gas-based solid-state reduction are the MIDREX and ENERGIRON processes. The future advancement of gas-based solid-state reduction faces two primary challenges: non-carbon heat source protection, and production efficiency improvement. The former can be solved by replacing carbon heat with electric energy, while hydrogen-rich reducing gas ($H_2$-CO instead of CO) and a higher reduction temperature help to further improve production efficiency. However, addressing the challenge of mineral phase sintering and iron phase bonding at high temperatures (above 900 °C) remains imperative [1].

Microwave irradiation, an electromagnetic heating method, is characterized by volume and selective heating [2–5]. Microwave irradiation significantly promotes chemical reactions [6]. Iron ore (powder or pellet) exhibits strong absorption characteristics [7]. The research on microwave reduction of iron-ore pellets and powder primarily utilizes the carbothermal reduction method [8–10]. Nonetheless, the literature reflects a few studies on microwave gas-based reduction. Applying the microwave field to the gas-based solid-state reduction process is expected to achieve the dual goals of heat source protection and reduction-rate improvement. It is of great importance to carry out fundamental and prospective studies on the metallurgical effect and mechanism of action of microwave irradiation, while exploring and studying various potential applications of microwave treatment in metallurgical processes.

The common properties of gas-based solid-state reduction of iron-ore powder or pellets are refined from basic metallurgical mechanics. The reduction process can be divided into the following component steps:

1. Gas phase diffusion. This includes gas phase volume diffusion, pore diffusion, and gas-film diffusion of reducing and product gas.
2. Solid phase diffusion. This includes the diffusion of the reduction process through the product layer and the internal diffusion of unreduced minerals.
3. Interface reaction. This includes interface physics, chemical adsorption, and chemical reaction.
4. Nucleation and growth. This includes the precipitation nucleation and growth of the new product phase.

Based on the relative resistance of the above steps in the reduction process, it becomes feasible to identify the primary limiting factors within the kinetics of the reduction process. The key to improving the metallurgical effect is mitigating the impedance caused by the constraining steps. In general, due to the high relative rate of the interfacial reaction in the reduction process, diffusion (gas or solid phase) can easily become a limiting factor in gas-based reduction. According to previous research, microwave irradiation can increase the solid phase diffusion rate [11]. However, whether microwave irradiation can improve the reduction rate when there is a gas phase diffusion rate limit has not been reported. In this study, a comparative analysis between microwave and conventional heating conditions was conducted, focusing on the microstructure and product morphology of the reduction process. CO has a lower diffusion capacity than $H_2$, and the diffusion resistance of the ore-powder layer is higher than that of the ore-pellet bed. Therefore, this study designed a material bed using magnetite concentrate powder as the raw material and CO gas as the reducing gas. Experimental research was carried out to verify that the reduction process occurred within the constraints of the gas phase diffusion limitation mechanism.

## 2. Experiment

Magnetite concentrate powder obtained from a steel plant production unit was heated and dehydrated at 180 °C for 4 h in a drying oven. The X-ray diffraction pattern (XRD) of the magnetite powder indicated its composition to be 95.7% magnetite with small amounts of hornblende and mica [12]. The particle size distribution of the magnetite concentrate powder was measured using a Mastersizer 2000 laser particle size analyzer, presenting values ranging from 10 to 200 μm; the median particle size ($D_{50}$) was 60.26 μm [12].

The magnetite powder (15 g) was placed in a corundum crucible (60 mm × 30 mm × 15 mm) without compaction. A tube furnace with conventional heating and microwave irradiation (1–4 kW, 2.45 GHz) was used in the experiment. The inner diameter (40 mm) and the length (100 mm) of the quartz tube. The heating rate was controlled at 20 °C/min. Once the reduction temperature was reached, it was maintained for a holding time of 40 min. High temperature and prolonged heating ensured that the temperature inside and outside the furnace tube remained consistent and uniform. A schematic diagram of the experimental device is shown in Figure 1.

Argon gas (purity 99.99% Ar) was used to protect the heating and cooling process. Different volumetric proportions of Ar-CO were introduced into the reduction process, and the total Ar-CO gas flow rate was 1.66 L/min. The gas flow was separately controlled by the mass flowmeter. A higher flow rate was used to avoid the experimental interference that can be caused by an inadequate CO flow rate. An online gas analyzer (Gasboard-3100), which was connected to the gas outlet of the tube-furnace reactor, was used to detect and record the real-time gas compositions. To facilitate prompt gas phase analysis, the quartz furnace tube was filled with a sealed quartz tube to shorten the residence time of the gas in the tube. After cooling, the magnetite was weighed to calculate the weight loss. The reduced products were subjected to chemical analysis, mineral phase composition analysis via XRD, and electron microscope morphology analysis via scanning electron microscopy (SEM). Details on the conventional tubular electric-heating furnace can be found in a previous study [13].

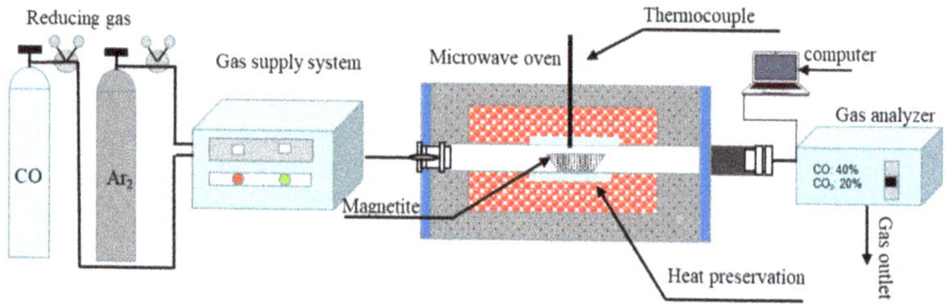


**Figure 1.** Microwave experiment system.

## 3. Results and Discussion

### 3.1. Reduction Indexes

Notably, the degree of metallization denotes the degree of conversion of iron oxide into metallic iron during reduction. It is defined as the percentage of the mass of metallic iron (*MFe*) divided by the total iron mass (*TFe*) in the reduced product ore powder, as shown in Equation (1).

$$Md = \frac{MFe}{TFe} \times 100\%. \quad (1)$$

During CO reduction of the iron-ore powder, oxygen in the iron oxide reactant is combined with the CO reducing gas to form $CO_2$, which decreases the reactant mass. Therefore, the reaction process can be analyzed by the changes of quality or by the amount of $CO_2$ released. The reduction ratio ($x$) is a key parameter adopted to characterize the reduction difficulty of reactants, that is, the difficulty of removing oxygen from reactants during the reaction process; moreover, it is an important fundamental parameter for assessing the rate-determining step. The calculation formula is as follows (Equation (2)):

$$x = \frac{\Delta W_t}{W_\infty}. \quad (2)$$

Here, $\Delta W_t$ is the mass of accumulated oxygen loss at time $t$, and $W_\infty$ is the initial mass of oxygen contained in the solid iron oxide.

Figure 2a,b illustrate the effects of temperature on metallization degree and reduction ratio, respectively. The experiment was conducted for 40 min under 60 vol.% CO-Ar reduction gas via microwave and conventional heating.

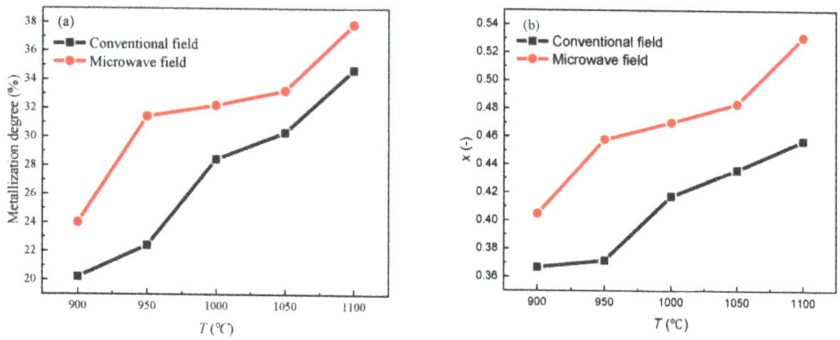


**Figure 2.** Variation in the metallization degree of magnetite under microwave or conventional heating at different temperatures: (**a**) metallization degree (**b**) reduction ratio.

Under the 60% CO-Ar atmosphere, both heating processes exhibited an increase in metallization degree and reduction ratio with increasing temperature, and the reduction processes were enhanced by microwave irradiation. The metallization degree and reduction ratio under microwave irradiation were higher than those under conventional heating. At 950 °C, the *Md* and *x* under the microwave field were 1.45 and 1.23 times higher, respectively, than those under the conventional field. As the temperature rose, the effect of microwave irradiation on enhancing reduction decreased. At 1100 °C, the metallization degree and reduction rate under the microwave field were 1.16 and 1.09 times those under the conventional field, respectively. However, under certain conditions, the reduction rate and metallization rate remained low, in accordance with the basic characteristics of diffusion limitation.

When the iron oxide is completely reduced, $x = Md/100$. As shown in Figure 2, there is a large differential value between $Md/100$ and $x$. This indicates that the reduction process has an intermediate (FeO) phase. This is due to the following two reactions:

$$Fe_3O_4 + CO = 3FeO + CO_2. \tag{3}$$

$$FeO + CO = Fe + CO_2. \tag{4}$$

According to $\%TFe$, $\%MFe$, $x$, etc., the oxygen loss rate corresponding to Fe and intermediate FeO in the reduced product can be calculated. The mass loss caused by the reduction process due to the removal of S and P can be disregarded, as it accounts for a negligible proportion of the total mass. The above difference represents the oxygen-loss rate of the intermediate phase FeO generated by the reduction of $Fe_3O_4$:

$$x = \frac{\Delta W_t}{W_\infty} = \frac{\Delta WO_{FeO} + \Delta WO_{Fe}}{W_\infty} = x_{FeO} + x_{Fe} = \frac{\left(\%FeO_t \frac{M_O}{3M_{FeO}} + \%MFe_t \frac{4M_O}{3M_{Fe}}\right)}{0.957 \frac{4M_O}{M_{Fe_3O_4}}} \cdot \frac{W_t}{W_0}. \tag{5}$$

Based on this, the ratio of oxygen lost in the Fe and FeO phases to the total oxygen loss can be obtained ($x_{Fe}/x$, $x_{FeO}/x$) as shown in Figure 3a. Further, the contents of Fe and FeO in the products were obtained (Figure 3b). Microwave irradiation can promote the conversion of $Fe_3O_4$ to FeO and the conversion of FeO to Fe. With the increase in temperature, the effect of the microwave field on the conversion of $Fe_3O_4$ to FeO becomes more apparent.

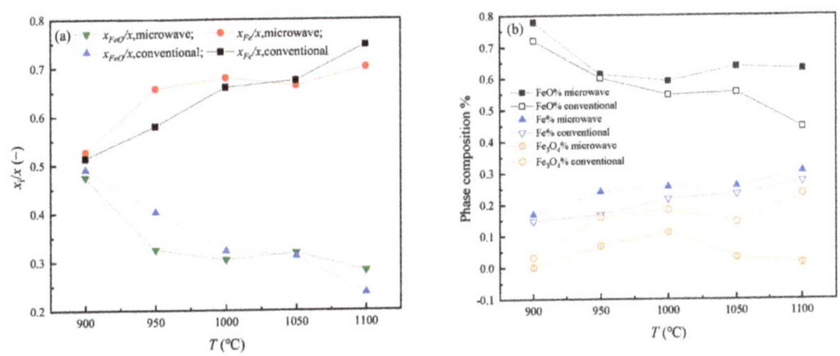


**Figure 3.** (**a**) The oxygen loss rate of each step. (**b**) Phase composition after reduction.

## *3.2. Phases and Microscopic Morphology*

The backscattered electron image obtained via SEM analysis of a magnetite particle is shown in Figure 4. A conductive mounting material was selected. The bright white in the figure is copper in the mounting material. Figure 4a shows the backscattered electron pattern of the cross section of the original particle, $Fe_3O_4$ in dark gray. Figure 4b shows the surface of the original fine ore particles, which is relatively smooth and dense, with hardly any pores observed.

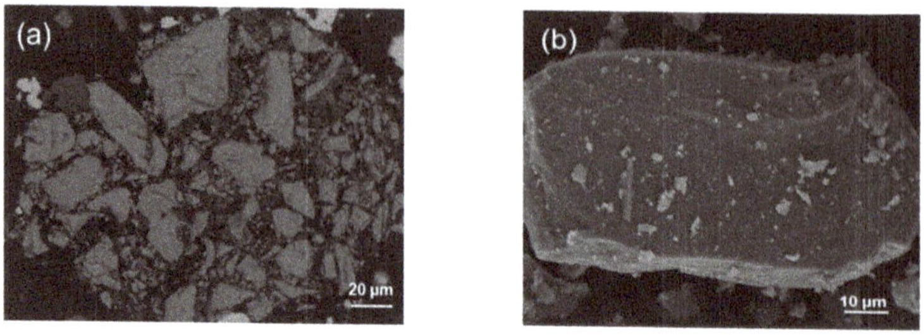


**Figure 4.** SEM images of an original magnetite particle: (**a**) cross section and (**b**) surface.

The micromorphology and structure after reduction under microwave irradiation are shown in Figures 5–7. Figure 5 presents the SEM-energy-dispersive X-ray spectroscopy (EDS) images and compositional data of key points of magnetite reduced at 1000 °C in 60% CO-Ar at different times by microwave irradiation. The atomic ratio O/Fe at Points 1 and 4 is 1:1; hence, the region corresponding to these points can be considered as FeO. The region corresponding to Point 2 exhibits an O/Fe atomic ratio of 4:3, which is attributed to $Fe_3O_4$. Finally, Point 3 indicates the presence of Fe (99.01 at. %).

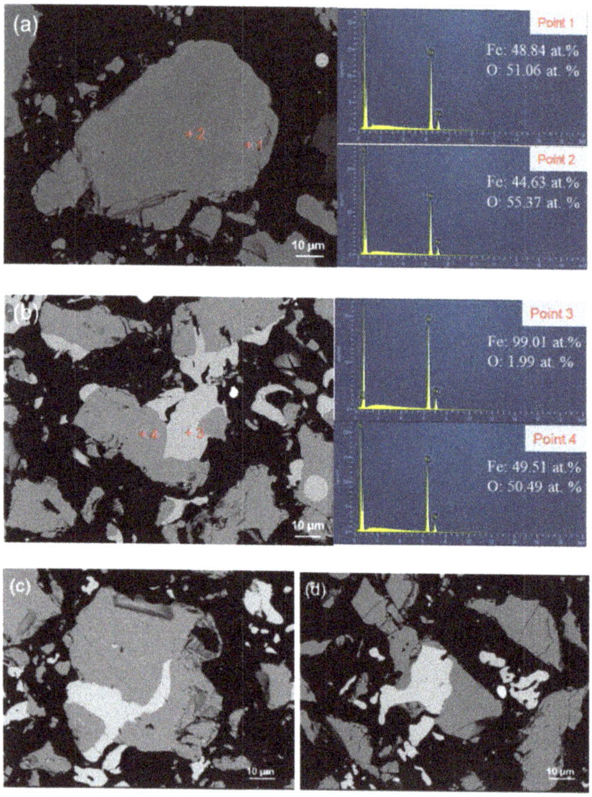


**Figure 5.** Cross-sectional SEM-EDS images of reduced magnetite particles obtained via microwave irradiation at various time intervals: (**a**) 10, (**b**) 20, (**c**) 30, and (**d**) 40 min.

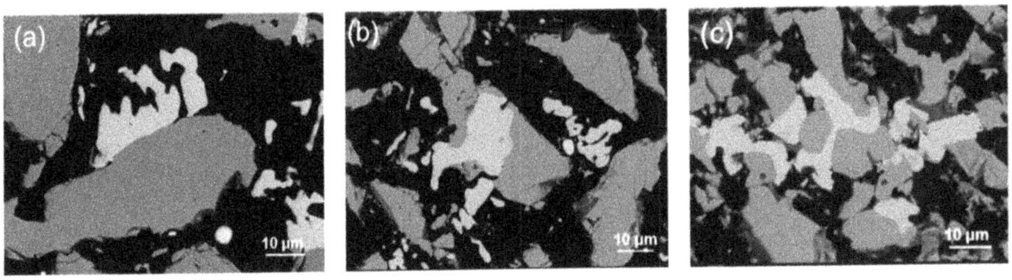


**Figure 6.** SEM images of the reduction at 40 min in a 60% CO-Ar atmosphere, at temperatures of (**a**) 900 °C; (**b**) 1000 °C; (**c**) 1100 °C.

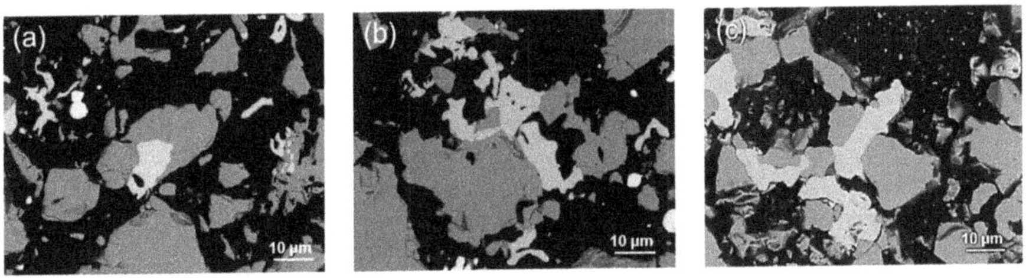


**Figure 7.** SEM images of the reduction at 40 min and 1000 °C, under reduction atmospheres of (**a**) 30%; (**b**) 40%; (**c**) 60% CO-Ar.

Figure 5a,b present the results of magnetite reduction via microwave irradiation at 10 min; here, the sample is taken from the top 1/3 of the material layer. The interior of the particles is composed of unreacted dense $Fe_3O_4$ (Point 2), and the exterior is composed of FeO (Point 1). The thickness of the FeO layer varies greatly at different particle positions, indicating that the reduction process of $Fe_3O_4 \rightarrow FeO$ is not a uniform process to the center of the particles. Figure 4a shows that the $Fe_3O_4$ phase is more compact, but some microcracks are generated at the edge position (Point $\alpha$ in Figure 5a). Some of the crack positions are restored (Point $\beta$ in Figure 5a). Some microcracks develop into large cracks (Point $\gamma$ in Figure 5a), and particle disintegration even occurs (Point $\delta$ in Figure 5a). This phenomenon shows that some cracks are produced before the iron phase. This may be related to the reverse temperature distribution during microwave irradiation (the relative high temperature inside the particles and the relative low temperature outside the particles). After 20 min progression of the reaction, the formation and aggregation of the dense metal iron phase was primarily concentrated within the particle cracks and certain segments of the outer surface. Crystallization occurring within a crack extends the crack into a fissure, leading the iron crystals to develop into strips as they traverse the gap between particles. This transformation ultimately results in the fragmentation of the particles into two distinct parts. However, these two parts isolated by the iron phase will not be further fragmented due to the bonding of the intermediate iron phase. By contrast, the selective reduction and crystallization of some areas on the outer surface should be affected by local CO partial pressure conditions and CO selective adsorption. Figure 5c,d shows the microstructure at 30 and 40 min of reduction, respectively, with the samples taken from the middle 1/3 and lower 1/3 of the material layer in the crucible, respectively. The figure shows that the iron phase produced by reduction mainly exists at the crack, and the reduction crystallization on the surface of the particles decreases. Compared with Figure 5b, the reduction process differs from the surface layer to the bottom layer inside the crucible, due to the deterioration of CO diffusion conditions. The upper region experiences a higher degree of reduction.

Figure 6 shows the cross-sectional morphology of magnetite particles reduced by 60% CO-Ar under a microwave field at different temperatures. Figure 7 shows the morphology of magnetite reduced by a microwave field at 1000 °C under different CO partial pressures. It can be seen that even under different conditions, the metal iron phase produced by reduction is dense and has no obvious porous structure. Higher temperature and higher CO partial pressure can obtain a higher metallization degree.

SEM images of the reduced samples at different time intervals, under conventional heating at 1000 °C and a reduction atmosphere of 60% CO-Ar, are shown in Figure 8. Metal iron forms under the conventional field at a significantly later stage than under the microwave field. The reduction process produces a large number of microcracks inside the unreduced matrix, resulting in the destruction of the original dense continuous structure. The iron phase is primarily formed at and infiltrated into the internal cracks in the particles, and there is no obvious iron phase on the outer surface of the particles. Therefore, under conventional heating at a higher reduction temperature, the unreduced phase undergoes sintering. By contrast, when under microwave irradiation, the sintering of the unreduced phase manifests as bridging and adhesion between iron phases. Microwave irradiation can effectively improve the phenomenon of mineral sintering. The expansion of the material bed is obvious in the reduction process under the microwave field, whereas its expansion is relatively small after conventional field reduction. The expansion phenomenon diminishes with the increase in temperature. Material-bed expansion provides a channel for gas diffusion and accelerates the reduction reaction.

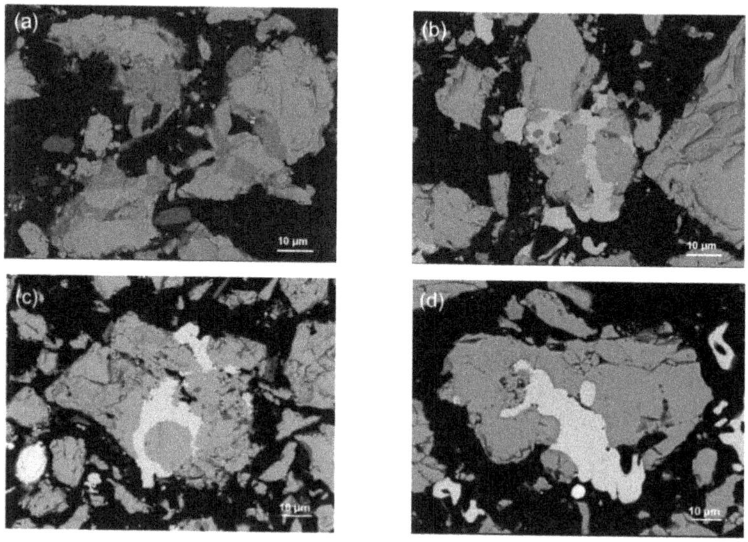


**Figure 8.** Cross-sectional SEM images of reduced magnetite particles obtained via conventional heating at various time intervals: (**a**) 10, (**b**) 20, (**c**) 30, and (**d**) 40 min.

There are generally three types of iron crystals produced by reduction [14,15]: dense punctate or layered, porous, and whisker fibrous. When the reduction potential of the reducing atmosphere is high, iron crystals are primarily precipitated in the first form. Therefore, the hydrogen-reduction iron-ore powder is generally porous [16]. In the process of CO fluidization reduction, iron whiskers are easily produced. Investigations [17] indicated that three distinct final product morphologies can be obtained on the reduction of wustite in $CO/CO_2$ or $H_2/H_2O$ gas mixtures: porous iron, porous wustite covered with dense iron, and dense wustite covered with dense iron. Matthew [18] reports that the porous structure is mainly controlled by atmospheric conditions and temperature. With the increase of the $CO/CO_2$ or $H_2/CO$ partial-pressure ratio, the iron phase morphology of the product develops from dense to bursting and then to honeycomb porous. Heating is beneficial for the formation of porous structures. At lower proportions of $CO_2$ or $H_2O$, however, despite the initial formation of a dense iron layer, the final structure comprises a porous iron morphology.

Under the conditions of this experiment, the iron crystals form dense bands across various time intervals, temperature levels, and CO partial pressures. The initial particles do not coat the outer surface, and there is no obvious porous structure under high CO partial pressure. Instead, they are selectively reduced and crystallized from the newly formed microcracks and gradually expanded along the normal direction of the ore powder length. Their in-depth growth forms a strip of iron crystal that penetrates the ore powder and expands outward width-ways as a result of the combined action of CO adsorption orientation [19] and mineral microdefects. This study shows that the prerequisite for the generation of porous structures is that the reduction takes places under a non-diffusion-limiting mechanism.

Figure 9 depicts the XRD patterns of reduced magnetite obtained using the microwave and conventional heating methods under a 60% CO-Ar atmosphere at 1000 °C. From the XRD analysis, the phase transition of magnetite during the reduction process was determined. Overall, in both heating methods, the diffraction peak of $Fe_3O_4$ steadily diminished, whereas the diffraction peak of Fe gradually increased. Under microwave irradiation, the FeO diffraction peak appeared after 10 min, the Fe peak appeared after 20 min, and the $Fe_3O_4$ diffraction peak completely disappeared after 40 min. Under conventional heating, the FeO diffraction peak appeared after 10 min, the Fe peak appeared after 30 min, and the $Fe_3O_4$ peak was still present after 40 min. The two reduction processes (CO reduction of $Fe_3O_4$ to $Fe_xO$, and $Fe_xO$ reduction to Fe) intersect. Owing to the high potential for CO reduction in the near-surface region of the ore powder, iron is produced before reduction completely eliminates the $Fe_3O_4$. The reduction of $Fe_3O_4$ by $H_2$ is divided into two independent processes: $Fe_3O_4 \rightarrow Fe_xO$, and $Fe_xO \rightarrow Fe$ [20].

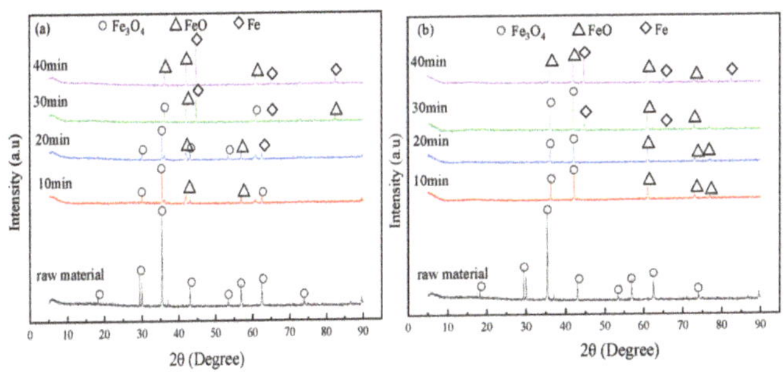


**Figure 9.** XRD diffraction patterns of reduced magnetite after (**a**) microwave and (**b**) conventional heating recorded at various time intervals (10, 20, 30, and 40 min).

### 3.3. Micro Kinetics

Detailed analysis of the microstructure of the reaction products shows that the iron crystals formed under the experimental conditions are dense punctate. However, porous iron crystals are formed during reduction under many other process conditions. Due to the change in material structure and mass transfer conditions in the reduction process, the rate-limiting steps will vary across different stages, thus undergoing different mechanisms of reduction. Therefore, it is necessary to conduct a kinetic analysis of the magnetite reduction process. Notably, understanding the rate-determining steps in complex metallurgical reactions is highly significant in the field of metallurgical dynamics theory, as well as in reactor design for practical industrial applications. For reactors in actual production, complex metallurgical reactions are involved. Figure 10a presents variation curves of $CO_2$ formation with time when magnetite is reduced under a 60% CO-Ar environment at 900–1100 °C. The experimental data were obtained using an online infrared gas analyzer, as previously described. The change in reaction rate with time was calculated in accordance with the change in $CO_2$ % generated by the reduction reaction (Figure 10b).

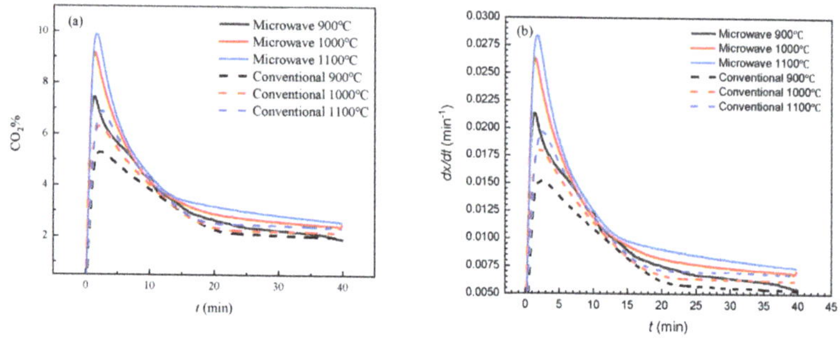


**Figure 10.** (**a**) Variation in the percentage of $CO_2$ generation from magnetite reduction via microwave and conventional heating. (**b**) The reduction ratio of microwave irradiation and conventional heating changes with time.

Figure 11 presents the reduction degree of the magnetite sample reduced at various temperatures (900, 1000, and 1100 °C) for 40 min using microwave and conventional heating. The slope of the curve represents the reduction reaction ratio, which increases with increasing reduction time.

A model-fitting method can be used to determine the rate control step of the reduction process [21–23]. The obtained reaction kinetics can be fitted with the gas–solid reaction models in Table 1. The corresponding expressions $y(x)$ are given on the left-hand sides of the equations in Table 1. If the expression $y(x)$ fits the experimental data well throughout the reaction process, plotting it under constant temperature conditions yields a straight line with a slope $k$ (rate constant). Therefore, fitting the data at different temperatures by analyzing $y(x)$ against time ($t$) produces a series of rate constant values. Tables 2 and 3 display the average values of the coefficient of determination ($R^2$) corresponding to different $y(x)$ function lines for the microwave and conventional heating processes, respectively. Figures 12 and 13 present the fitted curves of CO-reduced magnetite under microwave and conventional heating, respectively. When the reduction reaction gas flow begins, the CO concentration in the reaction gas rapidly increases from zero to the target concentration in the gas flow. Therefore, this time period is excluded from the study, and the kinetic analysis of microwave and conventional heating begins at 2 min and 2.7 min, respectively.

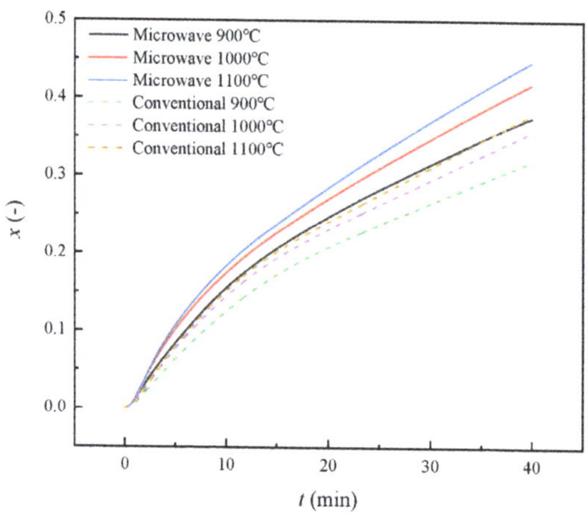


**Figure 11.** Variation in reduction ratio with time for microwave and conventional heating.

**Table 1.** Gas–solid reaction models.

| No. | Reduction Mechanism | Equation $y(x) = k_1 t$ |
|---|---|---|
| 1 | phase-boundary-controlled reaction (infinite slabs) | $x = k_1 t$ |
| 2 | phase-boundary-controlled reaction (contracting cylinder) | $1 - (1-x)^{1/2} = k_1 t$ |
| 3 | phase-boundary-controlled reaction (contracting sphere) | $1 - (1-x)^{1/3} = k_1 t$ |
| 4 | one-dimensional diffusion | $x^2 = kt$ |
| 5 | two-dimensional diffusion | $x + (1-x)\ln(1-x) = k_1 t$ |
| 6 | three-dimensional diffusion | $1 - 3(1-x)^{2/3} + 2(1-x) = k_1 t$ |
| 7 | first-order reaction (random nucleation) | $[-\ln(1-x)] = k_1 t$ |
| 8 | two-dimensional growth of nuclei (Avrami–Erofeyev equation) | $[-\ln(1-x)]^{1/2} = k_1 t$ |
| 9 | three-dimensional growth of nuclei (Avrami–Erofeyev equation) | $[-\ln(1-x)]^{1/3} = k_1 t$ |

**Table 2.** Correlation coefficients of $y(x)$ at different temperatures under microwave irradiation.

| | $R^2$ (900 °C) | $R^2$ (1000 °C) | $R^2$ (1100 °C) |
|---|---|---|---|
| 1 | 0.9653 | 0.9677 | 0.9809 |
| 2 | 0.9771 | 0.9796 | 0.9896 |
| 3 | 0.9804 | 0.9829 | 0.9919 |
| 4 | 0.9956 | 0.9947 | 0.9950 |
| 5 | 0.9915 | 0.9898 | 0.9897 |
| 6 | 0.9896 | 0.9876 | 0.9873 |
| 7 | 0.9863 | 0.9886 | 0.9953 |
| 8 | 0.8923 | 0.8959 | 0.9563 |
| 9 | 0.8195 | 0.8197 | 0.9314 |

**Table 3.** Correlation coefficients of $y(x)$ at different temperatures under conventional heating.

| | $R^2$ (900 °C) | $R^2$ (1000 °C) | $R^2$ (1100 °C) |
|---|---|---|---|
| 1 | 0.9719 | 0.9755 | 0.9784 |
| 2 | 0.9791 | 0.9829 | 0.9858 |
| 3 | 0.9812 | 0.9851 | 0.9879 |
| 4 | 0.9981 | 0.9973 | 0.9954 |
| 5 | 0.9963 | 0.9945 | 0.9916 |
| 6 | 0.9955 | 0.9933 | 0.9900 |
| 7 | 0.9852 | 0.9889 | 0.9915 |
| 8 | 0.9352 | 0.9442 | 0.9485 |
| 9 | 0.9087 | 0.9197 | 0.9236 |

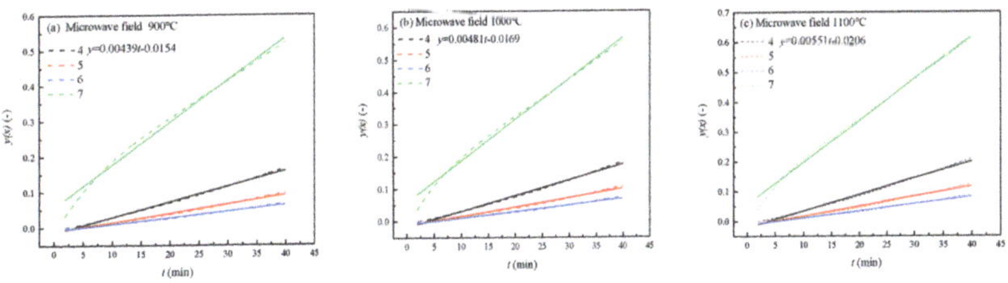


**Figure 12.** Fitted lines of the relationship between $y(x)$ and $t$ for CO reduction under microwave irradiation.

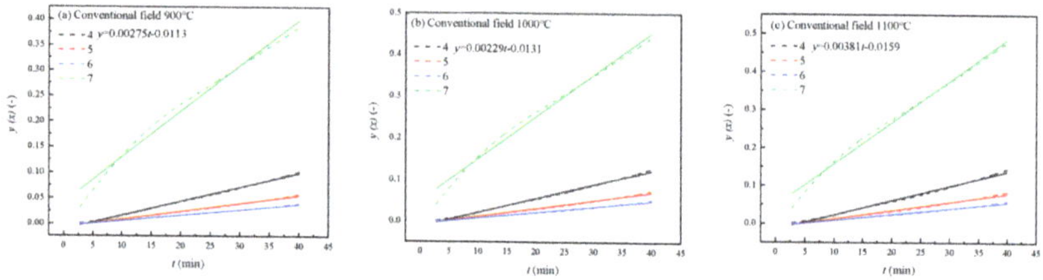


**Figure 13.** Fitted lines of the relationship between $y(x)$ and $t$ for CO reduction under conventional heating.

From Tables 2 and 3, under the microwave field, model 4 (diffusion control) has a higher fitting degree at 900 °C and 1000 °C. Models 7 and 4 (apparent first-order reaction and diffusion control, respectively) have a higher fitting degree at 1100 °C. Under the conventional field, model 4 (diffusion control) has a higher fitting degree at 900–1100 °C, and displays obvious one-dimensional characteristics. This shows that microwave irradiation does not significantly change the macroscopic restrictive step of magnetite powder reduction, but at a high temperature (1100 °C), a microwave field has a tendency to cause the apparent first-order reaction.

### 3.4. Microkinetics

The macroscopic kinetic analysis mentioned above, relying on the linear model, assumes that the restrictive steps of the reduction process remain constant. However, changes may occur to the restrictive steps of the reduction process due to changes in gas phase composition, mineral phase, and product structure. Therefore, it is necessary to study the microkinetic mechanism in the reduction process. In order to analyze the microscopic process kinetics of magnetite concentrate powder reduction, an approach combining the *lnln* method with the JMA (Johnson–Mehl–Avrami) model was used.

Based on the *lnln* method [21], the possible intervals of each limiting step are estimated, providing a reference for the fitting of segmentation and hybrid control.

According to the *lnln* method, the reduction mechanism model can be uniformly expressed as:

$$x = 1 - \exp\left[-(k_2 t)^m\right], \tag{6}$$

where $x$ is the degree of reduction at time $t$ (%), $t$ is reaction time (s), $k$ is the reduction rate, and $m$ is the Avrami exponent.

Therefore:

$$\frac{dx}{dt} = mk(k_2 t)^{m-1} \cdot \exp[-(k_2 t)^m], \tag{7}$$

$$\frac{dx}{dt} \cdot t = m(k_2 t)^m \cdot \exp\left[-(k_2 t)^m\right] = m[-\ln(1-x)] \cdot (1-x). \tag{8}$$

Therefore:

$$m = (\frac{dx}{dt} \cdot t)/\{[-\ln(1-x)] \cdot (1-x)\} = \frac{\frac{dx}{dt}}{(1-x)} \cdot \frac{t}{[-\ln(1-x)]}. \tag{9}$$

The first term on the right of the above formula is the relative specific rate (the specific rate per unit mass of the remaining unreacted phase). The second term is $k_c$ reduced by the apparent first-order reaction.

Therefore:

$$\frac{dx}{dt} = m \cdot [k_c(1-x)] = m \cdot (\frac{dx}{dt})_c, \tag{10}$$

where $k_c$ and $(\frac{dx}{dt})_c$ represent the converted rate constant and rate of the corresponding $t$ and $x$, respectively, converted according to the apparent first-order reaction (random nucleation) model. The above formula shows that the actual reduction transient rate $dx/dt$

is $m$ times the fitting rate of the first-order reaction (which can be physically understood as a multidimensional extension of the one-dimensional reaction).

In Figure 14, the control mechanism range corresponding to the $m$ value [24] is marked. The microwave field does not change the limiting step. The restrictive steps under the two heating methods are primarily characterized by diffusion. By using the calculation results of the $m$-$t$ relationship and the experimental results of the $x$-$t$ relationship, the change in $k$ with the reduction ratio $x$ can be calculated. The results are shown in Figure 15. In the first stage of reduction, the $k$ value decreases sharply, and in the second stage, it is almost flat. The $k$ value increases with the increase in temperature. Based on a $k$ value of 900 °C under the conventional field, the ratio $\beta$ of each temperature to 900 °C under the conventional field is calculated (Figure 16). The reduction rate constant at 900 °C under the microwave field is 1.11–1.57 times that under the conventional field. The reduction rate constant at 1100 °C under the microwave field is 1.47–2.29 times that at 900 °C under the conventional field. Microwave irradiation proves to be the most effective way to improve both the initial reduction rate and that exhibited during the primary iron crystal precipitation phase.

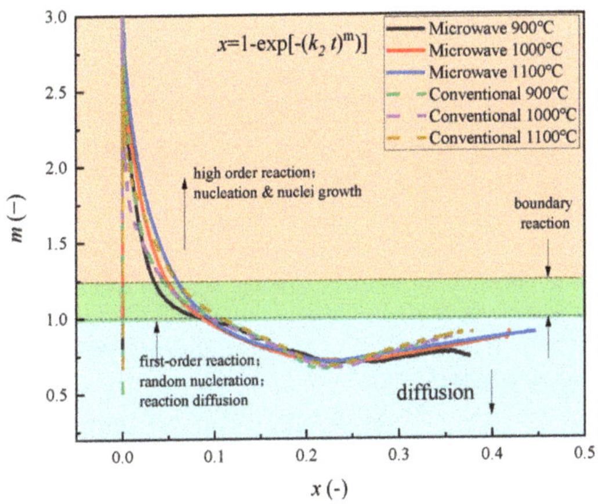


**Figure 14.** The range of control mechanism corresponding to $m$-value.

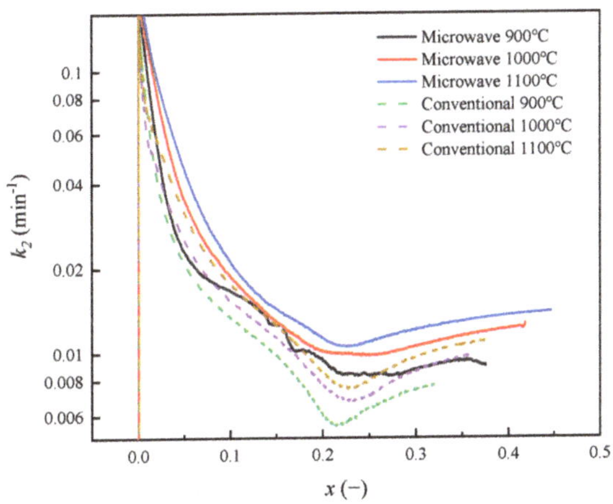


**Figure 15.** The rate constants under different conditions.

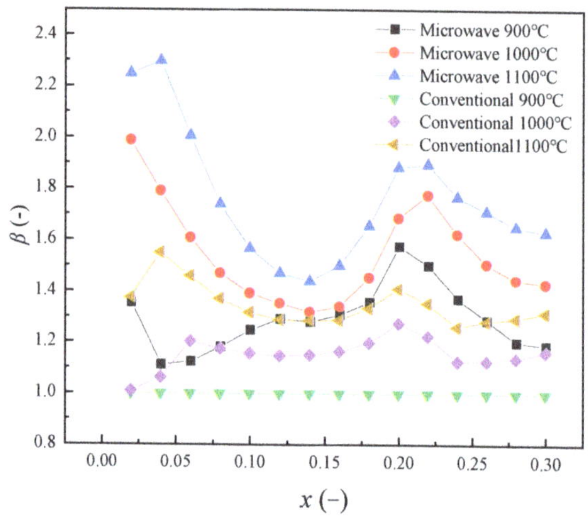


**Figure 16.** The ratio at each temperature based on the conventional 900 °C.

Jiang Chen [25] found that the formation of porous iron by CO reduction is related to CO partial pressure and reduction temperature. However, under the experimental conditions of the present study, the reduced iron is a dense punctate. The morphology of iron crystals may be essentially controlled by the ratio of diffusion rate to reaction rate [19]. It is also related to the adsorption force of CO and $H_2$ on precipitated iron, which is the root cause of the difference in the driving force of precipitated iron growth. Therefore, under the experimental conditions, when the diffusion rate is relatively low, it tends to produce a dense structure.

The avrami indexes $m$ and $k$ obtained from the above calculation are substituted into Equation (6). The obtained $x$-$t$ relationship completely coincides with the experimental results, indicating that the obtained $m$ and $k$ are correct. The JMA model can describe the reduction process under diffusion constraints.

### 3.5. Activation Energy

According to the reduction ratio displayed in Figures 12 and 13, the apparent activation energy of magnetite reduction to metallic iron can be calculated using the Arrhenius equation (Equation (11)):

$$k = \mathrm{A}exp\left(-\frac{E_a}{RT}\right). \tag{11}$$

Here, $k$ is the reaction rate constant, A is a pre-exponential factor, $R$ is the gas constant, $T$ is temperature, and $E_a$ is the apparent activation energy. Furthermore, taking the logarithm of both sides of Equation (11), a linearized Arrhenius equation can be obtained (Equation (12)).

$$\ln k = -\frac{E_a}{RT} + \ln A. \tag{12}$$

The activation energy under each experimental condition is obtained from the fitted linear equations displayed. The microkinetic activation energy was calculated (Figure 17). The activation energy under the microwave field was 15.106 kJ·mol$^{-1}$, and the activation energy under the conventional field was 21.949 kJ·mol$^{-1}$.

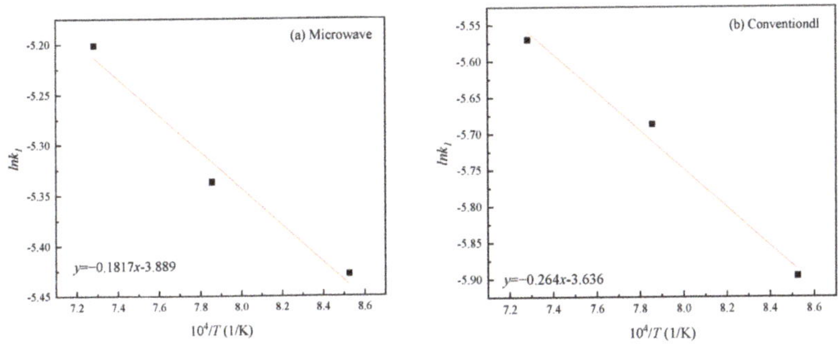


**Figure 17.** Microkinetic activation energy; (**a**) microwave field, (**b**) conventional field.

The apparent activation energy at different stages can be calculated according to the $k_2$ obtained from microscopic kinetic analysis (Figure 18). The activation energy under the microwave field is 11.889–21.283 kJ·mol$^{-1}$, and the activation energy under the conventional field is 18.374–23.196 kJ·mol$^{-1}$. The activation energy under the microwave field is lower than that under the conventional field because microwave irradiation intensifies the oscillation of molecules at a low temperature, thereby achieving the effect of high temperature. Microwave irradiation also accelerates the oscillation of CO, thereby accelerating the adsorption of CO on the lattice surface. Due to the oscillation, the lattice gap increases and the diffusion becomes easier, thereby accelerating the internal diffusion.

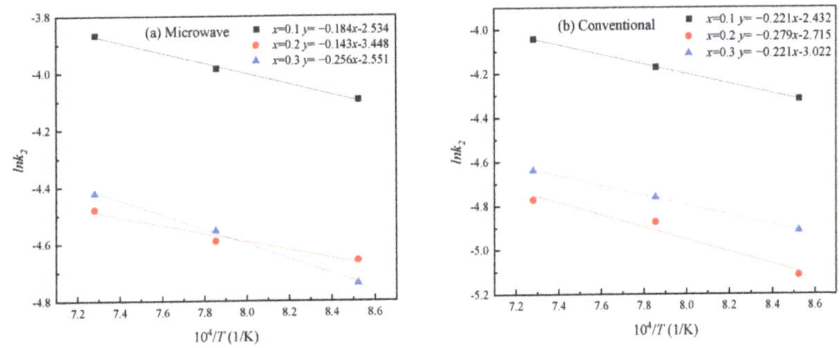


**Figure 18.** Microscopic kinetic activation energy; (**a**) microwave field, (**b**) conventional field.

## 4. Conclusions

The kinetics of CO reduction of magnetite under microwave irradiation and conventional heating conditions were studied. The impact of the microwave field on the reduction effect and process under the condition of gas phase diffusion as a restrictive step was clarified, and the morphology characteristics of the product layer were analyzed. The results lay a foundation for the kinetic study of CO + $H_2$ reduction. The primary conclusions are summarized as follows:

1. Under the 60% CO-Ar atmosphere at 900–1100 °C, the metallization rate of the microwave field is 1.4 times that of the conventional field. The reduction degree of the microwave field is 1.2 times that of the conventional field. This shows that the microwave field can promote gas phase diffusion, but microwave irradiation does not significantly change the restrictive step. The promoting effect of the microwave field diminishes at higher temperature.
2. Cross sections of magnetite particles were observed via SEM analysis. The morphologies of the reduced particles were similar for both microwave irradiation and conventional heating, with metallic iron forming a pointed-inward morphology causing the rupture of the particles. Under the condition of CO-limited reduction, the

iron crystals produced by the reduction process under the microwave field and the conventional field are dense structures; porous components cannot be detected. The microwave field does not significantly change the morphology of the iron crystal. In addition to the influence of a reduction temperature and gas phase composition, the morphology of iron crystal can also be controlled by the reduction limiting step. This is a useful supplement to the conditions for the formation of iron crystal morphology. The morphology of iron crystals may be essentially controlled by the ratio of diffusion rate to reaction rate. When the diffusion rate is relatively low, it tends to produce dense structures.

3. The microstructures of the reduction process under the microwave and conventional fields show that the reduction and crystallization of iron under the condition of CO reduction rate limiting is not like that of the $H_2$ reduction, which will first produce a structure-coating layer around the exterior of the mineral. Instead, it is selectively crystallized from the microcracks generated by the stress in the early stage of the unreduced phase, and gradually expanded along the normal direction of the length of the ore powder. The in-depth growth results in the formation of strips of iron crystals that penetrate the ore powder and then expand outward width-ways, which is the result of CO adsorption orientation and mineral microdefects.
4. Microwave irradiation can effectively improve the phenomenon of mineral sintering. The expansion of the material bed is obvious in the reduction process under the microwave field, whereas it is relatively small under the conventional field. The expansion phenomenon diminishes with the increase in temperature. The bed expansion provides a channel for gas diffusion and accelerates the reduction reaction.
5. By contrast with the $H_2$ reduction, the two reduction processes of CO reduction of $Fe_3O_4$ to $Fe_xO$, and $Fe_xO$ to Fe, overlap. Due to the high potential for CO reduction in the near-surface region of the ore powder, iron is produced before reduction eliminates the $Fe_3O_4$. The reduction of $Fe_3O_4$ by $H_2$ is divided into two independent cascade processes: $Fe_3O_4$ to $Fe_xO$, and $Fe_xO$ to Fe.
6. The JMA model was combined with the *lnln* method to determine the changes in the Avrami indexes $m$ and $k$ in the reduction process. It shows that the Avrami index $m$ can be used to analyze the restrictive steps of the reduction process and its transformation. Under the condition of CO gas diffusion rate limiting, the reaction rate constant at 900 °C under the microwave field is 1.11–1.57 times higher than that under the conventional field at 900 °C. Microwave irradiation proves the most effective way to improve both the initial reduction rate and the reduction rate when the iron phase precipitates.
7. Under the adopted experimental conditions, microwave irradiation reduced the activation energy of the reduction reaction. Microwave irradiation promoted the layer reduction phenomenon and directional reduction (the directional growth of the reduced iron phase) of the magnetite powder. Kinetic analysis of both the macroscopic linear model and the microscopic process shows that under the condition of gas diffusion as the limiting step, the activation energy of microwave irradiation reduction of magnetite concentrate powder is lower than that of conventional heating.

**Author Contributions:** Methodology, L.A.; Validation, L.H., C.S. and S.T.; Formal analysis, S.T.; Data curation, M.Z.; Writing—original draft, M.Z.; Writing—review & editing, L.A.; Visualization, L.H. All authors have read and agreed to the published version of the manuscript.

**Funding:** This study was supported by the Natural Science Foundation of Hebei Province (E2021209101; E2022209112); Science and Technology Research Projects of Higher Education Institutions in Hebei Province (ZD2022125); Tangshan Talent Funding Project (A20220212).

**Conflicts of Interest:** On behalf of all authors, the corresponding author states that there are no conflict of interest.

## References

1. Raymond, L.; Leiv, K. Iron Ore Reduction with CO and $H_2$ Gas Mixtures–Thermodynamic and Kinetic Modelling. In Proceedings of the 4th Ulcos Seminar, University of Wollongong. Available online: https://ro.uow.edu.au/engpapers/1260 (accessed on 1 October 2008).
2. Hunt, J.; Ferrari, A.; Lita, A.; Crosswhite, M.; Ashley, B.; Stiegman, A.E. Microwave-specific Enhancement of the Carbon-carbon Dioxide (Boudouard) Reaction. *J. Phys. Chem. C* **2013**, *117*, 26871–26880. [CrossRef]
3. Zhou, J.; Xu, W.; You, Z.; Wang, Z.; Luo, Y.; Gao, L.; Yin, C.; Peng, R.; Lan, L. A New Type of Power energy for Accelerating Chemical Reactions: The Nature of a Microwave-driving Force for Accelerating Chemical Reactions. *Sci. Rep.* **2016**, *6*, 25149. [CrossRef]
4. Ayvaz, S.I.; Aydin, I. Effect of the Microwave Heating on Diffusion Kinetics and Mechanical Properties of Borides in AISI 316L. *Indian Inst. Met.* **2020**, *73*, 2635–2644. [CrossRef]
5. Amini, A.; Ohno, K.; Maeda, T.; Kunitomo, K. Comparison between Microwave Heating and Conventional Heating of FeS-CaO Mixture during Hydrogen-Reduction. *Chem. Eng. J.* **2019**, *374*, 648–657. [CrossRef]
6. Binner, J.G.P.; Al-Dawery, I.A.H. Microwave Melt Texturing of Bulk YBCO Superconductors. *Supercond. Sci. Technol.* **1998**, *11*, 1230–1236. [CrossRef]
7. Chen, J.; Liu, L.; Zeng, J.Q. Experimental study on reduction of carbonaceous iron ore powder by microwave heating. *Steel* **2004**, *39*, 1–5. [CrossRef]
8. Hayashi, M.; Takeda, K.; Kashimura, K.; Watanabe, T.; Nagata, K. Carbothermic Reduction of Hematite Powders by Microwave Heating. *ISIJ Int.* **2013**, *53*, 1125–1130. [CrossRef]
9. Ishizaki, K.; Nagata, K.; Hayashi, T. Localized Heating and Reduction of Magnetite Ore with Coal in Composite Pellets Using Microwave Irradiation. *ISIJ Int.* **2007**, *47*, 817–822. [CrossRef]
10. Hara, K.; Hayashi, M.; Sato, M.; Nagata, K. Pig Iron Making by Focused Microwave Beams with 20 kW at 2.45 GHz. *Iron Steel Inst. Jpn.* **2012**, *52*, 2149–2157. [CrossRef]
11. Kashimura, K.; Sato, M.; Hotta, M.; Agrawal, D.K.; Nagata, K.; Hayashi, M.; Mitani, T.; Shinohara, N. Iron production from $Fe_3O_4$ and graphite by applying 915 MHz microwaves. *Mater. Sci. Eng. A* **2012**, *556*, 977–979. [CrossRef]
12. Zhou, M.; Ai, L.; Hong, L.; Sun, C.; Yuan, Y.; Tong, S. Comparison of Microstructures of Magnetite Reduced by $H_2$ and CO under Microwave Field. *Metals* **2023**, *13*, 1367. [CrossRef]
13. Sun, C.; Ai, L.; Hong, L.; Li, Y. Study on solid state steelmaking from thin cast iron sheets through decarburization in $H_2O$–$H_2$. *Ironmak. Steelmak.* **2019**, *47*, 1015–1021. [CrossRef]
14. Moujahid, S.E.; Rist, A. The Nucleation of Iron on Dense Wustite: A Morphological Study. *Metall. Mater. Trans. B* **1988**, *19*, 787–802. [CrossRef]
15. Amini, A.; Ohno, K.I.; Maeda, T.; Kunitomo, K. Effect of the Ratio of Magnetite Particle Size to Microwave Penetration Depth on Reduction Reaction Behaviour by $H_2$. *Sci. Rep. UK* **2018**, *8*, 15023. [CrossRef]
16. Kim, S.H.; Zhang, X.; Ma, Y.; Filho, I.R.S.; Schweinar, K.; Angenendt, K.; Vogel, D.; Stephenson, L.T.; El-Zoka, A.A.; Mianroodi, J.R.; et al. Influence of microstructure and atomic-scale chemistry on the direct reduction of iron ore with hydrogen at 700 °C. *Acta Mater.* **2021**, *212*, 116933. [CrossRef]
17. John, D.H.S.; Matthew, S.P.; Hayes, P.C. The breakdown of dense iron layers on wustite in $CO/CO_2$ and $H_2/H_2O$ systems. *Metall. Mater. Trans. B* **1984**, *15*, 701–708. [CrossRef]
18. Matthew, S.P.; Cho, T.R.; Hayes, P.C. Mechanisms of porous iron growth on wustite and magnetite during gaseous reduction. *Metall. Mater. Trans. B* **1990**, *21*, 733–741. [CrossRef]
19. Wei, J. Study on the Mechanism of the Difference in the Morphology of Iron Precipitated during the Reduction of FeO by CO and $H_2$. Ph.D. Thesis, Chongqing University, Chongqing, China, 2018.
20. Kuila, S.K.; Chatterjee, R.; Ghosh, D. Kinetics of Hydrogen Reduction of Magnetite Ore Fines. *Int. J. Hydrogen Energy* **2016**, *41*, 9256–9266. [CrossRef]
21. Abolpoura, B.; Afsahia, M.; Azizkarimi, M. Reduction kinetics of magnetite concentrate particles by carbon monoxide. *Miner. Process Extr. Metall.* **2018**, *127*, 29–39. [CrossRef]
22. Stir, M.; Ishizaki, K.; Vaucher, S.; Nicula, R. Mechanism and kinetics of the reduction of magnetite to iron during heating in a microwave E-field maximum. *J. Appl. Phys.* **2009**, *105*, 124901. [CrossRef]
23. Janković, B. Kinetic analysis of the no isothermal decomposition of potassium metabisulfite using the model-fitting and is conversional (model-free) methods. *Chem. Eng. J.* **2008**, *139*, 128–135. [CrossRef]
24. Hancock, J.D.; Sharp, J.H. Method of Comparing Solid-State Kinetic Data and Its Application to the Decomposition of Kaolinite, Brucite, and $BaCO_3$. *J. Am. Ceram. Soc.* **2010**, *55*, 74–77. [CrossRef]
25. Chen, J.; Zhang, R.; Simmonds, T.; Hayes, P.C. Microstructural Changes and Kinetics of Reduction of Hematite to Magnetite in $CO/CO_2$ Gas Atmospheres. *Metall. Trans. B* **2019**, *50*, 2612–2622. [CrossRef]

processes

MDPI

Article

# Study on Efficient Removal Method of Fine Particulate Dust in Green Metallurgy Process

**Haiying Li, Hairui Xue *, Junya Zhang and Guijie Zhang ***

College of Metallurgy and Energy, North China University of Science and Technology, Tangshan 063000, China; hblglhy@sina.com (H.L.); zhangjy2366@163.com (J.Z.)
* Correspondence: xuehairui93@163.com (H.X.); zhguijie@vip.sina.com (G.Z.)

**Abstract:** In order to solve the problem of the low removal efficiency of fine particles in the flue gases of the metallurgy process, a chemical agglomeration pretreatment method was studied. The coagulant solution of xanthan gum, konjac gum, and their mixtures was selected to research the reunion effects of and the efficiency of gravitational dust removal of fine dust in the gas of the converter flue using a self-built experimental platform. Moreover, the effects of wetting agent type, dust concentration, pressure, and flue gas velocity on the fine grain removal efficiency were investigated. The results showed that the mixed solution of 1 g/L mixed gum and 0.5 g/L SDS had the most obvious effect on the particle size increasing of fine dust particles and the best removal effect when the flue gas velocity was 10 m/s. There was a peak particle size of 85.32 μm increased about eight times larger, and the removal efficiencies reached 51.46% for $PM_{2.5}$ and 53.13% for $PM_{10}$. The Box–Behnken experimental design combined with a response surface analysis method was used to optimize the parameters of the mixed gum concentration, pressure, and flue gas velocity. The optimal removal conditions were 1 g/L, 0.4 MPa, and 10 m/s. The results of this study can provide efficient methods and technical support for pre-processing and efficient removal of fine particles in heavy-polluting industries such as steel making. This will promote the green development of the metallurgical industry.

**Keywords:** green metallurgy; chemical agglomeration; fine particulate dust

**Citation:** Li, H.; Xue, H.; Zhang, J.; Zhang, G. Study on Efficient Removal Method of Fine Particulate Dust in Green Metallurgy Process. *Processes* **2023**, *11*, 2573. https://doi.org/10.3390/pr11092573

Academic Editors: Xin Yao and Huaqing Xie

Received: 31 July 2023
Revised: 19 August 2023
Accepted: 25 August 2023
Published: 28 August 2023



## 1. Introduction

Fine particulate dust is one of the main pollutants causing atmospheric pollution [1]. Since particles with small particle sizes can stay in the atmosphere for a longer period of time, they are thus the cause of hazy weather and reduced visibility. In addition, it poses a serious risk to public health, such as lung cancer and cardiovascular diseases [2]. According to the previous research, fine particulates in China mainly come from industrial production processes such as iron and steel. Currently, industries mainly use bag filters and electrostatic precipitators to remove particulate matter from dust [3], and the comprehensive removal efficiency is more than 99%. However, the collection performance of fine particles, such as $PM_{2.5}$ and $PM_{10}$, is still unsatisfactory with the removal efficiencies being below 50% [4,5]. Therefore, how to improve the fine particle removal effect is the key challenge currently.

To improve the removal efficiency of fine particulate dust, researchers have proposed coagulation and agglomeration to increase larger particle size through physical and chemical action and then removing them by dust collectors, which has achieved certain results in practice [6]. The main current agglomeration techniques are acoustic, turbulent, electrical, and chemical [7–10]. Among them, chemical agglomeration technology was a very promising pretreatment means, and researchers have studied enhanced dust removal from coal-fired power plants [11] and sintering processes [12]. Liu [13] analyzed the effects of the operating parameters, such as the species and concentration of the agglomeration solution, flue gas temperature, pH value of the agglomeration solution, and diameter of the spray droplets, on the fine particle removal efficiency. The results show that increasing the agglomerate concentration facilitates agglomerate size growth. However, as the concentration increases, the viscosity also increases. At the same time, this leads to the deterioration of atomization performance and negatively affects the removal of fine particles. Therefore, the addition of wetting agents is necessary to improve the agglomerate removal effect. The main function of the wetting agent is to reduce the surface tension of the agglomerate; at this time, the agglomerate is easier to flow, so it also plays a role in reducing viscosity. In order to reduce the fugitive emission of fine particles from open-pit lignite mines, Yang [10] developed a lignite dust suppressant by using a mixture of 0.7% guar gum and 0.1% Triton X-100. At present, researchers have extensively studied the wetting performance of coal dust in coal-fired power plants [14–16] and found that wetting agents can effectively reduce surface tension and settling time, which in turn improves the wetting effect of coal dust. Ding [14] investigated the effect of different surfactants on dust suppression efficiency, and the results showed that the addition of anionic and nonionic surfactants effectively improved the average suppression efficiency to 50.07%–70.57%. Zhou [15] conducted a chemical agglomeration experiment on a simulated coal-fired furnace and discussed the effect of air pressure on the chemical agglomeration performance. The results showed that increasing the air pressure was beneficial to produce small droplets, and the collision and collection of droplets on fine particles could be greatly improved. However, the effect of coupling agglomeration on the removal of fine particles in flue gas in the iron and steel industry has not been studied.

In this paper, fine particulate dust in the converter flue gas was selected to study the reduction in $PM_{2.5}$ and $PM_{10}$ emission based on the preparation of coupled agglomerator solution by combining agglomerator and wetting agent.

## 2. Materials and Methods

### 2.1. Experimental Setup

A gas-liquid two-phase flow nozzle atomization was used to construct the chemical agglomeration apparatus test bench shown in Figure 1. The experimental setup included the simulated flue gas system, the nozzle atomization system, the dust agglomeration system, the sampling and measurement system, and the waste liquid collection and treatment system. The simulated flue gas system consists of three parts: an air preheater, a blower, and a powder feeder. The dust agglomeration system mainly consisted of nozzles and agglomeration chambers. Among them, the nozzle adopts a gas–liquid two-phase flow nozzle, in which one side is introduced into high-pressure nitrogen, and the other side is introduced with water or agglomerate solution. The two fluids collide at the nozzle, crushing the agglomerate solution and atomizing it into droplets. The reunification chamber consists of four 10 × 100 × 0.5 cm high-temperature acrylic plates flanked by conical stainless steel inlets and outlets. The sampling device is located at the end of the entire apparatus. A fiberglass membrane is used to collect dust at the end. Finally, the waste liquid flows into a special collection unit for recycling so as not to harm the environment. The test was repeated three times to ensure accuracy.

### 2.2. Experimental Reagent

In the chemical agglomeration process, agglomerating agents can increase the adhesion between particles, and wetting agents can increase the wettability of dust. Under the action of agglomerating and wetting agents, fine particles agglomerate to form large particles. The non-toxic and non-hazardous xanthan gum, konjac gum two agglomerating agents, anionic sodium dodecyl sulfate (SDS) and sodium dodecyl benzene sulfonate (SDBS), nonionic Tween 20 (TW-20) and Tralatone 100 (TX-100) four wetting agents, and deionized water were selected during the experiment.

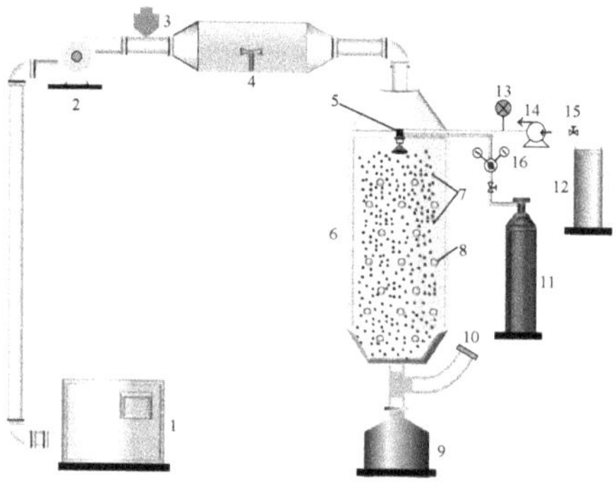


**Figure 1.** Diagram of chemical agglomeration experimental device. 1—air preheating chamber; 2—blower; 3—electromagnetic vibration powder feeder; 4—buffer premixing chamber; 5—gas–liquid two-phase flow nozzle; 6—chemical agglomeration chamber; 7—dust particles; 8—agglomerated liquid droplets; 9—waste liquid collection device; 10—sampling device; 11—nitrogen bottle; 12—agglomerate solution pool; 13—liquid flow meter; 14—centrifugal pump; 15—rotary screw valve; and 16—pressure gauge.

### 2.3. Experimental Method

During the experiment, the viscosity of the agglomerating agents was measured using the viscometer (Shanghai Lichen, NDJ-5S type, Shanghai, China). The contact angle of the agglomerate containing the wetting agent on the dry dust was measured using the contact angle meter (Shanghai Yinuo Precision Instruments, CA 100B type, Shanghai, China), and the filming opportunity was immediately analyzed and measured in the automatic image analysis system by computer to obtain the contact angle. The zeta potential was measured using a zeta potential meter (Malvern, UK, Zeta sizer Mano zs9003030810 type), and the selected experimental objects were consistent with the contact angle experiments conducted. To illustrate the effect of composite agglomerate on dust, the composite agglomerate solution was characterized by a viscometer, contact angle meter, and zeta potential meter.

The dust particles before and after agglomeration, and analysis of the volume distribution, were examined by the laser particle size analyzer (Mastersizer 2000). The agglomeration efficiency was expressed by $\eta$:

$$\eta = \frac{\varphi_{in} - \varphi_{out}}{\varphi_{in}} \times 100\% \tag{1}$$

where $\varphi_{in}$ is the volume fraction of particles before agglomeration, and $\varphi_{out}$ is the volume fraction of particles after agglomeration.

### 2.4. Size Distribution of Raw Dust

The experimental dust samples were obtained from the converter process at a steel plant in Hebei Province, China. The particle size distribution was examined by a laser particle size analyzer. The results are shown in Figure 2. From the results shown, the dust particle size ranges from 0 to 60 μm and the trend has a single-peak distribution. There is a peak at 10.74 μm and it accounts for 5.24%. At this point, at the peak granularity position, the cumulative distribution reaches nearly 70%. The volume fractions of $PM_{2.5}$ and $PM_{10}$ are 22.54% and 61.1%, respectively. The results show that the dust collector is not effective in capturing the fine particles in the flue gas.

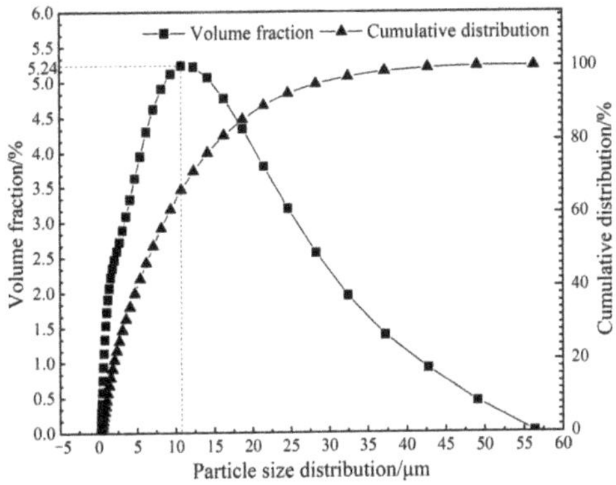


**Figure 2.** Particle size distribution of raw dust.

### *2.5. Response Surface Test Design*

Based on the results of the single-factor test, mixed gum concentration (A), nitrogen pressure (B), and flue gas flow rate (C) were selected as independent variables, and $PM_{2.5}$ and $PM_{10}$ removal rates ($\eta_1$ and $\eta_2$) were used as response values. The Box–Behnken 3-factor, 3-level response surface design was performed using Design-Expert 8.0.6 software. The results are presented in Table 1. The response surface test investigated the effect of each factor on the response values, and the data were analyzed to optimize the determination of the optimal process conditions.

**Table 1.** Box–Behnken test factors and levels.

| Level | Factor | | |
|---|---|---|---|
| | A, Mixed Gel Concentration /(g/L) | B, Nitrogen Pressure /MPa | C, Flue Gas Flow Velocity /(m/s) |
| 1 | 0.75 | 0.2 | 8 |
| 0 | 1 | 0.4 | 10 |
| −1 | 1.25 | 0.6 | 12 |

## 3. Results and Discussion

### *3.1. Effect of the Agglomerator Solutions on Fine Dust Particles*

Konjac gum, xanthan gum, and konjac gum and xanthan gum mixed solution with deionized water were selected during the agglomeration experiment. The experimental results of raw dust (agglomeration-less process), deionized water (agglutinant-free), and agglomerating agent are shown in Figure 3. The deionized water and agglomerating agent are sprayed using a two-phase flow nozzle. Fine particles adhere and agglomerate through the combined effect of van der Waals and liquid bridge forces. This improves cohesion between particles and enables fine particles to grow into larger ones, resulting in an increase in peak particle size. As seen in Figure 3, when the concentration is kept constant, the agglomeration effect of dust is mixed gum > xanthan gum > konjac gum > deionized water, and mixed gum has the best effect. This is due to the increased force of liquid bridge formation between the particles. Under the action of mixed gum agglomerating agent, the peak particle size of the grown dust is 49.1 μm, and the removal efficiencies of $PM_{2.5}$ and $PM_{10}$ were 42.6% and 46.43%, respectively.

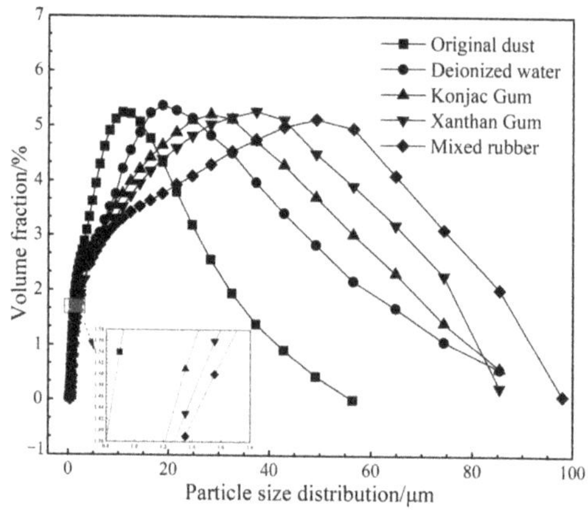


**Figure 3.** Effect of agglomerating agent on dust.

From Figure 4, the viscosity of the gum-containing solution is higher than that of deionized water. As the temperature increases, the viscosity of the agglomerate solution tends to decrease. The reason is that as the temperature increases, the molecular spacing of the liquid becomes larger, the activity of the polymer chain is increased, and the ability to restrain the molecular motion is weakened, so the viscosity decreases. The viscosity of the mixed gum is highest at 30 °C and is more than six times higher than that of deionized water. Combined with Figure 3, it is shown that the higher viscosity of the mixed gum liquid is more conducive to the adhesion and growth of microfine dust and higher dust removal efficiency.

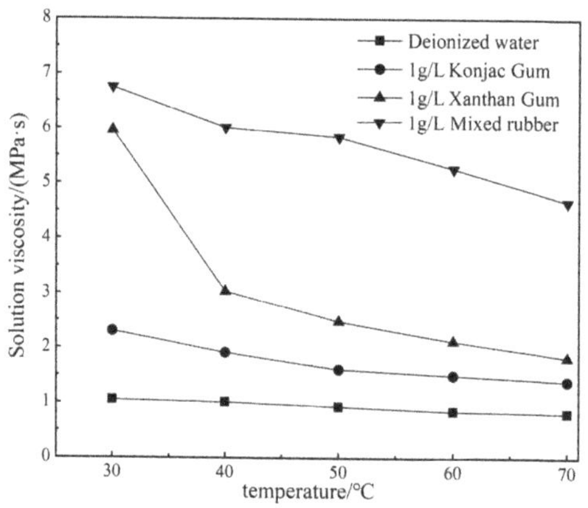


**Figure 4.** Effect of temperature on solution viscosity.

### 3.2. *Effect of Different Coupled Agglomerates on Fine Dust Particles*

The agglomerating agent will make the surface tension of the solution too high and reduce the wetting effect on dust, which in turn will affect the dust removal effect. Therefore, a combination of agglomerating agent and wetting agent was used for the experiment. As shown in Figure 5, the particle size distribution shifted to the right after the wetting agent was added. This indicates that the agglomerated droplets are more likely to adhere to the fine particles after wetting, and the adsorption reaction between the fine particles leads to further growth of the particles. The results are consistent with the study of Zhou [10]. Wetting agents improve the wetting ability of dust. Comparing the effect of the four wetting agents on particle growth, SDS has the largest right shift and the best effect. The particles grow significantly after agglomeration, with a peak particle size of 85.32 μm, which is about eight times larger. The wetting agents contributed to the particle growth, and the removal rates of $PM_{2.5}$ and $PM_{10}$ were improved to 51.46% and 53.13%, respectively.

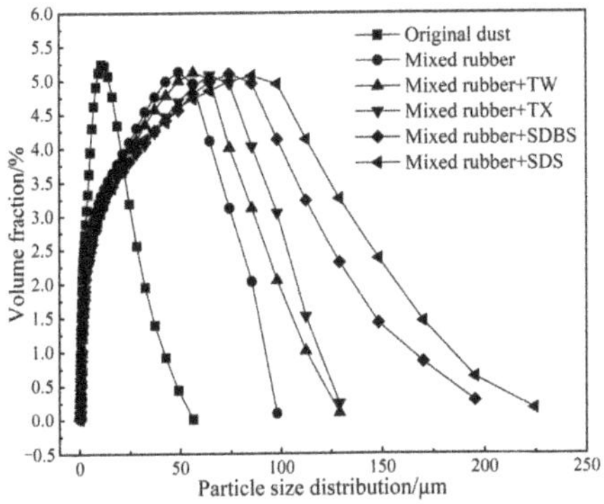


**Figure 5.** Effect of different coupling agglomerates.

The coupling agglomerating agent promotes fine dust with obvious effect under the action of the wetting agent. The wetting agent can reduce the surface tension of the droplets, which makes it easier for the droplets to unfold on the solid surface and the particles to agglomerate and condense [13]. On the other hand, the wetting agent adsorbed on the dust surface changes the charge state of the dust and increases the electrostatic force between the fine dust, which promotes the agglomeration of dust particles. As shown in Figure 6, the contact angle was reduced when the wetting agent was added to the agglomerate. In addition, the absolute value of the zeta potential increased when the nonionic wetting agent was added and decreased when the anionic wetting agent was added. The lower contact angle indicates that coupling agglomerate solution with dust wetting performance has the best SDS wetting effect. Zeta potential evaluates the strength of the interaction between particles. The higher the absolute value, the more stable the particle solution and the less likely aggregation is to occur. It can be inferred that the suspensions with lower zeta potential exhibited have better fine particle removal efficiency [17]. Comparing the zeta potential of the mixed gum solution of SDBS and SDS, it was found that the absolute value of the mixed gum solution of SDS was the smallest; therefore, the mixed gum solution of SDS was the easiest to coalesce.

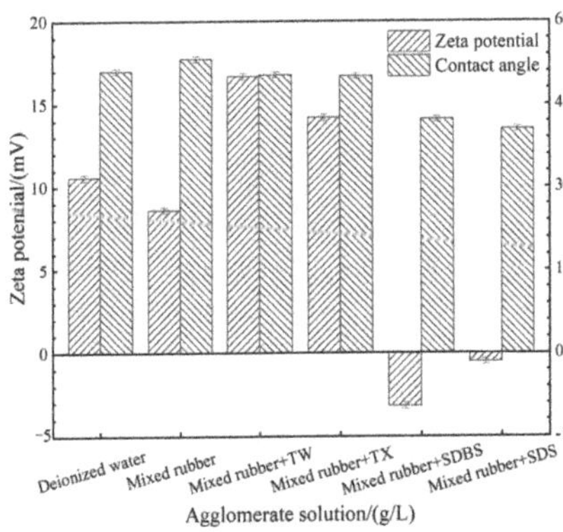


**Figure 6.** Zeta potential and contact angle of the solution.

### 3.3. Effect of Mixed Gel Concentration on Fine Dust Particles

The effect of mixed gel concentration on fine dust particles is shown in Figure 7. As the agglomerate concentration increases, the dust removal efficiency shows an increase and then a decrease. This is because as the agglomerate concentration increases, the number of agglomerate molecules contained in the agglomerate droplets increases and the adhesion capacity to the dust increases, while the collision probability between the droplets and the dust also increases, which promotes dust agglomeration and thus improves the dust removal efficiency [15]. When the content of mixed gum is too high, it leads to an increase in the diameter of the agglomerate atomized droplets and thus a decrease in the number and difficulty in the expansion of macromolecular chains after adsorption on the particles [11]. This also leads to a low number of surface adsorption sites and thus a weakening of the positive effect of liquid bridge force formation, affecting fine particle agglomeration [18]. In conclusion, the optimal concentration of mixed gum solution is 1.0 g/L, corresponding to the highest efficiency of fine dust removal.

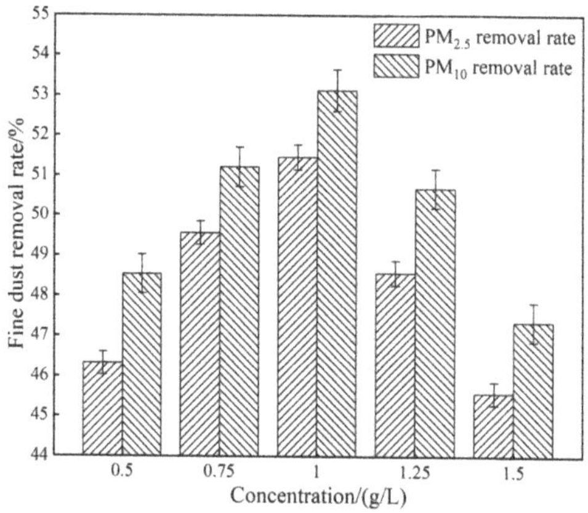


**Figure 7.** Dust removal rate of different solution concentrations.

### 3.4. Effect of Gas Pressure on Fine Dust Particles

The interaction force between nitrogen and the agglomerate solution was utilized to break the solution into fine droplets using a two-phase flow nozzle. Therefore, the nitrogen pressure affects the atomization effect of the fluid and also plays a crucial role in the removal of fine dust. It can be seen in Figure 8 that as the pressure increases from 0.1 to 0.5 MPa, the removal rate of fine dust also increases, among which the removal rate of $PM_{2.5}$ and $PM_{10}$ increases from 43.16% to 51.46% and from 45.31% to 53.13%, respectively. However, when the pressure is higher than 0.5 MPa, the fine dust removal rate decreases significantly. The reason for this is that the increase in pressure is conducive to the generation of fine droplets, the possibility of collision and collection of droplets on fine particles is greatly enhanced, and the particle growth effect is more obvious. However, when the pressure reaches a critical value or is too high, on the one hand, the gas strongly cuts the solution, leading to the deterioration of the nozzle atomization droplets; on the other hand, it changes the agglomeration chamber disturbance velocity, resulting in an increased disturbance effect unfavorable to the adhesion between particles [15].

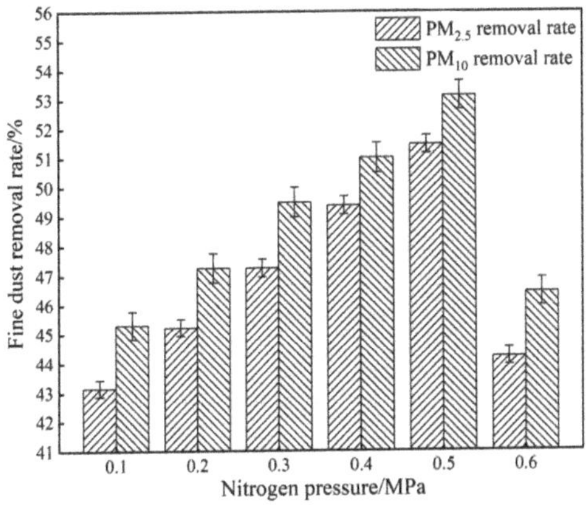


**Figure 8.** Dust removal rates for different nitrogen pressures.

### 3.5. *Effect of Flue Gas Flow Rate on Fine Dust Particles*

From Figure 9, the fine dust removal rate increases and then decreases with the increase in the flue gas velocity, and the dust removal effect is generally poor on both sides. The reason is that as the flue gas velocity increases, the disturbance effect in the agglomeration chamber increases, leading to an increase in the probability of particle–droplet collision. When the flue gas velocity is higher than 10 m/s, the velocity of dust and agglomerate droplets decreases their reaction time in the agglomeration chamber [13], which leads to the decrease in particle capture by droplets. The highest dust removal rate was achieved when the flue gas flow rate was 10 m/s. Finally, the removal rates of $PM_{2.5}$ and $PM_{10}$ reached 51.46% and 53.13%, respectively.

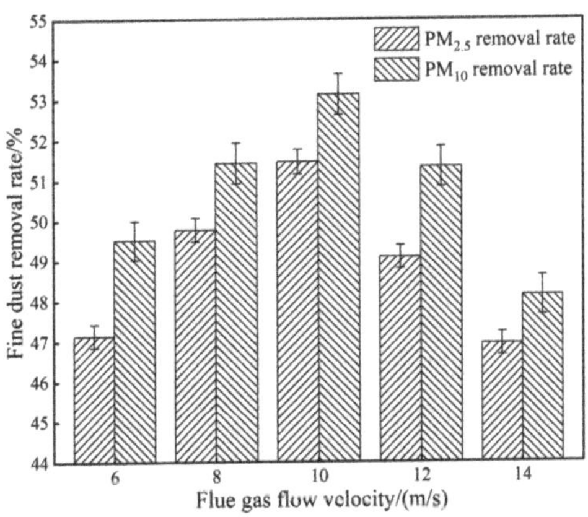


**Figure 9.** Dust removal rate with different flue gas flow rates.

### 3.6. *Response Surface Experiments*

In order to find the optimal process conditions, the response surface experiments were conducted. Based on the experimental results, a three-factor, three-level response surface experiment was designed using Design-Expert. The mixed gel concentration (A), nitrogen pressure (B), and flue gas flow rate (C) were selected as independent variables by single-factor experiments, and $PM_{2.5}$ and $PM_{10}$ removal rates ($\eta_1$ and $\eta_2$) were used as response values, and their experimental protocols and results are shown in Table 2.

**Table 2.** Box–Behnken test protocol and response values.

| Serial Number | A, Mixed Gum Concentration/(g/L) | B, Nitrogen Pressure/MPa | C, Flue Gas Flow Velocity/(m/s) | $\eta_1$/% | $\eta_2$/% |
|---|---|---|---|---|---|
| 1 | 0.75 | 0.2 | 10 | 46.61 | 48.92 |
| 2 | 1.25 | 0.2 | 10 | 46.49 | 48.73 |
| 3 | 0.75 | 0.6 | 10 | 47.14 | 49.53 |
| 4 | 1.25 | 0.6 | 10 | 46.91 | 49.31 |
| 5 | 0.75 | 0.4 | 8 | 46.51 | 49.97 |
| 6 | 1.25 | 0.4 | 8 | 47.52 | 49.37 |
| 7 | 0.75 | 0.4 | 12 | 48.63 | 49.43 |
| 8 | 1.25 | 0.4 | 12 | 46.59 | 48.64 |
| 9 | 1 | 0.2 | 8 | 47.15 | 48.63 |
| 10 | 1 | 0.6 | 8 | 47.54 | 49.52 |
| 11 | 1 | 0.2 | 12 | 47.25 | 49.24 |
| 12 | 1 | 0.6 | 12 | 47.44 | 49.53 |
| 13 | 1 | 0.4 | 10 | 50.37 | 52.41 |
| 14 | 1 | 0.4 | 10 | 50.61 | 52.76 |
| 15 | 1 | 0.4 | 10 | 50.69 | 52.97 |
| 16 | 1 | 0.4 | 10 | 50.68 | 52.37 |
| 17 | 1 | 0.4 | 10 | 50.73 | 52.75 |

The results of the mathematical and statistical analysis of the experimental results using Design-Expert 8.0.6 software are shown in Tables 3 and 4. After fitting, the regression equation with the dust removal rate as the response value was obtained as

$$\eta_1 = 50.62 - 0.17 \times A + 0.19 \times B + 0.15 \times C - 0.028 \times A \times B - 0.76 \times A \times C - 0.050 \times B \times C - 1.93 \times A^2 - 1.90 \times B^2 - 1.37 \times C^2$$

$$\eta_2 = 52.65 - 0.22 \times A + 0.30 \times B - 0.081 \times C - 7.500 \times 10^{-3} \times A \times B - 0.047 \times A \times C - 0.15 \times B \times C - 1.70 \times A^2 - 1.83 \times B^2 - 1.60 \times C^2$$

As shown in Table 3, the $PM_{2.5}$ model F = 107.41, $p < 0.0001$, indicating that the model is highly significant. The out-of-fit term F = 4.04, $p = 0.1054$ (>0.05), indicating that the regression equation is a good fit for the actual. In this case, B, AC, $A^2$, $B^2$, and $C^2$ are the significant model terms. From Table 4, the $PM_{10}$ model F = 40.04, $p < 0.0001$, indicating that the model is highly significant. The out-of-fit term F = 2.5, $p = 0.1651$ (>0.05), indicating that the regression equation is a good fit for the actual. In this case, B, $A^2$, $B^2$, and $C^2$ are the significant model terms. From the data F-values in the table, it is known that the order of the intensity of each response factor on $PM_{2.5}$ and $PM_{10}$ removal efficiency is B (nitrogen pressure) > A (mixed gel concentration) > C (flue gas flow rate).

**Table 3.** Analysis of $PM_{2.5}$ regression equation variance.

| Variance Source | Sum of Squares | Degree of Freedom | Mean Square | F-Value | Prob > F | Significance |
|---|---|---|---|---|---|---|
| Model | 46.27 | 9 | 5.14 | 107.41 | <0.0001 | Significant |
| A | 0.24 | 1 | 0.24 | 4.97 | 0.0610 | |
| B | 0.29 | 1 | 0.29 | 6.11 | 0.0427 | Significant |
| C | 0.18 | 1 | 0.18 | 3.70 | 0.0959 | |
| AB | $3.025 \times 10^{-3}$ | 1 | $3.025 \times 10^{-3}$ | 0.063 | 0.8087 | |
| AC | 2.33 | 1 | 2.33 | 48.59 | 0.0002 | Significant |
| BC | 0.010 | 1 | 0.010 | 0.21 | 0.6615 | |
| $A^2$ | 15.69 | 1 | 15.69 | 327.85 | <0.0001 | Significant |
| $B^2$ | 15.17 | 1 | 15.17 | 316.90 | <0.0001 | Significant |
| $C^2$ | 7.94 | 1 | 7.94 | 165.83 | <0.0001 | Significant |
| Residual | 0.34 | 7 | 0.048 | | | |
| Loss of proposed project | 0.25 | 3 | 0.084 | 4.04 | 0.1054 | Insignificant |

**Table 4.** Analysis of $PM_{10}$ regression equation variance.

| Variance Source | Sum of Squares | Degree of Freedom | Mean Square | F-Value | Prob > F | Significance |
|---|---|---|---|---|---|---|
| Model | 42.57 | 9 | 4.73 | 40.04 | <0.0001 | Significant |
| A | 0.41 | 1 | 0.41 | 3.43 | 0.1065 | |
| B | 0.70 | 1 | 0.70 | 5.94 | 0.0449 | Significant |
| C | 0.053 | 1 | 0.053 | 0.45 | 0.5252 | |
| AB | $2.250 \times 10^{-4}$ | 1 | $2.250 \times 10^{-4}$ | $1.90 \times 10^{-3}$ | 0.9664 | |
| AC | $9.025 \times 10^{-3}$ | 1 | $9.025 \times 10^{-3}$ | 0.076 | 0.7902 | |
| BC | 0.09 | 1 | 0.09 | 0.76 | 0.4117 | |
| $A^2$ | 12.22 | 1 | 12.22 | 103.42 | <0.0001 | Significant |
| $B^2$ | 14.04 | 1 | 14.04 | 118.82 | <0.0001 | Significant |
| $C^2$ | 10.73 | 1 | 10.73 | 90.77 | <0.0001 | Significant |
| Residual | 0.83 | 7 | 0.12 | | | |
| Loss of proposed project | 0.57 | 3 | 0.19 | 2.90 | 0.1651 | Insignificant |

The response surface is a three-dimensional surface plot of the response values against each factor (A, B, C), in which the interaction of mixed gel concentration and nitrogen pressure has the most significant effect on $PM_{2.5}$ and $PM_{10}$ removal efficiency, as shown in Figures 10 and 11. Figure 10 shows that the $PM_{2.5}$ removal efficiency increases with the increase in mixed gel concentration and nitrogen pressure, and the best point changes to the mixed gel concentration of 0.99 g/L and nitrogen pressure of 0.41 MPa and reaches the maximum around this point. The $PM_{2.5}$ removal efficiency starts to decrease when the parameters are increased again. Figure 11 shows that the $PM_{10}$ removal efficiency increases with the increase in mixed gel concentration and nitrogen pressure, and the optimum point is near the mixed gel concentration of 0.98 g/L and nitrogen pressure of 0.42 MPa and reaches the maximum around this point. Then, the $PM_{10}$ removal efficiency starts to decrease with the increase in parameter.

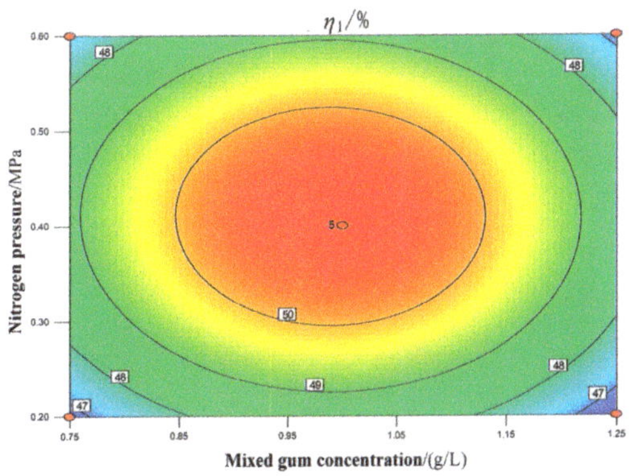


(a)

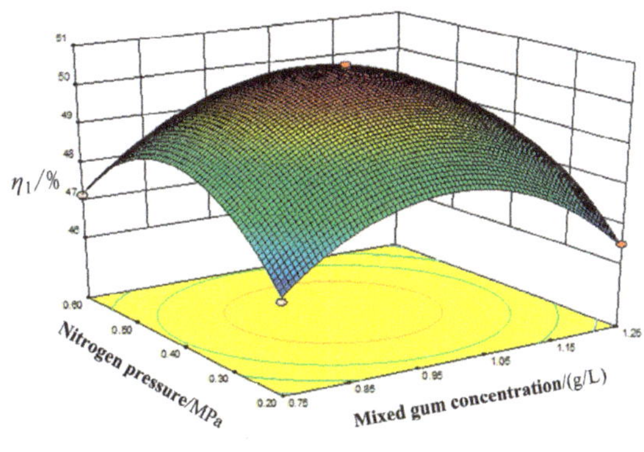


(b)

**Figure 10.** (**a**) Contour plot of $PM_{2.5}$ where mixed glue concentration and nitrogen pressure interact with each other; (**b**) response surface diagram of $PM_{2.5}$ where mixed glue concentration and nitrogen pressure interact with each other.

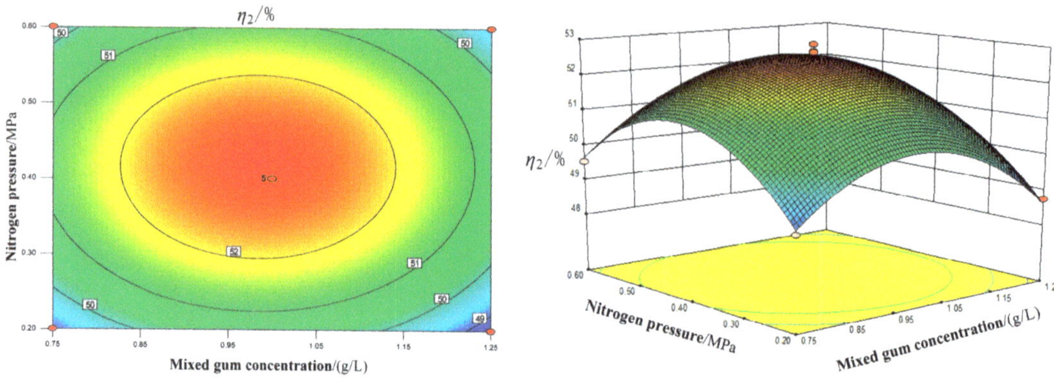


(a) (b)

**Figure 11.** (**a**) Contour diagram of $PM_{10}$ where mixed glue concentration and nitrogen pressure affect each other; (**b**) response surface diagram of $PM_{10}$ where mixed glue concentration and nitrogen pressure affect each other.

The optimal process parameters were obtained by Design-Expert 8.0.6 software, and the optimal process parameters were mixed gel concentration of 0.99 g/L, nitrogen pressure of 0.41 MPa, flue gas flow rate of 10.03 m/s, and the predicted $PM_{2.5}$ and $PM_{10}$ removal efficiencies of 50.63% and 52.67%. To verify the applicability of this model equation, the flue gas flow rate was set to 10 m/s, and the nitrogen pressure was set to 1 g/L and 0.40 MPa. To verify the applicability of this model equation, verification tests were conducted according to the optimized process parameters, and after three verification tests, we found that the optimized process parameters achieved 50.68% and 52.73% $PM_{2.5}$ and $PM_{10}$ removal efficiency. The experimental results are very close to the expected results, indicating that the applicability of the model equation is reliable and feasible.

## 4. Conclusions

To enhance the removal effect of chemical agglomeration on fine particulate matter in converter flue gas, the impact of different factors on the removal efficiency of $PM_{2.5}$ and $PM_{10}$ was studied during the experiment. The findings of the study are as follows:

(1) The best results in single chemical agglomeration experiments were achieved by using a mixed gum solution of xanthan gum and konjac gum, which produced a peak particle size of 49.1 μm, over twice as large as deionized water.

(2) In the wetting agent coupling screening experiment, a combination of 0.5 g/L SDS and 1 g/L mixed gum solution was the best agglomerator concentration since it provided the most effective dust wetting, thus promoting dust agglomeration and growth. The maximum particle size is 85.32 μm, about eight times greater. During this time, the $PM_{2.5}$ and $PM_{10}$ removal efficiency reached 51.46% and 53.13%, respectively.

(3) The Box–Behnken model was utilized to optimize the effects of the coupling agglomerator. The primary factors that influenced the $PM_{2.5}$ and $PM_{10}$ removal rates were identified. Additionally, the optimal process conditions were obtained with a mixed gel concentration of 1 g/L, nitrogen pressure of 0.4 MPa, and flue gas flow rate of 10 m/s. The model was tested and found to agree well with reality.

**Author Contributions:** Conceptualization, H.L. and H.X.; methodology, H.L.; software, H.X.; validation, H.L., G.Z., H.X. and J.Z.; formal analysis, H.X.; investigation, J.Z.; resources, H.L.; data curation, H.X.; writing—original draft preparation, H.X.; writing—review and editing, H.L.; supervision, G.Z.; project administration, H.L.; funding acquisition, H.L. All authors have read and agreed to the published version of the manuscript.

**Funding:** This research was supported by the Natural Science Foundation of Hebei Province (E2022209138) and Tangshan Science and Technology Plan Project (No.: 21130231C).

**Data Availability Statement:** The data that support the findings of this study are available from the corresponding author, upon reasonable request.

**Acknowledgments:** Thanks to the editors and reviewers of the journal for their suggestions on this article.

**Conflicts of Interest:** The authors declare no conflict of interest.

## References

1. Atkinson, R.W.; Mills, I.C.; Walton, H.A.; Anderson, H.R. Fine Particle Components and Health-a Systematic Review and Meta-Analysis of Epidemiological Time Series Studies of Daily Mortality and Hospital Admissions. *J. Expo. Sci. Environ. Epidemiol.* **2015**, *25*, 208–214. [CrossRef] [PubMed]
2. Knaapen, A.M.; Borm, P.J.A.; Albrecht, C.; Schins, R.P.F. Inhaled Particles and Lung Cancer. Part A: Mechanisms. *Int. J. Cancer* **2004**, *109*, 799–809. [CrossRef] [PubMed]
3. Chen, D.L.; Wu, K.; Mi, J.C. Experimental Investigation of Aerodynamic Agglomeration of Fine Ash Particles from a 330 Mw Pc-Fired Boiler. *Fuel* **2016**, *165*, 86–93. [CrossRef]
4. Asanakham, A.; Kiatsiriroat, T. Use of Bluff Body for Enhancing Submicron Particle Agglomeration in Plasma Field. *Part. Sci. Technol.* **2018**, *36*, 123–130. [CrossRef]
5. Yang, Z.D.; Cai, Y.X.; Li, Q.Y.; Li, H.Q.; Jiang, Y.; Lin, R.Y.; Zheng, C.H.; Sun, D.S.; Gao, X. Predicting Particle Collection Performance of a Wet Electrostatic Precipitator under Varied Conditions with Artificial Neural Networks. *Powder Technol.* **2021**, *377*, 632–639. [CrossRef]
6. Wang, Y.; Yu, J.L.; Wang, Z.H.; Liu, Y.X.; Zhao, Y.C. A Review on Arsenic Removal from Coal Combustion: Advances, Challenges and Opportunities. *Chem. Eng. J.* **2021**, *414*, 128785. [CrossRef]
7. Zu, K.; Yao, Y.; Cai, M.; Zhao, F.; Cheng, D.L. Modeling and Experimental Study on Acoustic Agglomeration for Dust Particle Removal. *J. Aerosol Sci.* **2017**, *114*, 62–76. [CrossRef]
8. Sun, Z.K.; Yang, L.J.; Chen, S.; Wu, X. Promoting the Removal of Fine Particles by Turbulent Agglomeration with the Coupling of Different-Scale Vortexes. *Powder Technol.* **2020**, *367*, 399–410. [CrossRef]
9. Luo, Z.Y.; Chen, H.; Wang, T.; Zhou, D.; Lu, M.S.; He, M.C.; Fang, M.X.; Cen, K.F. Agglomeration and Capture of Fine Particles in the Coupling Effect of Pulsed Corona Discharge and Acoustic Wave Enhanced by Spray Droplets. *Powder Technol.* **2017**, *312*, 21–28. [CrossRef]
10. Zhou, L.; Liu, Y.; Luo, L.Y.; Yuan, Z.L.; Yang, L.J.; Wu, H. Improving the Removal of Fine Particles by Chemical Agglomeration During the Limestone-Gypsum Wet Flue Gas Desulfurization Process. *J. Environ. Sci.* **2019**, *80*, 35–44. [CrossRef] [PubMed]
11. Shen, X.K.; Sun, D.S.; Liu, Y.L.; Yu, Y.; Yan, P.L. Agglomeration and removal of coal-fired particulate matter under the action of chemical agglomeration agent. *Environ. Sci. Technol.* **2017**, *40*, 111–116.
12. Dai, M.B.; Ruan, Z.Y.; Geng, Y.; Li, W.C.; Mu, G.T.; Chun, T.J. Study on removal of PM10/2.5/1.0 based on chemical agglomerated iron ore sintering process. *J. Iron Steel Res.* **2021**, *33*, 385–393.
13. Liu, Y.; Hu, B.; Zhou, L.; Jiang, Y.Z.; Yang, L.J. Improving the Removal of Fine Particles with an Electrostatic Precipitator by Chemical Agglomeration. *Energy Fuels* **2016**, *30*, 8441–8447. [CrossRef]
14. Ding, X.H.; Wang, D.F.; Luo, Z.M.; Wang, T.; Deng, J. Multi-Scale Study on the Agglomerating and Wetting Behaviour of Binary Electrolyte-Surfactant Coal Dust Suppression Based on Multiple Light Scattering. *Fuel* **2022**, *311*, 122515. [CrossRef]
15. Zhou, L.; Chen, W.; Wu, H.; Shen, A.; Yuan, Z.L.; Yang, L.J. Investigation on the Relationship of Droplet Atomization Performance and Fine Particle Abatement During the Chemical Agglomeration Process. *Fuel* **2019**, *245*, 65–77. [CrossRef]
16. Shi, G.Q.; Qi, J.M.; Wang, Y.M.; Shen, H.Y. Synergistic Influence of Noncationic Surfactants on the Wettability and Functional Groups of Coal. *Powder Technol.* **2021**, *385*, 92–105. [CrossRef]
17. Guo, Y.Q.; Zhao, Y.C.; Wang, S.L.; Jiang, C.; Zhang, J.Y. Relationship between the Zeta Potential and the Chemical Agglomeration Efficiency of Fine Particles in Flue Gas During Coal Combustion. *Fuel* **2018**, *215*, 756–765. [CrossRef]
18. Guo, Y.Q.; Zhang, J.Y.; Zhao, Y.C.; Wang, S.L.; Jiang, C.; Zheng, C.G. Chemical Agglomeration of Fine Particles in Coal Combustion Flue Gas: Experimental Evaluation. *Fuel* **2017**, *203*, 557–569. [CrossRef]

processes

MDPI

Article

# Research on Metallurgical Saw Blade Surface Defect Detection Algorithm Based on SC-YOLOv5

Lili Meng [1,*], Xi Cui [1], Ran Liu [2,*], Zhi Zheng [1], Hongli Shao [3], Jinxiang Liu [1], Yao Peng [1] and Lei Zheng [1]

1 College of Mechanical Engineering, North China University of Science and Technology, Tangshan 063210, China
2 College of Metallurgy & Energy, North China University of Science and Technology, Tangshan 063210, China
3 Tangshan Metallurgical Saw Blade Co., Ltd., Tangshan 063000, China
* Correspondence: mengll@ncst.edu.cn (L.M.); liuran@ncst.edu.cn (R.L.)

**Abstract:** Under the background of intelligent manufacturing, in order to solve the complex problems of manual detection of metallurgical saw blade defects in enterprises, such as real-time detection, false detection, and the detection model being too large to deploy, a study on a metallurgical saw blade surface defect detection algorithm based on SC-YOLOv5 is proposed. Firstly, the SC network is built by integrating coordinate attention (CA) into the Shufflenet-V2 network, and the backbone network of YOLOv5 is replaced by the SC network to improve detection accuracy. Then, the SIOU loss function is used in the YOLOv5 prediction layer to solve the angle problem between the prediction frame and the real frame. Finally, in order to ensure both accuracy and speed, lightweight convolution (GSConv) is used to replace the ordinary convolution module. The experimental results show that the mAP@0.5 of the improved YOLOv5 model is 88.5%, and the parameter is 31.1M. Compared with the original YOLOv5 model, the calculation amount is reduced by 56.36%, and the map value is increased by 0.021. In addition, the overall performance of the improved SC-YOLOv5 model is better than that of the SSD and YOLOv3 target detection models. This method not only ensures the high detection rate of the model, but also significantly reduces the complexity of the model and the amount of parameter calculation. It meets the needs of deploying mobile terminals and provides an effective reference direction for applications in enterprises.

**Keywords:** YOLOv5; metallurgical saw blade; lightweight; deep learning; defect detecting

**Citation:** Meng, L.; Cui, X.; Liu, R.; Zheng, Z.; Shao, H.; Liu, J.; Peng, Y.; Zheng, L. Research on Metallurgical Saw Blade Surface Defect Detection Algorithm Based on SC-YOLOv5. *Processes* **2023**, *11*, 2564. https://doi.org/10.3390/pr11092564

Academic Editors: Xin Yao and Huaqing Xie

Received: 9 August 2023
Revised: 21 August 2023
Accepted: 25 August 2023
Published: 27 August 2023



## 1. Introduction

In the context of intelligent manufacturing, machine vision [1,2] has become an important research direction for artificial intelligence, and same-target defect detection has become the focus of research. In recent years, with the rapid development of China's industrial level, the demand for metallurgical saw blades has increased significantly. While ensuring the output of metallurgical saw blades, the quality requirements have also become an important issue. As a multi-blade tool [3], which is essentially a consumable, the saw blade will still have defects even though metallurgical technology and the manufacturing process of saw blades have been continuously improved in recent years. These defects, such as cracks, pitting, and scarring, greatly affect the service life of a saw blade. The real-time performance of manual detection [4,5] cannot be guaranteed. Therefore, a good surface defect detection technology for metallurgical saw blades can not only ensure real-time performance, but also ensure surface quality, which is of great significance to service life.

With the development of industrial intelligence, defect detection is broadly divided into two directions [6–8]. One is the direction based on traditional image processing, and the other is the direction based on deep learning image detection. In traditional image processing, Liu et al. [9] proposed a two-branch balanced saliency model based on discriminant features for fabric defect detection. This method can be used for accurate fabric defect detection and even surface defect detection for other industrial products. Joung et al. [10] used infrared imaging to detect defects in pipelines. Zhang et al. [11] proposed an algorithm combining local binary patterns (LBPs) and a gray-level co-occurrence matrix (GLCM). The local feature information and overall texture information of the defect image are extracted using LBPs and the GLCM, respectively. There are many ways of feature extraction. However, for specific traditional image processing methods, real-time processing and accuracy should be considered to meet the demand. In image detection methods based on deep learning, faster RCNN, YOLO and other series of networks are mainly used. Faster RCNN is a typical representative of two-stage target detection model, but the detection and training process is still relatively complex. Sun et al. [12] proposed a new face detection scheme based on deep learning and obtained advanced face detection performance by improving the faster RCNN framework. Shou et al. [13] proposed an improved region-based convolutional neural network (RCNN) fast detection method for airport detection in large-scale remote sensing images. Multi-scale training is applied to faster RCNN to enhance the robustness of the network for airport detection of different scales. The core idea of YOLO is to transform the target detection into a regression problem [14,15], using the whole map as the input of the network, just after a neural network, so that YOLO uses the whole graph as the input to the network, and just goes through a neural network to obtain the position of the bounding box and its category. Wang et al. [16] investigated the improved YOLOv4 algorithm using a shallow feature enhancement mechanism for the problems of insensitivity to small objects and low detection accuracy in traffic light detection and recognition. Xian et al. [17] used a triple loss function in YOT-Net in order to improve defect detection accuracy for copper elbows. Image similarity was used to enhance the feature extraction capability. The YOT-Net method for copper elbow surface defects was proposed. Wang et al. [18] proposed an optimized micro YOLOv3 algorithm with less computation and higher accuracy to solve the problem of insufficient accuracy of the original micro YOLOv3 algorithm in target detection in a lawn environment. Xue et al. [19] proposed an improved forest fire small-target detection model based on YOLOv5 for the problems of high unpredictability and strong destructiveness of forest fires. The model improves the backbone layer of YOLOv5 and adds an attention mechanism module to improve the identifiability of small forest fire targets.

This paper focuses on a defect detection algorithm based on deep learning, which can have high transmission efficiency while ensuring high detection accuracy, and the model is lightweight enough to meet the needs of enterprise applications. Taking the metallurgical saw blade of a metallurgical saw blade factory in Tangshan as the research object, the main network Shuffenet-V2 [20] was replaced in the YOLOv5 model, and the coordinate attention mechanism CA [21–23] was added to effectively improve the ability to extract metallurgical saw blade features. The lightweight convolution GSConv module was used to reduce the volume of the model, accelerate the speed of model reasoning, and realize the lightweight nature of the model. The regression loss function SIoU was introduced to accelerate the fitting of the data [24,25]. At the same time, the application of the improved lightweight identification model was studied.

## 2. Improved YOLOv5 Methods

### *2.1. YOLOv5s Target Detection Models*

YOLOv5 is a classical algorithm for single-stage target detection. The YOLOv5s architecture comprises the input, backbone network, neck network, and prediction head [26]. The backbone is responsible for feature extraction. The neck is responsible for feature fusion. The head contains three detection heads, which are responsible for outputting detection information. The YOLOv5s network is the network with the smallest depth and the smallest width of the feature map in the YOLOv5 series. The YOLOv5 network structure is shown in Figure 1 below.

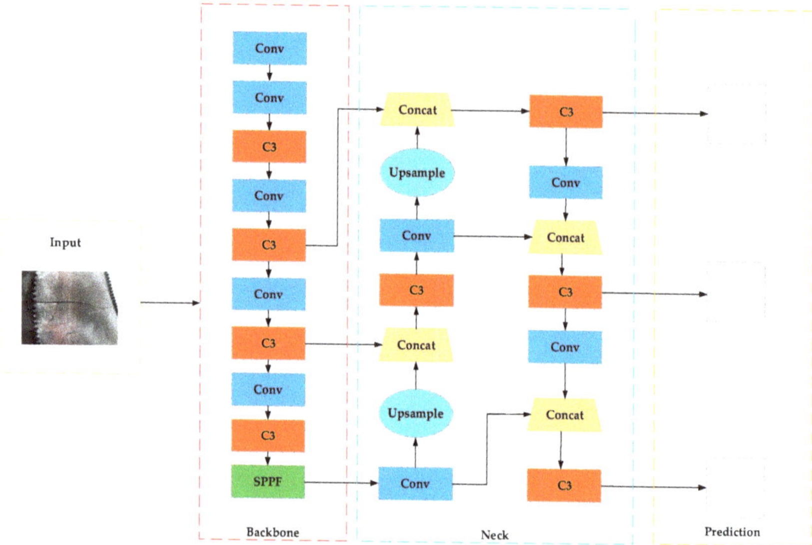


**Figure 1.** YOLOv5 network architecture.

*2.2. SC-YOLOv5 Improved Model*

To meet the needs of low recognition accuracy and a large number of parameters, this study improves the YOLOv5 model. While improving the recognition accuracy, the parameters of the model are reduced and the model inference speed is accelerated. In this study, the coordinate attention mechanism was introduced into the Shuffenet-V2 network (a lightweight network) to construct the SC network, and the SC network structure was used as the lightweight backbone network of the YOLOv5 model. The lightweight convolution module (GSConv) was used to replace the convolution module, and SIoU was introduced to redefine the loss function so that the lightweight degree of the backbone network was effectively improved. The improved SC-YOLOv5 model network structure based on YOLOv5 is shown in Figure 2.

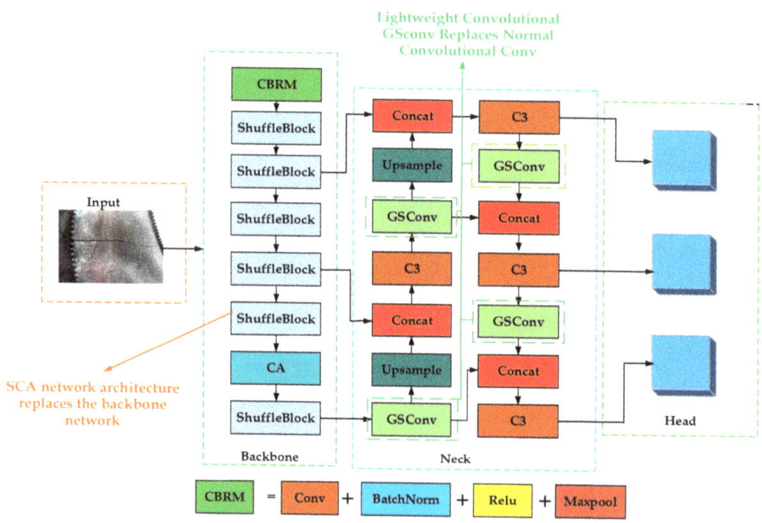


**Figure 2.** SC-YOLOv5 network structure.

In Figure 2, the SC-YOLOv5 network structure is divided into four parts, in which Input inputs and processes the dataset. Backbone extracts the features of the input metallurgical saw blade images and uses the SC network module as a new backbone network to improve the feature extraction ability of the model for key information. The acquired feature map is fused by Neck, and the lightweight convolution module is used instead of the ordinary convolution module. Head performs regression prediction, introduces the SIoU function to calculate the regression loss, and improves the convergence ability of the model.

### 2.3. *Shuffenet-V2 Architecture*

The most important part of the Shufflenet V2 network structure is the basic residual unit (block), which has two branching structures. As shown in Figure 3, the first structure performs a channel division operation at the input and divides the input feature map into two branches [27]; the primary branch contains three convolution operations the secondary branch does not perform any operation, and the input and output channels of each branch remain the same. The second structure splits the feature map into two branches, with three convolution operations in the primary branch and one depth-divisible convolution and one pointwise convolution in the secondary branch. The residual unit merges the output feature maps of the two branches by splicing at the output and further performs channel blending on the merged feature maps. Different subgroups are randomly extracted for rearrangement into new feature maps so that the next group convolution can fuse the input feature information from different groups, improving the information flow between channel groups and ensuring that the input and output channels are correlated.

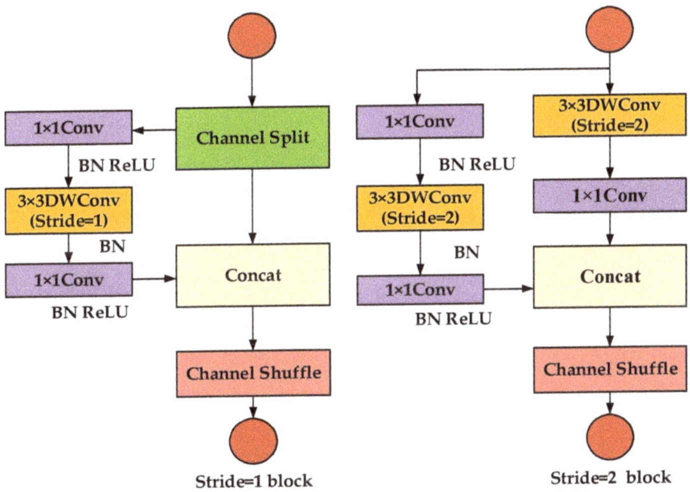


**Figure 3.** Basic residual unit network structure.

The ShuffleNetV2 network structure is mainly composed of the Conv1 layer, MaxPool layer, Stage 2 layer, Stage 3 layer, Stage 4 layer, Conv5 layer, and FC layer. The Stage 2 layer, Stage 3 layer, and Stage 4 layer are composed of the superposition of basic residual units. Among them, the Stage 2 layer and the Stage 4 layer are superimposed with a total of 4 basic residual units, and the Stage 3 layer is superimposed with a total of 8 basic residual units. The step size of the first basic residual unit in each Stage is 2, the main purpose is for downsampling, and the step size of other basic residual units is 1. A network with different complexity can be designed by changing the number of output channels in each structural layer of the network structure. Based on the ShuffleNet V2 1 ×, the number of output channels in the Conv1 layer, MaxPool layer, Stage 2 layer, Stage 3 layer, Stage

4 layer, Conv5 layer, and FC layer is 24, 24, 116, 232, 464, 1024, and 1000, respectively, as shown in Table 1.

**Table 1.** ShuffleNetV2 network architecture.

| Layer | Output Size | KSize | Stride | Repeat | Output Channels |
|---|---|---|---|---|---|
| Image | 224 × 224 | | | | 3 |
| Conv1 | 112 × 112 | 3 × 3 | 2 | 1 | 24 |
| MaxPool | 56 × 56 | 3 × 3 | 2 | | 24 |
| Stage 2 | 28 × 28 | | 2 | 1 | 116 |
| | 28 × 28 | | 1 | 3 | |
| Stage 3 | 14 × 14 | | 2 | 1 | 232 |
| | 14 × 14 | | 1 | 7 | |
| Stage 4 | 7 × 7 | | 2 | 1 | 464 |
| | 7 × 7 | | 1 | 3 | |
| Conv 5 | 7 × 7 | 1 × 1 | | 1 | 1024 |
| FC | | | | | 1000 |

### 2.4. Coordinate Attention Mechanisms

Most previous attention mechanisms for lightweight networks used SE modules [28,29], which only consider interchannel information but ignore positional information. Although later CBAM modules [30] tried to extract positional attention information by convolution after reducing the number of channels, convolution can only extract local relations and lacks the ability to extract long-range relations. Therefore, a new efficient attention mechanism, coordinate attention (CA), is proposed, which is able to encode horizontal and vertical location information into channel attention, allowing mobile networks to focus on a wide range of location information without incurring too much computational cost. It is shown in Figure 4, where C is the number of channels, W is the width, H is the height, and r is the dimension.

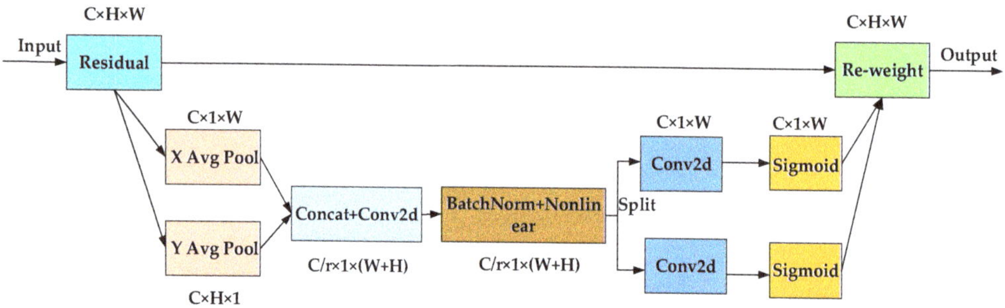


**Figure 4.** Coordinate attention network structure.

The advantages of the location attention module are mainly as follows:

1. It captures not only interchannel information, but also direction-dependent positional information, which helps the model to better locate and identify the target.
2. It is flexible and lightweight enough to be easily inserted into the core structure of a mobile network.
3. It can be used as a pre-trained model for a variety of tasks such as detection and segmentation, both of which show good performance improvement.

### 2.5. Redefined SIoU Loss Functions

YOLOv5s uses the CIoU loss function [31,32], although the length–width ratio of the predicted bounding box to the real bounding box is introduced on the basis of the DIoU loss, which makes the loss function pay more attention to the shape of the bounding box.

However, the calculation of CIoU loss is relatively complex, which may lead to a large computational overhead in the training process. In order to solve the above problems, the loss function SIoU is introduced into the improved network model, which not only considers the overlapping area, distance, length, and width, but also considers the angle between the prediction box and the real box. The SIoU loss function consists of four cost functions: angle cost, distance cost, shape cost, and IoU cost.

The SIOU loss function is as follows:

$$L_{\text{box}} = 1 - IOU + \frac{\Delta + \Omega}{2} \tag{1}$$

The overall loss function is as follows:

$$L = W_{box} L_{box} + W_{cls} L_{cls} \tag{2}$$

In the formulas above, $IOU$ denotes conventional regression loss, $\Delta$ denotes distance loss, $\Omega$ denotes shape loss, $W_{\text{box}}$ denotes frame loss model volume, $L_{\text{box}}$ denotes regression loss, $W_{\text{cls}}$ denotes categorical loss model volume, and $L_{\text{cls}}$ denotes focal loss.

### 2.6. Lightweight Convolutional GSConv

In the lightweight model design, the deep neural network has only used deep separated convolution (DSC) from start to finish. Although the number of parameters is reduced and the separation of channels and regions is achieved, DSC does not effectively use the feature information of different channels at the same spatial location. In order to make the output of the DSC as close to the SC as possible, a new method, GSConv, was introduced, which, as shown in Figure 5, introduces the SC-generated information (dense convolution operation) into each part of the DSC-generated information using shuffle. This method allows the information from SC to be completely blended into the output of DSC. GSConv first downsamples the inputs for normal convolution and then performs deep convolution using DSC; the results from SC and DSC are stitched together, and finally, a shuffle operation is performed.

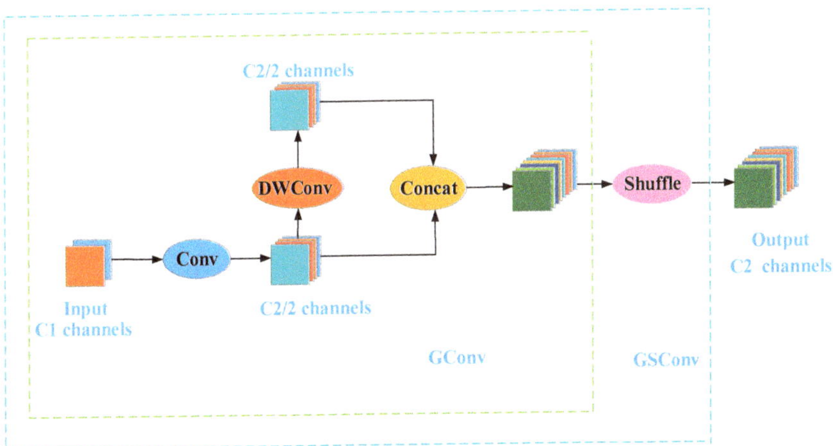


**Figure 5.** GSConv network structure.

## 3. Metallurgical Saw Blade Dataset

### 3.1. Collection of Datasets

The dataset used industrial cameras mainly collected from a metallurgical saw blade factory in Tangshan. In order to increase the diversity of metallurgical saw blades and improve the generalization ability of the recognition model, natural image data from different angles and different environments were selected when collecting datasets. Through random noise, Gaussian blur, random cropping, random rotation, random translation, and other data amplification methods of the original image, three kinds of metallurgical saw blades with cracks, pitting corrosion, and scarring were collected. The metallurgical saw blade defect part of the picture is shown in Table 2.

**Table 2.** Metallurgical saw blade partial data sheet.

| Photos | Description |
| --- | --- |
|  | Collected images of some metallurgical saw blade defects, numbered 000001–000004. |
| 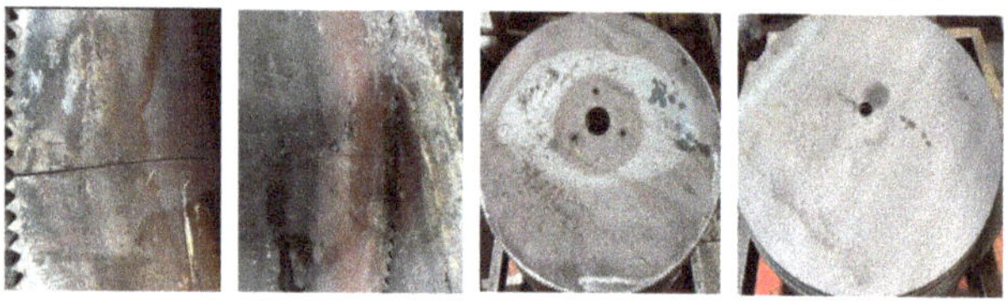 | Collected images of some metallurgical saw blade defects, numbered 000005–000008. |
| 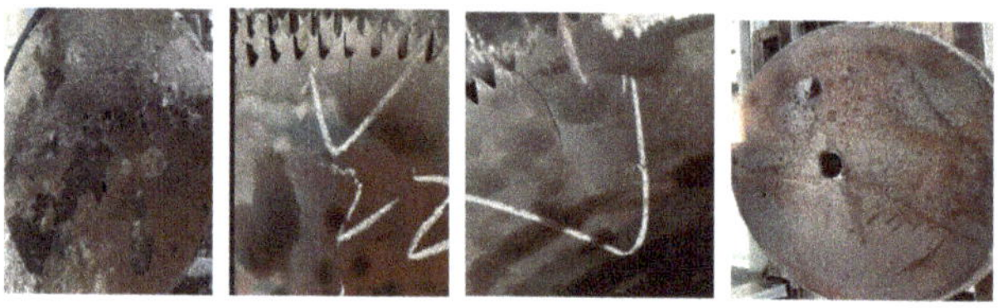 | Collected images of some metallurgical saw blade defects, numbered 000009–000012. |
| 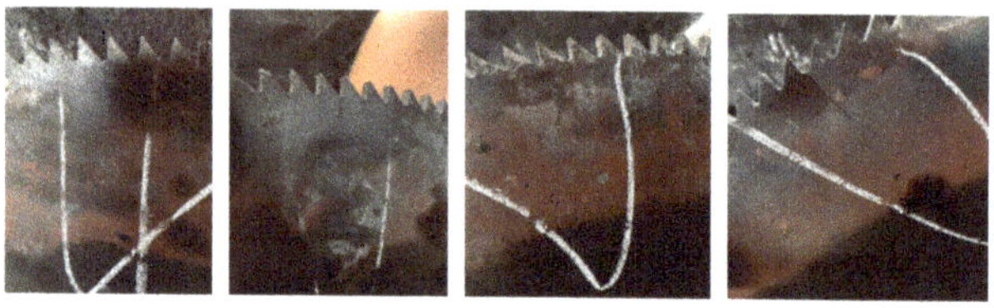 | Collected images of some metallurgical saw blade defects, numbered 000013–000016. |
| 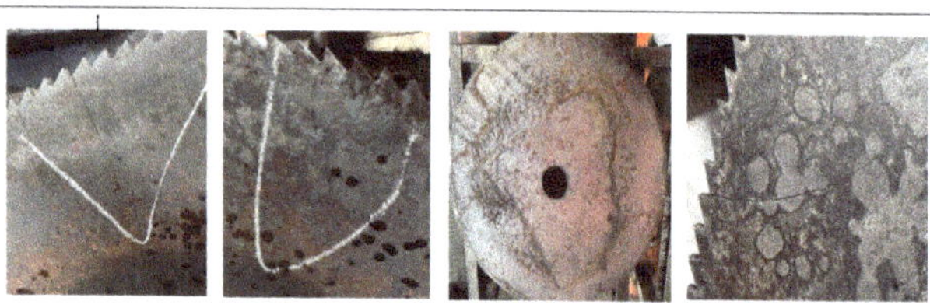 | Collected images of some metallurgical saw blade defects, numbered 000017–000020. |

### 3.2. Dataset Processing

The collected dataset needed to be pre-processed, including data labeling, label conversion, and data storage.

1. Use labelImg software (https://gitcode.net/mirrors/tzutalin/labelimg?utm_source=csdn_github_accelerator, accessed on 8 August 2023) to label the dataset information,

use the labeling software to frame the position of defects in the image, and save it as an xml format file, as shown in Table 3.

**Table 3.** Multiple defect labeling data.

| Photos | Description |
| --- | --- |
| 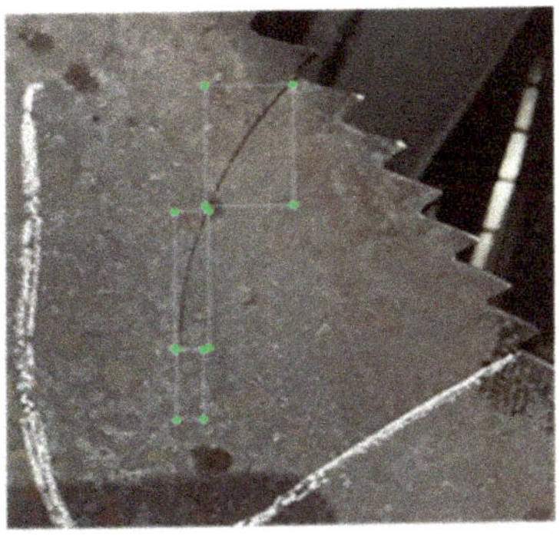 | The picture on the left shows the crack defects of metallurgical saw blades, which are easily caused by the large temperature difference during the quenching process. |
| 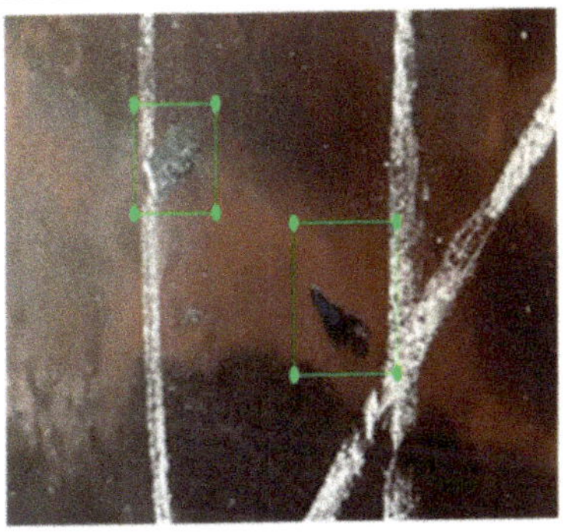 | The picture on the left shows a metallurgical saw blade scarring defect, which is easily caused by poor surface cleaning when cutting steel. |
| 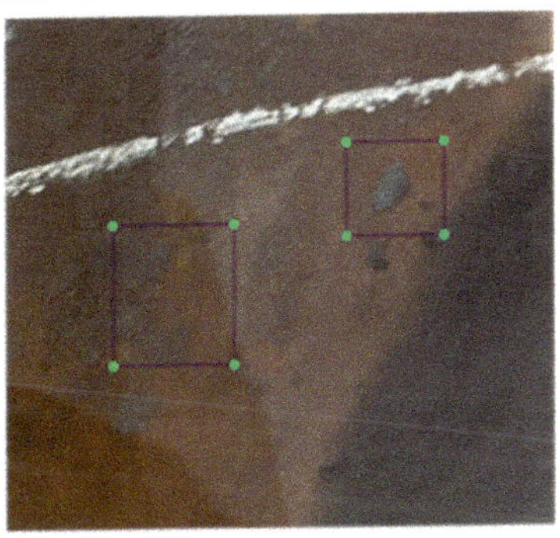 | The figure on the left represents the pitting corrosion defects of metallurgical saw blades. When there are localized damages on the surface of metallurgical saw blades, such as small holes, microcracks, and scratches, corrosive media and oxides are easily gathered and thus form pitting corrosion. |

2. The annotation information required by the Yolo algorithm is the coordinates of the center of the target frame in the sample image and the width and height of the target frame. Therefore, the xml annotation file needs to be converted. The specific steps are as follows:

- Calculate the YOLO type annotation data: set the coordinates of the center point of the target frame on the sample image as $(x, y)$ and the width and height of the target frame as $w$ and $h$, respectively. According to the known data of <*xmin*>, <*xmax*>, <*ymin*>, and <*ymax*>, they can be represented as follows (3):

$$\begin{aligned} w &= (x\text{max}) - (x\text{min}) \\ h &= (y\text{max}) - (y\text{min}) \\ x &= (x\text{min}) + \tfrac{w}{2} \\ y &= (y\text{min}) + \tfrac{h}{2} \end{aligned} \tag{3}$$

- Normalization: Let the width normalization coefficient be $dw$ and the height normalization coefficient be $dh$, as shown in Equation (4):

$$dw = \frac{1}{width} dh = \frac{1}{height} \tag{4}$$

- Data normalization requires multiplying the $x$ coordinate of the center point of the target frame and the width $w$ of the target frame by the width normalization coefficient $dw$, and multiplying the coordinate y of the center point of the target frame and the height $h$ of the target frame by the height normalization coefficient $dw$, as shown in Equation (5):

$$\begin{aligned} \overset{\wedge}{x} &= x \bullet dx \\ \overset{\wedge}{y} &= y \bullet dh \\ \overset{\wedge}{w} &= y \bullet dw \\ \overset{\wedge}{h} &= h \bullet dh \end{aligned} \tag{5}$$

3. Data storage: The YOLO-type txt annotation file contains a row of data, followed by category information, the $x$ coordinate of the center point of the target box, the $y$ coordinate of the center point of the target box, the width $w$ of the target box, and the height $h$ of the target box.

### *3.3. Experimental Environment*

This experiment was carried out using a Win11 operating system and NVIDIA GeForce GTX 105 graphics card. The model was built, trained, and validated using the Pytorch framework. The parameters of the YOLO model were initialized, and during the training process, the applied network model was designed with adaptive anchor frames, with initial anchor frame sizes set to [10, 13, 16, 23, 30, 33], [30, 45, 59, 61, 62, 119], and [90, 116, 156, 198, 326, 373]. To perform network training, the learning rate was set to 0.01, and the number of training epochs was set to 200. The test platform is shown in Table 4.

**Table 4.** Experimental platform.

| Designation | Configure |
|---|---|
| CPU | Intel Core i11400H 2.7 GHZ |
| Graphics Board | NVIDIA GeForce GTX 105 |
| CUDA | CUDA 11.7 |
| Operating System | W11 Operating System |
| Experimental Platform | Python 3.8, Pytorch 3.8 |

## 4. Improved Model Evaluation and Comparative Ablation Experiments

### *4.1. Model Performance Evaluation*

To accurately evaluate the improved SC-YOLOv5 network model, the model evaluation indicators of this experiment included the average accuracy mean mAP@0.5, confusion matrix, PR curve, and loss fitting diagram to evaluate the improved model. The PR curve reflects the relationship between precision and recall. The relationship between precision and recall is reciprocal. If the classifier only predicts the samples with high probability as positive samples, many positive samples with relatively low probability but still satisfied will be missed, resulting in a decrease in recall. The PR curve of the improved model reflects the average accuracy mAP@0.5 of the three defects in the model evaluation standard, and the mAP@0.5 values of cracks, pitting corrosion, and scarring are 0.866, 0.897, and 0.893. The average mAP@0.5 value of the three defects reached 0.885, as shown in Figure 6; the model has relatively high accuracy and good performance.

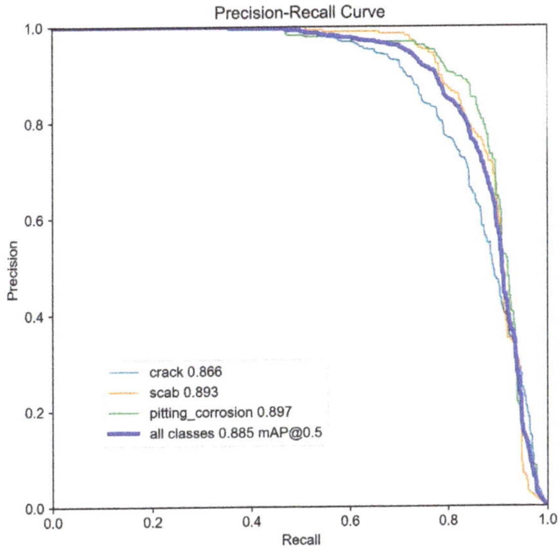


**Figure 6.** Model PR curve.

The confusion matrix is a summary of the prediction results of the defect types. The count value is used to summarize the number of correct and incorrect predictions by classification and is broken down by defect type, showing which part of the classification model is confused when making predictions. From Figure 7 it can be seen that the actual value is close to the predicted value, and the positive sample ratios for cracks, pitting, and scars are 0.89, 0.91, and 0.92, respectively.

The mAP is one of the metrics used to evaluate the detection performance of the improved model, combining the performance of the precision and recall metrics, taking into account the performance at different confidence levels. Specifically, mAP is obtained by averaging the average precision (AP) over all defect categories, and AP is the size of the area under the precision–recall curve. The precision and recall are shown below, where $TP$ denotes data predicted to be defective and actually defective, $FP$ denotes data predicted to be defective but actually not defective, $TN$ denotes data predicted not to be defective and actually not defective, and $FN$ denotes data predicted not to be defective but actually defective.

$$\text{Precision} = \frac{TP}{TP + FP} \tag{6}$$

$$\text{Recall} = \frac{TP}{TP + FN} \tag{7}$$

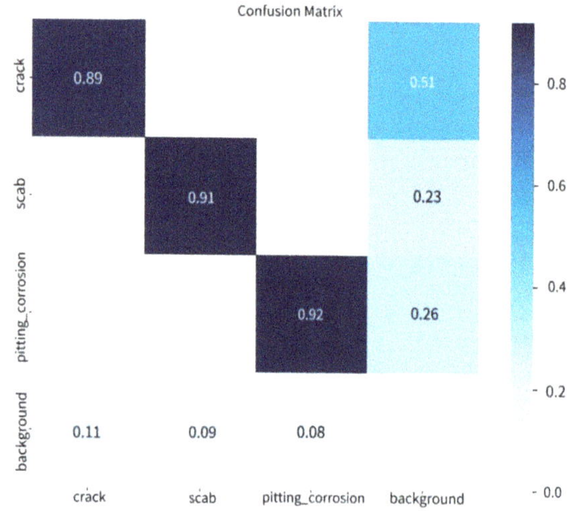


**Figure 7.** Model confusion matrix.

Precision indicates the accuracy of the model in predicting correctly; the higher the value, the less the model misclassifies as a positive case. Recall reflects the ability of the model to identify correct samples; the higher it is, the lower the risk of the model under-reporting (predicting a positive case as a false case). As shown in Table 5, the smaller the mean value of the loss fit plot, the more accurate the detection. Six metrics can be ideally fitted after 200 epochs of training, as expected.

**Table 5.** Multiple indicator fit charts.

| Photos | | Description |
|---|---|---|
| 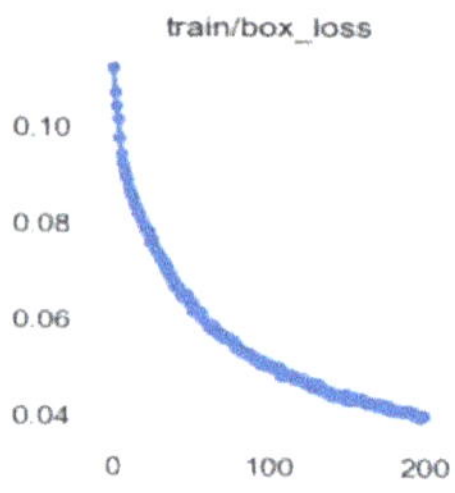  | 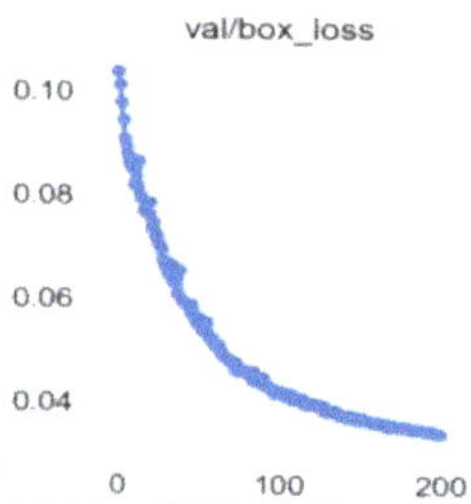  | The box_loss plots indicate the error between the prediction box and the calibration box; the smaller it is, the more accurate the localization is. The error in the graph is less than 0.04 and its defects are more accurately localized. |
| 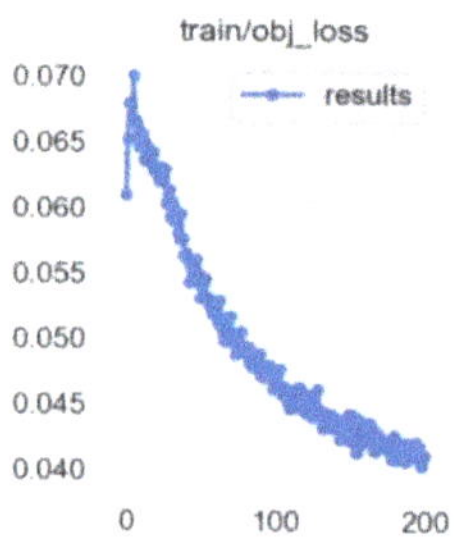  | 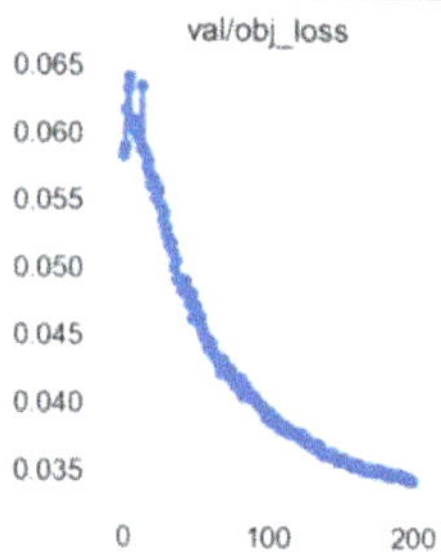  | The obj_loss plots indicate the confidence level of the computational network; the smaller it is, the more accurate the ability to judge it as a target. The higher the confidence level of the graphs are all below 0.04, the higher their ability to be judged as defective targets. |

**Table 5.** *Cont.*

| Photos | | Description |
|---|---|---|
| 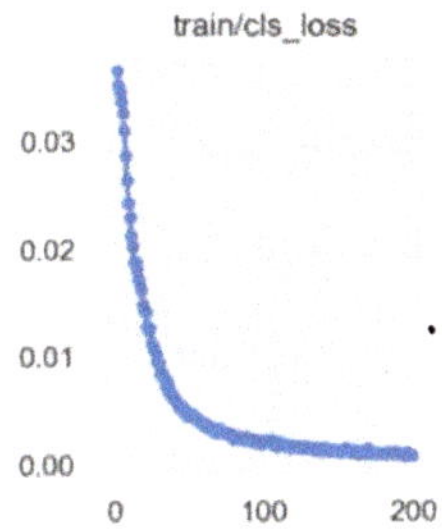 | 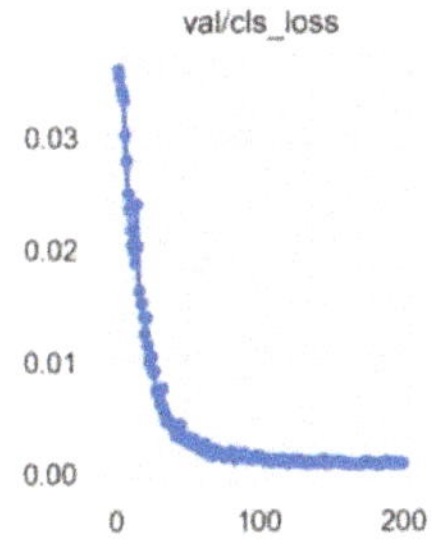 | The cls_loss plots indicate whether the computational anchor box is correctly categorized with the corresponding calibration; the smaller it is, the more accurate the categorization is. The values in the graph are all below 0.01 and their defects are categorized more accurately. |
| 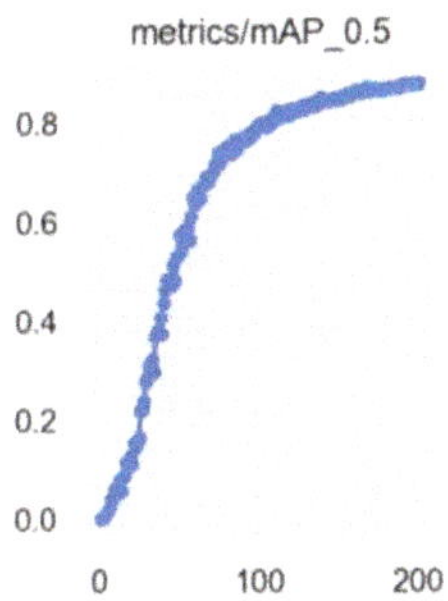 | 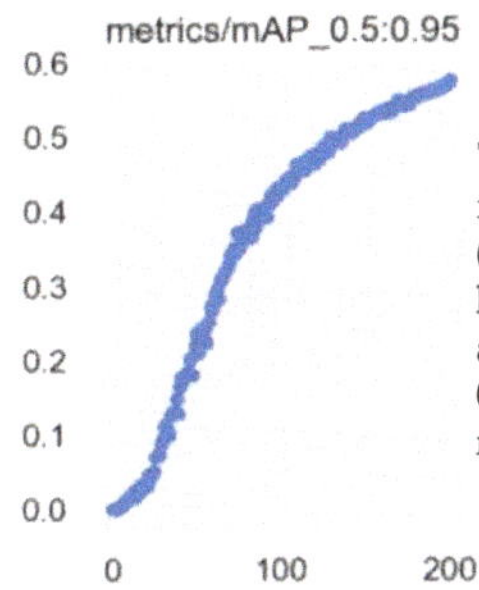 | The mAP plots represent the average mAP over different IoU thresholds (from 0.5 to 0.95 in steps of 0.05), with higher mAP values being more accurate. In the graph mAP@0.5 Above 0.8, mAP@0.9 Above 0.6, its accuracy is relatively accurate. |
| 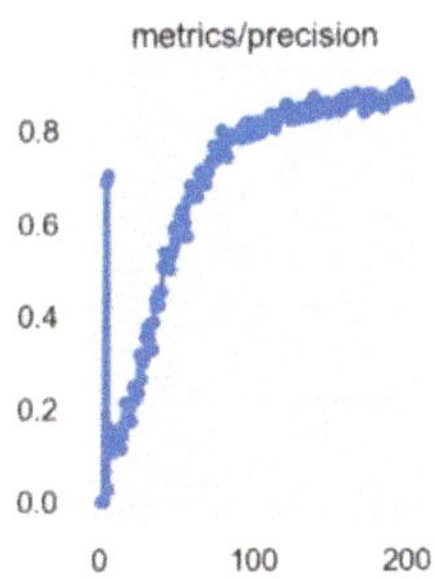 | 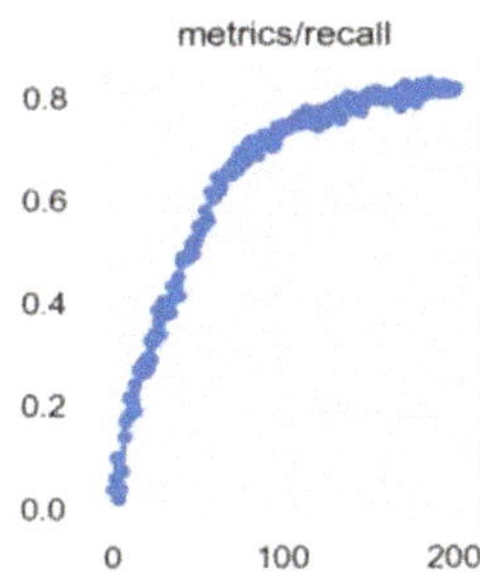 | The precision plot represents the proportion of truly defective samples out of all samples that were predicted as defective. The recall plot represents the proportion of samples that are correctly predicted as defective out of all samples that are truly defective. Both the predicted value and the recall in the figure are higher than 0.8, which gives better precision results. |

## *4.2. Ablation Contrast Experiment*

### 4.2.1. GSConv Improves the Performance of the Model

To prove that the lightweight convolution module GSConv has a better effect on the YOLOv5 network structure and to obtain a better performance of the feature network SC-YOLOv5, comparative experiments were carried out for this paper. First, the original YOLOv5 network structure was experimentally analyzed using the lightweight convolutional module GSConv and the ordinary convolutional module Conv. The experiments show that by using the lightweight convolution module GSConv in the original YOLOv5 network structure, the model parameters are reduced by 6.4%, the F1 value is increased by 0.03, and the map value is increased by 0.04, so the network lightweighting is improved and the accuracy is effectively improved, as shown in Figure 8.

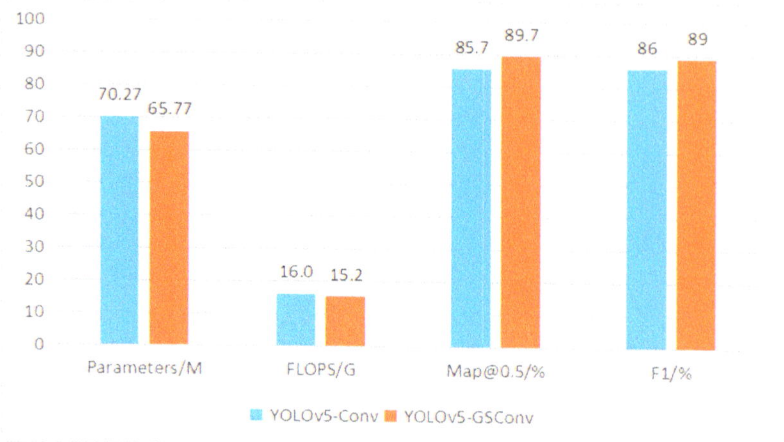


**Figure 8.** GSconv lightweight module ablation experiment.

4.2.2. Network Model Performance Comparison Experiment

In order to objectively evaluate the performance of the model proposed in this paper, the self-made metallurgical saw blade defect dataset was selected to test SC-YOLOv5 and compare it with SSD, YOLO v3-tiny, and YOLOv5. The comparison results are shown in Table 6.

**Table 6.** Comparison of detection performance of models.

| Network Models | Model Size/MB | mAP@0.5/% | Parameters/MB |
|---|---|---|---|
| SSD | 91.68 | 0.629 | 138.12 |
| YOLOv3-tiny | 16.6 | 0.883 | 86.74 |
| YOLOv5 | 27.8 | 0.864 | 71.27 |
| SC-YOLOv5 | 5.78 | 0.885 | 31.10 |

From Table 3, it can be seen that this study further compares SC-YOLOv5 with SSD, YOLOv3-tiny, and YOLOv5 models on the metallurgical saw blade dataset. It can be seen from Table 3 that the mAP@0.5 of the SC-YOLOv5 model is 88.5%, and the memory size of the model is 5.78 MB. Compared with the SSD, YOLOv3-tiny, and YOLOv5 models, the mAP@0.5 of the SC-YOLOv5 model is increased by 25.6, 0.2, and 2.1 percentage points, respectively. The memory size of the model is reduced by 93.7%, 65.1%, and 79.2%, respectively. It can be seen that compared with the SSD, YOLOv3-tiny, and YOLOv5 models, the SC-YOLOv5 model has the best overall performance on metallurgical saw blade defects, and the model complexity is the lowest, which is conducive to the deployment of an improved metallurgical saw blade defect detection model in low-power equipment.

4.2.3. Improvement of SC Network Structure and GSConv on Model Performance

The SC network structure is used to replace the backbone network part in the original YOLOv5, SIoU is redefined as the loss function, and the light convolution module GSConv is replaced to improve the detection effect and further lighten the model parameters. For this paper, the SIoU loss function is first replaced by CIoU, and the SE, CBAM attention mechanism, and CA attention mechanism are added at the same position of the end layer of the YOLOv5 backbone network for ablation experiments. At the same time, the lightweight convolution module is replaced to complete the final improvement of the model, as shown in Table 7. In order to effectively evaluate the performance of the improved model SC-YOLOv5, other models were selected for comparison with SC-YOLOv5. The mAP training curves of the improved model and other models are shown in Figure 9.

**Table 7.** YOLOv5 model ablation contrast experiment.

| Model | ShuffleNetv2 | SIOU | SE | CBAM | CA | GSConv | Parameters/MB | Map@0.5/% | FLOPS/G |
|---|---|---|---|---|---|---|---|---|---|
| 1 | × | × | × | × | × | × | 71.27 | 0.864 | 16.0 |
| 2 | √ | × | × | × | × | × | 31.12 | 0.738 | 6.7 |
| 3 | √ | √ | × | × | × | × | 31.12 | 0.728 | 6.7 |
| 4 | √ | √ | √ | × | × | × | 32.91 | 0.833 | 7.0 |
| 5 | √ | √ | × | √ | × | × | 32.95 | 0.848 | 7.0 |
| 6 | √ | √ | × | × | √ | × | 32.89 | 0.867 | 7.9 |
| 7 | √ | √ | √ | √ | √ | √ | 31.10 | 0.885 | 6.7 |

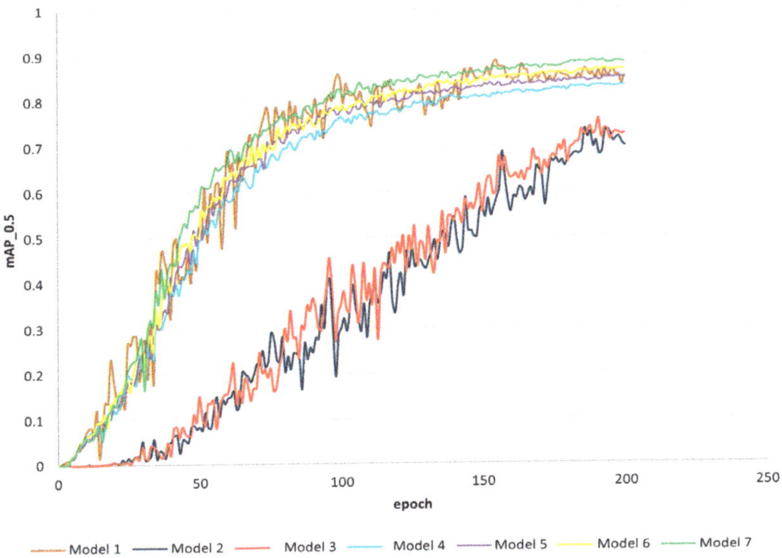


**Figure 9.** Variation curves of mAP for different models.

Experiments show that after the ShuffleNetV2 network structure replaces the original YOLOv5 backbone network structure, its model parameters are greatly reduced, and the lightweight degree of the network is greatly improved, as shown in Table 6. Compared with Model 6, the model parameters of SC-YOLOv5 (Model 7) in this paper are reduced by 2.5%, the map value is increased by 0.018, and the overall lightweight degree is improved. Compared with Model 5, the model parameters are reduced by 5.9%, and the map value is increased by 0.037. Compared with Model 4, the model parameters are reduced by 5.8%, and the mAP value is increased by 0.052. Compared with Model 1, the model parameters are reduced by 56.36%, and the lightweight degree of the network is greatly improved, which significantly improves the long-term transmission efficiency in industrial production, and the map value is slightly improved.

In Figure 9, compared with other models, after 200 rounds of full training, it is obvious that the improved Model 7 has a better convergence effect and better average accuracy.

In order to verify the effectiveness of the model improvement, a random mixed test diagram of multiple defects in the test set was selected, and SC-YOLOv5 was used for testing. The test results are shown in Figure 10. It can be seen from the figure that for metallurgical saw blades with various defects, the improved model can effectively identify the types of defects and locate the target.

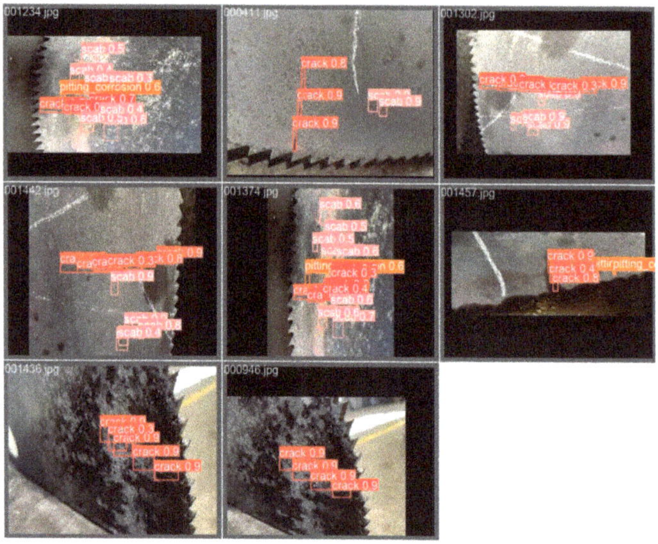

**Figure 10.** Randomized mixed test plot of multiple defects.

## 5. Conclusions

A metallurgical saw blade is a kind of consumable tool, and saw blade defect detection is particularly important. The surface defects of metallurgical saw blades are numerous and complex, and the small size leads to low accuracy and poor real-time performance. An improved YOLOv5 algorithm is proposed to overcome these problems. Verification experiments were carried out on the collected metallurgical saw blade dataset, and the conclusions of this investigation are as follows:

1. The ShuffleNetV2 structure and coordinate attention module were used to form a new network structure SC, which improves the network detection of small targets and multi-target detection ability, greatly reduces the number of model parameters, and improves the degree of network lightweighting.
2. The lightweight convolution module GSConv and loss function SIOU were used to improve YOLOv5, which ensures the accuracy and speed of SC-YOLOv5.
3. The improved YOLOv5 algorithm has low model complexity while ensuring accuracy. It can quickly and accurately detect small-size defect images on the surface of metallurgical saw blades, and it meets the requirements of lightweight model deployment for mobile devices.
4. This paper focuses on the detection of metallurgical saw blade surface defects. In the future, we will continue to optimize the YOLO algorithm, further improve the detection accuracy, and reduce the complexity of the model. In addition, a variety of defect categories for metallurgical saw blades will be added to meet the needs of the industry.

**Author Contributions:** Conceptualization, X.C. and L.M.; methodology, X.C. and R.L.; data curation, X.C. and H.S.; software, L.M.; formal analysis, R.L.; investigation, H.S.; writing—original draft preparation, L.M., X.C., J.L., Y.P. and L.Z.; writing—review and editing, L.M., Z.Z. and R.L.; supervision, L.M. and R.L.; project administration, R.L.; funding acquisition, L.M. and R.L. All authors have read and agreed to the published version of the manuscript.

**Funding:** This study was supported by the Natural Science Foundation of Hebei Province (E2019209314) and the Natural Science Foundation of Hebei Province (E2022209086).

**Data Availability Statement:** The data presented in this study are available on demand from the first author (at mengll@ncst.edu.cn).

**Acknowledgments:** The authors would like to express their sincere gratitude to their schools and colleges and for the funding of the project. We sincerely appreciate all the support and help. In addition, we sincerely thank the editors and reviewers of this paper for their work.

**Conflicts of Interest:** The authors declare no conflict of interest.

## References

1. Feng, L.; Tang, D.; Chen, M. Development of a measuring system for the diameter of a smooth plug gauge based on machine vision technology. *Meas. Control* **2023**, *56*, 966–974. [CrossRef]
2. Sarker, A.; Grift, T. Monitoring Postharvest Color Changes and Damage Progression of Cucumbers Using Machine Vision. *J. Food Res.* **2023**, *12*, 37–50. [CrossRef]
3. Pavlopoulos, G.A.; Hooper, S.D.; Sifrim, A.; Schneider, R.; Aerts, J. Medusa: A tool for exploring and clustering biological networks. *BMC Res. Notes* **2011**, *4*, 384. [CrossRef]
4. Benfenati, A.; Causin, P.; Oberti, R.; Stefanello, G. Unsupervised deep learning techniques for automatic detection of plant diseases: Reducing the need of manual labelling of plant images. *J. Math. Ind.* **2023**, *13*, 5. [CrossRef]
5. Li, J.; Li, S.; Li, X.; Miao, S.; Dong, C.; Gao, C.; Liu, X.; Hao, D.; Xu, W.; Huang, M. Primary bone tumor detection and classification in full-field bone radiographs via YOLO deep learning model. *Eur. Radiol.* **2023**, *33*, 4237–4248. [CrossRef]
6. Wang, L. Retraction Note: Application of deep learning to detect defects on the surface of steel balls in an IoT environment. *J. Supercomput.* **2023**, *79*, 13907. [CrossRef]
7. Hou, X.; Liu, M.; Zhang, S.; Wei, P.; Chen, B. CANet: Contextual Information and Spatial Attention Based Network for Detecting Small Defects in Manufacturing Industry. *Pattern Recognit.* **2023**, *140*, 109558. [CrossRef]
8. Abagiu, M.M.; Cojocaru, D.; Manta, F.; Mariniuc, A. Detecting Machining Defects inside Engine Piston Chamber with Computer Vision and Machine Learning. *Sensors* **2023**, *23*, 785. [CrossRef]
9. Liu, Z.; Wang, M.; Li, C.; Ding, S.; Li, B. A dual-branch balance saliency model based on discriminative feature for fabric defect detection. *Int. J. Cloth. Sci. Technol.* **2022**, *34*, 451–466. [CrossRef]
10. Joung, O.J.; Kim, Y.H. Application of an IR thermographic device for the detection of a simulated defect in a pipe. *Sensors* **2006**, *6*, 1199–1208. [CrossRef]
11. Zhang, L.; Jing, J.; Zhang, H. Fabric defect classification based on LBP and GLCM. *J. Fiber Bioeng. Inform.* **2015**, *8*, 81–89. [CrossRef]
12. Sun, X.; Wu, P. Face detection using deep learning: An improved faster RCNN approach. *Neuro Comput.* **2018**, *299*, 42–50. [CrossRef]
13. Yin, S.; Li, H.; Teng, L. Airport detection based on improved faster RCNN in large scale remote sensing images. *Sens. Imaging* **2020**, *21*, 49. [CrossRef]
14. Yin, H.; Xu, J.; Wang, Y.; Hu, D.; Yi, W. A Novel Method of Situ Measurement Algorithm for Oudemansiella raphanipies Caps Based on YOLO v4 and Distance Filtering. *Agronomy* **2022**, *13*, 134. [CrossRef]
15. Li, Z.; Zhang, J.; Wen, Y.; Dong, Y.; Xu, W. A multiscale fusion YOLOV3-based model for human abnormal behavior detection in special scenarios. *J. Transp. Eng.* **2023**, *149*, 04022150. [CrossRef]
16. Wang, Q.; Zhang, Q.; Liang, X.; Wang, Y.; Zhou, C. Traffic lights detection and recognition method based on the improved YOLOv4 algorithm. *Sensors* **2021**, *22*, 200. [CrossRef] [PubMed]
17. Xian, Y.; Liu, G.; Fan, J.; Yu, Y.; Wang, Z. YOT-Net: YOLOv3 combined triplet loss network for copper elbow surface defect detection. *Sensors* **2021**, *21*, 7260. [CrossRef] [PubMed]
18. Wang, X.; Lv, F.; Li, L.; Yi, Z.; Jiang, Q. A novel optimized tiny YOLOv3 algorithm for the identification of objects in the lawn environment. *Sci. Rep.* **2022**, *12*, 15124. [CrossRef]
19. Xue, Z.; Lin, H.; Wang, F. A small target forest fire detection model based on YOLOv5 improvement. *Forests* **2022**, *13*, 1332. [CrossRef]
20. Li, H.; Zeng, X.; Li, Y.; Zhou, S.; Wang, J. Convolutional neural networks based indoor Wi-Fi localization with a novel kind of CSI images. *China Commun.* **2019**, *16*, 250–260. [CrossRef]
21. Zha, M.; Qian, W.; Yi, W.; Hua, J. A lightweight YOLOv4-Based forestry pest detection method using coordinate attention and feature fusion. *Entropy* **2021**, *23*, 1587. [CrossRef]
22. Zhang, Y.; Zhou, A.; Zhao, F.; Wu, H. A lightweight vehicle-pedestrian detection algorithm based on attention mechanism in traffic scenarios. *Sensors* **2022**, *22*, 8480. [CrossRef]
23. Li, Y.; Han, B.; Zeng, S.; Xu, S.; Yuan, C. End-to-End Point Cloud Completion Network with Attention Mechanism. *Sensors* **2022**, *22*, 6439. [CrossRef] [PubMed]
24. Du, S.; Zhang, B.; Zhang, P. Scale-sensitive IOU loss: An improved regression loss function in remote sensing object detection. *IEEE Access* **2021**, *9*, 141258–141272. [CrossRef]
25. Han, G.; Li, T.; Li, Q.; Zhao, F.; Zhang, M.; Wang, R.; Yuan, Q.; Liu, K.; Qin, L. Improved algorithm for insulator and its defect detection based on YOLOX. *Sensors* **2022**, *22*, 6186. [CrossRef]

26. Wang, L.; Zheng, H.; Yin, C.; Wang, Y.; Bai, Z.; Fu, W. Dense Papaya Target Detection in Natural Environment Based on Improved YOLOv5s. *Agronomy* **2023**, *13*, 2019. [CrossRef]
27. Kırık, M.; Hamamreh, J.M. A novel interference signal superposition algorithm for providing secrecy to subcarrier number modulation-based orthogonal frequency division multiplexing systems. *Trans. Emerg. Telecommun. Technol.* **2023**, *34*, 4678. [CrossRef]
28. Bai, Y.; Yan, Z.; Li, C. An attention mechanism-based deep regression approach with a sequence decomposition-granularity reconstruction-integration model for urban daily water supply forecasting. *J. Hydrol.* **2023**, *617*, 129032. [CrossRef]
29. Xu, B.; Cui, X.; Ji, W.; Yuan, H.; Wang, J. Apple grading method design and implementation for automatic grader based on improved YOLOv5. *Agriculture* **2023**, *13*, 124. [CrossRef]
30. Li, L.; Fang, B.; Zhu, J. Performance Analysis of the YOLOv4 Algorithm for Pavement Damage Image Detection with Different Embedding Positions of CBAM Modules. *Appl. Sci.* **2022**, *12*, 10180. [CrossRef]
31. He, X.; Cheng, R.; Zheng, Z.; Wang, Z. Small object detection in traffic scenes based on YOLO-MXANet. *Sensors* **2021**, *21*, 7422. [CrossRef] [PubMed]
32. Luo, Q.; Jiang, W.; Su, J.; Ai, J.; Yang, C. Smoothing Complete Feature Pyramid Networks for Roll Mark Detection of Steel Strips. *Sensors* **2021**, *21*, 7264. [CrossRef] [PubMed]

processes MDPI

Article

# Dynamic Characteristics Analysis of Metallurgical Waste Heat Radiative Drying of Thin Layers of Sewage Sludge

**Hongyan Guo [1], Zhiwei Tan [1], Haiying Li [1], Yue Long [1], Aimin Ji [2,*] and Liangxu Liu [1,*]**

1 School of Metallurgy and Energy, North China University of Science and Technology, Tangshan 063210, China
2 Ocean College, Tangshan Normal University, Tangshan 063210, China
* Correspondence: jiaimin2022@126.com (A.J.); liuliangxu@126.com (L.L.); Tel.: +86-152-3055-9182 (A.J.); +86-183-2270-3883 (L.L.)

**Abstract:** The utilization of metallurgical waste heat for urban sludge drying and dewatering not only affects the subsequent cost of sludge treatment but also provides a pathway for the rational utilization of metallurgical waste heat. The influence of different experimental conditions on sludge drying characteristics, such as drying temperature and thickness, was analyzed based on metallurgical waste heat. Based on the analysis and evaluation of the drying kinetics parameters of commonly used drying mathematical models, a modified Midilli drying kinetic model is proposed. The kinetic parameters and effective diffusivity of sludge drying were analyzed in three stages of sludge drying: rising rate, constant rate, and falling rate. By utilizing the Arrhenius equation, the relationship between the effective diffusion coefficient and thermodynamic temperature is established, revealing the apparent activation energies for the three stages of urban sludge drying as 29.772 $kJ \cdot mol^{-1}$, 37.129 $kJ \cdot mol^{-1}$, and 39.202 $kJ \cdot mol^{-1}$, respectively. This is closely related to the migration, diffusion, and mass transfer resistance of sludge moisture, indicating that the thickness of sludge accumulation affects the drying time of sludge during the treatment of municipal sludge.

**Keywords:** municipal sludge; dry model; effective diffusion coefficient; activation energy; metallurgical waste heat

**Citation:** Guo, H.; Tan, Z.; Li, H.; Long, Y.; Ji, A.; Liu, L. Dynamic Characteristics Analysis of Metallurgical Waste Heat Radiative Drying of Thin Layers of Sewage Sludge. *Processes* **2023**, *11*, 2535. https://doi.org/10.3390/pr11092535

Academic Editor: Carlos Sierra Fernández

Received: 12 July 2023
Revised: 9 August 2023
Accepted: 17 August 2023
Published: 24 August 2023



## 1. Introduction

With the advancement of industrialization and the rapid growth of the urbanization population, the output of municipal sludge is increasing day by day. According to statistics, the average sludge yield during the operation of domestic sewage treatment plants is 1.62 $tDS/10^4\ m^3$ [1]. Conventional drying methods are inadequate to address the current situation, and the drying technology for sludge in China lags significantly behind. Proper treatment and timely follow-up disposal of sludge are very difficult. Therefore, scholars must study the drying theory and develop treatment technology to solve the situation of sludge increasing efficiently. Currently, China produces over 750 million tons of metallurgical slag annually, with the discharged slag temperature typically exceeding 1200 °C and carrying a substantial amount of heat [2]. The recovery and utilization of this immense metallurgical waste heat is of utmost importance.

Sludge is a typical solid sediment, which is mainly obtained after the treatment of sewage produced by human production and life. It is extremely complex in composition and is present in large amounts, making it very difficult to treat [3]. To facilitate the transportation and logistics of sludge during the treatment process, sludge generally needs to be thickened or mechanically dewatered, which can reduce the moisture content of sludge to approximately 75% [4]. It will not only cause great damage to the natural environment but also endanger the physical and mental health of human beings and the survival of other animals and plants. Traditional municipal sludge disposal methods, such as landfill, composting, sea reclamation, incineration, etc. [5–7], can play a certain role in the harmless treatment of municipal sludge. But these traditional disposal methods cause secondary pollution and big energy consumption problems. At the same time, gradual low-temperature drying not only solves the issues of poisonous gases emanating from the pyrolysis of organic pollutants in sewage sludge and big issues such as energy consumption but also reduces the moisture content of sludge, reduces sludge volume capacity, and solves the problems of sludge and of its difficult transportation [8–10].

The most critical step in wastewater sludge treatment is sludge drying, which can be commonly categorized into mechanical drying, natural drying, and thermal drying methods. Mechanical drying involves using mechanical equipment such as centrifugation, filtration, or evaporation to reduce the moisture content of the sludge [11]. Natural drying refers to methods where moisture is allowed to evaporate naturally or gravity is utilized to separate water, with sludge drying beds being a typical example, where sludge is spread over sand beds until it dries [12]. Thermal drying refers to the utilization of thermal energy to conduct heat to the sludge, resulting in the evaporation of moisture [13]. Thermal drying can be further classified into convection drying, conduction drying, and radiation drying based on the modes of heat transfer [14]. Commonly used heat sources for drying are often low-grade waste heat. After the metallurgical blast furnace slag undergoes heat recovery using the air quenching method, the resulting granulated slag still maintains temperatures above 150 °C [15], which aligns with the requirements for thermal drying of sludge.

Currently, the commonly used drying models in research are mostly empirical or semi-empirical models, which rely on experimental data and may not necessarily be applicable to drying under different conditions. The simple Lewis model is easy to apply and has clear physical meaning [16]. The empirical Logarithmic model exhibits higher correlation coefficients and lower root mean square errors [17]. The semi-empirical Page model, compared to the Lewis model, better captures the descending rate period and fits well with the experimental data [18]. The Modified Page model allows for improved accuracy by adjusting the exponent [19]. The two-term exponential model describes the constant-rate and falling-rate periods with higher precision [20]. The Midilli model captures both the constant-rate and falling-rate periods and exhibits a high degree of fit with experimental data [21]. The Henderson and Pabis models can describe the falling rate period and are characterized by their simplicity [22]. Each of the seven models mentioned above has its advantages, but it is difficult to determine which model is more suitable for municipal sludge drying.

The main objective of this study is to conduct preliminary research for industrial projects involving the drying of metallurgical low-grade waste heat. To achieve this, experimental equipment was employed to simulate metallurgical waste heat, and drying experiments were performed on urban sludge samples with varying thicknesses and temperatures. The simulation of urban sludge thin-layer drying curves was carried out using the aforementioned seven models. Based on the goodness-of-fit, the most suitable drying model was selected, and further improvements were made. The kinetic parameters and effective diffusion coefficients of sludge drying in three different stages are analyzed, and the activation energy is calculated and analyzed to characterize the drying characteristics of urban sludge under different conditions and scenarios.

## 2. Experimental Materials and Methods

### 2.1. Experimental Materials

The sludge samples were collected from the wastewater treatment plant in Tangshan City. Before the experiments, solid particles and residual garbage were removed from the wet sludge, which was then stored in a refrigerator at 0–5 °C for 24 h to prevent fermentation. Prior to the experiments, a suitable amount of sludge was taken out and allowed to reach a temperature of 20–25 °C in the environment for two hours.

As the moisture content of the collected sludge was initially unknown in this study, a 105 °C electric blast drying oven was used to heat and dry the sludge. The weight was measured every two hours until a constant weight state was achieved. Eventually, the moisture content of the sludge in the experiment was determined to be approximately 80.26%. The fundamental characteristics of sludge are presented in Table 1.

**Table 1.** Fundamental characteristics of sludge.

| Industrial Analysis (ar [1])/% | | | | Elementary Analysis (d [2])/% | | | | | $Q_{net.ad}$ [3]/MJ·kg$^{-1}$ |
|---|---|---|---|---|---|---|---|---|---|
| M | A | V | FC | C | H | O | N | S | |
| 80.26 | 4.65 | 12.79 | 2.30 | 31.45 | 5.27 | 26.40 | 3.58 | 0.77 | 12.78 |

[1] as received; [2] dry basis; [3] air dry basis.

## 2.2. Experimental Instrument

DHS-16A halogen moisture tester (Changzhou Heng Positron Instrument Co. Ltd., Changzhou, China): This instrument has a maximum weighing capacity of 100 g, with a display readability of 1 mg and an error of ±1 mg. It uses a halogen lamp as the heating source, with a maximum temperature of 160 °C. The time control range is 0–90 min.

DHG-9070A Electric Blower drying oven (Shaoxing Yicheng Instrument Manufacturing Co. Ltd., Shaoxing, China): This equipment has a temperature control range of RT + 10 °C to 250 °C. The temperature stability is ±1.0 °C, and the temperature resolution is 0.1 °C. At the test point of 100 °C, the temperature uniformity is within ±3%. The volume of the oven is 80 L.

FA-2204 high precision electronic balance (Hangzhou Jingfei Instrument Technology Co. Ltd., Hangzhou, China): This balance has a maximum weighing capacity of 220 g, with a readability of 0.1 mg and a repeatability of ±0.0002 g. The linear error is ±0.0003 g.

## 2.3. Experimental Methods

The preparation of sludge cakes involves using two glass plates, each with a layer of 100-mesh sieve net on it. The sludge is then pressed between the glass plates to ensure flatness. After the sludge cake is formed, the glass plates are opened, the sieve net is removed, and a small knife is used to modify the sludge cake to meet the experimental conditions. DHS-16A halogen moisture tester was used to complete the drying experiment. The drying temperatures were 45 °C, 55 °C, 75 °C, 105 °C, and 135 °C, and the thicknesses were 0.5 mm, 1 mm, 2 mm, and 4 mm. The drying equipment incorporated a self-made balance for real-time measurement, continuously weighing the sludge cake during the moisture evaporation process. Three experiments were conducted for each sludge cake thickness, and the average value was obtained. During the experiment, the height of the radiant heat source was maintained at 1 cm and no hot air interference was ensured. The radiation temperature error was ±2 °C. Data were recorded once every minute after startup.

# 3. Analysis of Experimental Results

## 3.1. Influence of Temperature on Sludge Drying

Taking a thin municipal sludge layer with a thickness of 2 mm as an example, its moisture content changes over time at 45 °C, 55 °C, 75 °C, 105 °C, and 135 °C, as shown in the figure below.

As can be seen from Figure 1, when the thickness of municipal sludge is the same, the drying time of sludge is significantly shortened by increasing the drying temperature. The reason is that by increasing the drying temperature, that is, enhancing the radiation intensity on the surface of the sludge, the adsorbed water on the surface of the sludge evaporates rapidly, the surface humidity is significantly reduced, and there is a temperature gradient in the internal and external water. Under the driving force of the temperature difference in heat transfer, the water in the sludge migrates outward at a faster rate. Therefore, when the temperature rises from 45 °C to 135 °C, the drying time for a 2 mm thick municipal thin sludge layer to equilibrium water content is reduced from 245 min to about 20 min.

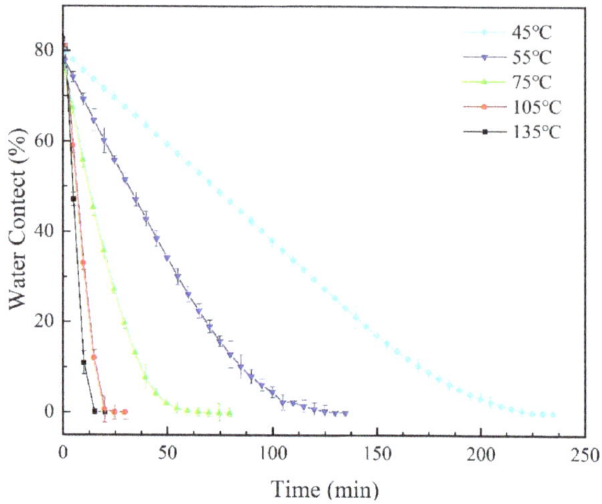


**Figure 1.** The relationship between sludge moisture content and temperature with a thickness of δ = 2 mm.

### 3.2. *Effect of Thickness on Sludge Drying*

Taking the thin sludge layer with a drying temperature of 75 °C as an example, its moisture content changes over time under the four thicknesses of 0.5 mm, 1 mm, 2 mm, and 4 mm, as shown in the figure below.

As can be seen from Figure 2, when the drying temperature is constant, the smaller the thickness of the thin sludge layer, the shorter the drying time of the thin layer. Because the thickness of the sludge is small, the internal water of the sludge is more likely to migrate outward, and the mass transfer resistance is smaller; therefore, the drying time of the sludge is shorter, and the water on the surface evaporates faster. When the drying temperature is constant and the thickness of the thin sludge layer decreases from 4 mm to 1 mm, the drying time to equilibrium moisture content is reduced from about 130 min to nearly 15 min.

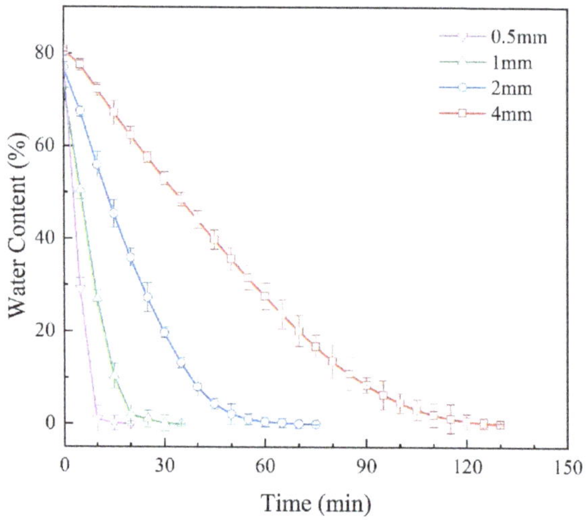


**Figure 2.** The relationship between moisture content and thickness of sludge at temperature T = 75 °C.

### 3.3. Effect of Temperature on Sludge Drying Rate

Taking the 2 mm thin sludge layer as an example, its drying rate changes over time at 45 °C, 55 °C, 75 °C, 105 °C, and 135 °C, as shown in the figure below.

As can be seen from Figure 3, when the thickness of municipal sludge is the same and the drying temperature is increased, the total drying time of the sludge is shorter. When the temperature rises from 45 °C to 135 °C, 2 mm increases obviously with the drying rate peak value of thin sludge layer, and the range of isometric drying stage of sludge decreases obviously, and the time of descending drying becomes shorter.

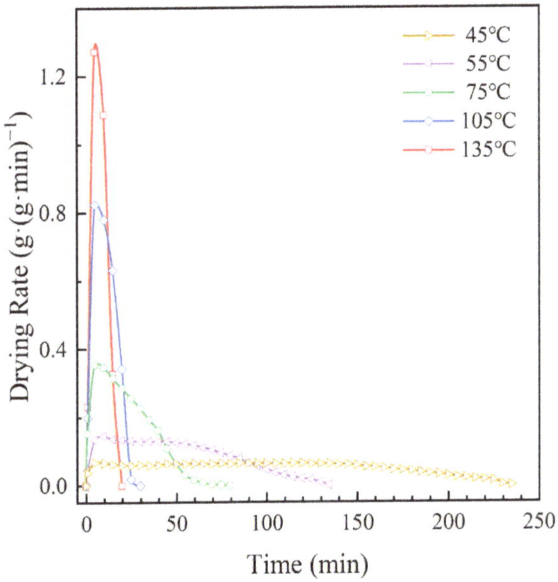


**Figure 3.** The relationship between sludge drying rate and temperature at a thickness of δ = 2 mm.

### 3.4. Effect of Thickness on Sludge Drying Rate

Taking the thin sludge layer at a drying temperature of 75 °C as an example, its moisture content changes over time under the four thicknesses of 0.5 mm, 1 mm, 2 mm, and 4 mm, as shown in the figure below.

It can be seen from Figure 4 that, under the same municipal sludge temperature, the thicker the thin sludge layer, the longer the total drying time. The increasing drying time of sludge has no obvious change, the peak value of municipal sludge drying rate increases from 0.0514 g·(g·min)$^{-1}$ to 0.2645 g·(g·min)$^{-1}$, the constant drying time also shows little change, and the decreasing drying time increases with the increase in thickness. The thickness of sludge increased from 0.5 mm to 4 mm, and the desiccating time of sludge decreased from 10 min to about 120 min. From this, it can be observed that, under the same thickness, higher drying temperatures result in a faster change in drying rate and a shorter total drying time. The higher the peak point of the drying rate during the up-rate drying stage, the more negligible the time in the constant rate drying stage, leading to a shortened time in the down-rate drying stage. Likewise, at the same temperature, thinner sludge exhibits a faster change in drying rate and a shorter total drying time. However, there is no significant change in the time of rise-rate drying and iso-rate drying, but the time of down-rate drying is noticeably shortened. Consequently, in the municipal sludge treatment process, measures such as timing agitation should be adopted during the deceleration stage to enhance the drying rate, conserve energy, and reduce costs.

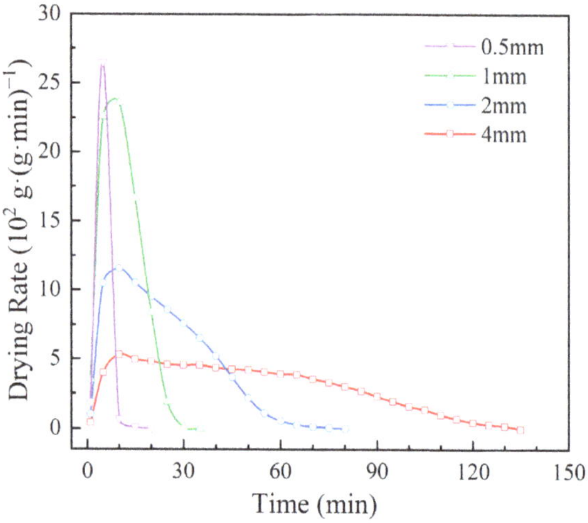


**Figure 4.** The relationship between sludge drying rate and thickness at temperature T = 75 °C.

## 4. Model Analysis and Goodness Evaluation

The drying process is a complicated heat and mass transfer process. Through numerous domestic and international scientific research, workers experiment with different materials to uncover the mechanisms and characteristics of the drying process. They summarize the experience of many incomplete experiences and theories, experiment with a material drying mathematical model, and, based on the characteristics of sludge and on tests on the working condition of municipal sludge, choose several common mathematical models of sludge drying. These are used to quantitatively describe the drying law of municipal sludge and predict and verify the accuracy and reliability of the test data.

The following table uses seven commonly used thin-layer drying models to simulate and analyze the drying curve of the thin municipal sludge layer. The moisture content measured in the drying test of 2 mm thin layer of municipal sludge at five temperatures (45 °C, 55 °C, 75 °C, 105 °C and 135 °C) is taken as the sample value. Origin software was used to fit the thin-layer drying model of sludge at different temperatures, and each parameter value of the drying model was obtained. The measured data obtained from the municipal sludge drying test was fitted and verified by the commonly used drying mathematical model. Standard deviation ($S_D$), residual sum of squares ($R_{SS}$), Chi-square ($\chi^2$), and correlation coefficient $R^2$ were used to evaluate the goodness of fit of the model.

The commonly used drying mathematical model is evaluated using fitting parameter values and the Origin software, as shown in Table 2 below.

**Table 2.** Fitting parameter values and accuracy evaluation of each model.

| Model Name | Value | 2 mm | | | | | Reference |
|---|---|---|---|---|---|---|---|
| | | 135 °C | 105 °C | 75 °C | 55 °C | 45 °C | |
| Lewis<br>MR = exp(−kt) | $k$ | 0.2195 | 0.1238 | −0.02133 | −0.0168 | −0.0114 | [23] |
| | $R^2$ | 0.9402 | 0.9493 | 0.3204 | −1.1088 | −1.4925 | |
| | $S_D$ | 0.0147 | 0.0059 | 0.1672 | 0.0144 | 0.0047 | |
| | $\chi^2$ | 0.0048 | 0.0042 | 1245.8521 | 1548.067 | 1693.049 | |
| | $R_{SS}$ | 0.0678 | 0.0967 | 21,072.4211 | 43,345.866 | 79,573.280 | |
| Logarithmic<br>MR = aexp(−kt) + b | $a$ | 1.1630 | 1.2923 | 90.6649 | 106.5904 | 161.6415 | [24] |
| | $b$ | −0.3160 | −0.4061 | −8.9063 | −25.4111 | −78.8582 | |
| | $k$ | 0.1048 | 0.0566 | 0.0378 | 0.0120 | 0.0033 | |
| | $R^2$ | 0.9907 | 0.9909 | 0.9895 | 0.9833 | 0.9852 | |
| | $S_D$ | 0.1704 | 0.1686 | 5.1016 | 7.7434 | 21.7998 | |
| | $\chi^2$ | 0.0008 | 0.0008 | 7.2975 | 4.9380 | 3.2936 | |
| | $R_{SS}$ | 0.0091 | 0.0159 | 94.8676 | 128.3873 | 148.2111 | |

**Table 2.** *Cont.*

| Model Name | Value | 2 mm | | | | | Reference |
|---|---|---|---|---|---|---|---|
| | | 135 °C | 105 °C | 75 °C | 55 °C | 45 °C | |
| Page MR = exp(−kt^n) | $k$ | 0.1789 | 0.0779 | −0.0293 | −4.6016 | −5.5922 | [25] |
| | $n$ | 1.1163 | 1.2015 | 0.9250 | −0.0862 | −0.1073 | |
| | $R^2$ | 0.9443 | 0.9586 | −0.6839 | 0.2637 | 0.4063 | |
| | $S_D$ | 0.1682 | 0.1053 | 0.5681 | 0.2415 | 0.3173 | |
| | $\chi^2$ | 0.0486 | 0.0034 | 1174.7973 | 540.5235 | 403.3151 | |
| | $R_{SS}$ | 0.0632 | 0.0755 | 17,621.9600 | 14,594.1360 | 18,552.495 | |
| Modified Page MR = exp(−(kt)^n) | $k$ | 0.2140 | 0.1195 | −0.0230 | −0.0120 | −0.0114 | [26] |
| | $n$ | 1.1180 | 1.2051 | 1.0000 | 1.0000 | 1.0000 | |
| | $R^2$ | 0.9401 | 0.9586 | −0.8066 | −1.1952 | −1.4925 | |
| | $S_D$ | 0.1409 | 0.0930 | 0.2276 | 0.1320 | 0.0465 | |
| | $\chi^2$ | 0.0049 | 0.0034 | 1240.4390 | 1611.4903 | 1729.8539 | |
| | $R_{SS}$ | 0.0632 | 0.0755 | 17,646.1500 | 43,510.2380 | 79,573.2800 | |
| Henderson and Pabis MR = aexp(−kt) | $a$ | 0.8969 | 0.9508 | 84.3280 | 85.3485 | 90.7964 | [27] |
| | $k$ | 0.1982 | 0.1180 | 0.0493 | 0.0211 | 0.0103 | |
| | $R^2$ | 0.9544 | 0.9506 | 0.9751 | 0.9652 | 0.9450 | |
| | $S_D$ | 0.0620 | 0.0478 | 3.3148 | 2.6224 | 2.7258 | |
| | $\chi^2$ | 0.0037 | 0.0041 | 17.3562 | 25.5843 | 37.3831 | |
| | $R_{SS}$ | 0.0480 | 0.0900 | 242.9874 | 690.7748 | 1719.6243 | |
| Two-term exponential MR = aexp(−kt) + bexp(−$k_1$at) | $a$ | 0.3899 | 0.4129 | 37.6105 | 7.0574 | 93.2083 | [28] |
| | $b$ | 0.5070 | 0.5379 | 46.7151 | 78.2889 | −13.0433 | |
| | $k$ | 0.1982 | 0.1179 | 0.0493 | 0.0211 | 0.0106 | |
| | $k_1$ | 0.5084 | 0.2858 | 0.0013 | 0.0030 | 1.0001 | |
| | $R^2$ | 0.9461 | 0.9456 | 0.9732 | 0.9624 | 0.9471 | |
| | $S_D$ | 0.7100 | 0.7543 | 249.0900 | 489.7758 | 87.3237 | |
| | $\chi^2$ | 0.0044 | 0.0045 | 18.6914 | 27.6310 | 35.9266 | |
| | $R_{SS}$ | 0.0480 | 0.0900 | 242.9877 | 690.7754 | 1580.7692 | |
| Midilli MR = aexp(−kt^n) + bt | $a$ | 0.7098 | 0.7543 | 76.8406 | 76.7693 | 77.5928 | [29] |
| | $k$ | −0.0036 | −0.0401 | 0.0114 | 0.0028 | 0.0006 | |
| | $n$ | $4.58 \times 10^{-9}$ | $6.41 \times 10^{-11}$ | 1.4097 | 1.4445 | 1.5195 | |
| | $b$ | −0.0596 | −0.0397 | −0.0221 | −0.0335 | −0.0441 | |
| | $R^2$ | 0.9229 | 0.9503 | 0.9981 | 0.9985 | 0.9985 | |
| | $S_D$ | 30.5821 | 1.5786 | 1.1165 | 0.6864 | 0.6008 | |
| | $\chi^2$ | 0.0054 | 0.0041 | 1.3399 | 1.1407 | 1.0414 | |
| | $R_{SS}$ | 0.0598 | 0.0824 | 16.0783 | 28.5171 | 45.8205 | |

Where $k$ and $k_1$ are the empirical coefficients of the model; $a$, $b$, and $n$ are empirical constants of the model.

It can be seen from Table 2 above that the correlation coefficients $R^2$ of the Logarithmic Model all exceed 0.99 under the medium temperature drying condition of 105–135 °C. The standard deviation ($S_D$), Chi-square ($\chi^2$), and residual square ($R_{SS}$) were also the smallest in all models. It shows that the Logarithmic model is more suitable to describe the change of drying water content and time prediction of thin municipal sludge layer in the range of 105–135 °C. Under low-temperature drying conditions (45~75 °C), $R^2$ of the Midilli model was more than 0.99, and standard deviation ($S_D$), Chi-square ($\chi^2$), and residual square ($R_{SS}$) were the smallest among all models. The results show that the Midilli model is more suitable to describe the variation of drying water content and the prediction of drying time of thin municipal sludge layer at 45–75 °C.

*Modeling*

Through fitting and verification of the above seven mathematical models and test values, it can be seen that Temperature conditions in the Logarithmic model are suitable for the municipal sludge drying process, and the Midilli model is more suitable for the low-temperature condition of the municipal sludge drying process. To unify the municipal sludge in the moderate temperature drying mathematical model, we simply introduce a Logarithmic model of a zero dimension number n to eliminate the equations describing the drying process in early and late moisture prediction error [30]. After considering the influence of water content change at the initial moment of municipal sludge, a linear term is introduced to modify Logarithmic model into Midilli model. Finally, to improve the description of the single diffusion model of the equation, an additional empirical constant term C is added to obtain a new model, namely the modified model Midilli, whose expression is MR = aexp(−kt^n) + bt + c. The fitting comparison between this model and the measured values of municipal sludge tests at different temperatures is shown below (Figure 5).

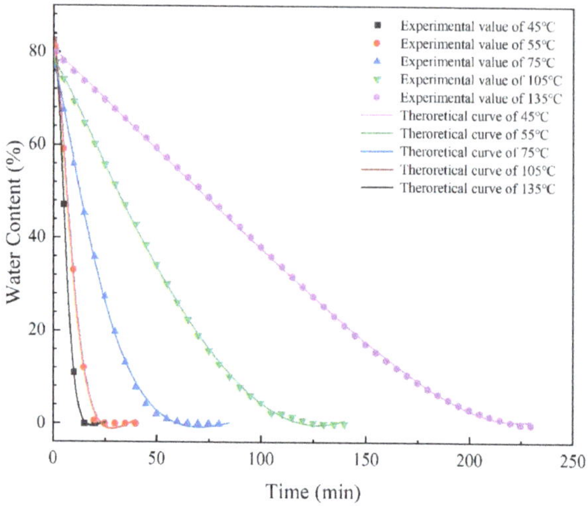


**Figure 5.** Fitting comparison between experimental and theoretical values.

The goodness of fit of the model was verified by taking the fitting results under five working conditions of 2 mm thickness and 45~135 °C drying temperature as an example, as shown in Table 3 below.

**Table 3.** Parameter values and evaluation excellence of modified Midilli model fitting.

| Model Name | Value | 135 °C | 105 °C | 75 °C | 55 °C | 45 °C |
|---|---|---|---|---|---|---|
| Modified Midilli MR = aexp(−kt$^n$) + bt + c | $a$ | 84.0776 | 87.9204 | 96.3962 | 60.6362 | 0.0006 |
| | $k$ | 0.0283 | 0.0224 | 0.0143 | −0.0307 | −0.4738 |
| | $n$ | 1.8481 | 1.6068 | 1.2798 | 0.7834 | 0.5681 |
| | $b$ | 0.0405 | 0.1383 | 0.2061 | −2.0067 | −0.4305 |
| | $c$ | −1.2315 | −5.5717 | −18.7135 | 17.0443 | 80.5924 |
| | $R^2$ | 0.9998 | 0.9983 | 0.9992 | 0.9996 | 0.9998 |
| | $S_D$ | 2.2793 | 4.5312 | 6.7908 | 0.8545 | 0.1309 |
| | $\chi^2$ | 0.2509 | 1.9180 | 0.6295 | 0.2722 | 0.1757 |
| | $R_{SS}$ | 1.0036 | 11.5079 | 9.4428 | 7.3486 | 8.2559 |

The comparison between the fitting value of the modified Midilli model in Table 3 and the fitting value of the traditional model in Table 2 shows that the correlation coefficient $R^2 > 0.998$, and the standard deviation $S_D$, Chi-square $\chi^2$, residual square, and $R_{SS}$ are also smaller. The accuracy of the fitting degree of this evaluation is better than that of the traditional drying model. It shows that the modified Midilli model can simulate municipal sludge drying well at 45~135 °C. Because the new model can accurately describe the variation trend of the water content of municipal sludge, it has a certain guiding significance for the subsequent treatment and disposal of sludge.

## 5. Kinetic Analysis of Sludge Water Transfer

Based on Fick's second law and Arrhenius law of liquid diffusion theory, kinetic characteristics of sludge drying were analyzed from two aspects of the effective water diffusion coefficient and sludge apparent activation energy.

### 5.1. Effective Water Diffusion Coefficient Calculation

Water transfer during sludge drying is a complex process of heat and mass transfer, which is reflected in Fick's Second Law. The integral expression of Fick's second law is as follows [31]:

$$MR = \frac{8}{\pi^2}\sum_{n=0}^{\infty}\frac{1}{(2n+1)^2}\exp\left(-(2n+1)^2\frac{\pi^2 D_{eff}t}{4\delta^2}\right) \tag{1}$$

where $D_{eff}$ is the effective water diffusion coefficient of the material in $m^2 \cdot s^{-1}$; $t$ is the drying time in the unit s; $\delta$ is the thickness of material drying in the unit m; $n$ is the number of experimental samples.

When the drying time is long, only the first term of the above equation [32] can be taken, and then logarithmic deformation can be taken on both sides of the equation at the same time, so that:

$$\ln MR = \ln\frac{8}{\pi^2} - \frac{\pi^2 D_{eff}}{4\delta^2}t \tag{2}$$

It can be seen from Formula (2) above that there is a linear relationship between the logarithm of the effective water diffusion coefficient and drying time. Therefore, this line can be fitted according to the relationship between water content measured in the test and time, and its slope can be obtained from the fitted $\ln D_{eff}$-$t$ function graph, and then the effective water diffusion coefficient $D_{eff}$ can be obtained according to the formula below.

$$D_{eff} = -\frac{4k\delta^2}{\pi^2} \tag{3}$$

Considering the different degrees of water migration in the sludge, the sludge drying rate curve should be divided into three main parts, namely, the ascending stage, the constant stage and the descending stage, and the effective diffusion coefficient of each stage can be calculated, respectively. In the up-rate drying stage, the adsorbed water on the surface of sludge is mainly removed. In the constant rate drying stage, the free water needs a certain amount of heat as a driving force to diffuse to the surface before evaporation and removal. In the down-rate drying stage, part of the bound water in the sludge is removed, which requires more energy to destroy the molecular structure. Taking the drying temperature of 75 °C and the test value of drying conditions of municipal sludge with different thicknesses as an example, the effective water diffusion coefficient can be calculated as $2.63 \times 10^{-9}$ $m^2 \cdot s^{-1}$ according to the Expression (3). Similarly, the effective water diffusion coefficient $\ln D_{eff}$ of municipal sludge with other thicknesses can be calculated. The results are shown in Table 4.

**Table 4.** Effective water diffusivity of the thin sludge layer with different thicknesses at a drying temperature of 75 °C.

| Thickness | Slope | $D_{eff}$ | $R^2$ |
|---|---|---|---|
| 0.5 mm | $-7.43 \times 10^{-3}$ | $7.53 \times 10^{-10}$ | 0.86467 |
| 1 mm | $-4.33 \times 10^{-3}$ | $1.75 \times 10^{-9}$ | 0.92945 |
| 2 mm | $-1.62 \times 10^{-3}$ | $2.63 \times 10^{-9}$ | 0.82413 |
| 4 mm | $-6.55 \times 10^{-4}$ | $4.25 \times 10^{-9}$ | 0.96312 |

It can be seen from Table 4 above that in the sludge drying process at 75 °C, when the thickness decreases from 4 mm to 0.5 mm, the effective diffusion coefficient of municipal sludge moisture increases from $-7.43 \times 10^{-3}$ $m^2 \cdot s^{-1}$ to $-6.55 \times 10^{-4}$ $m^2 \cdot s^{-1}$. Therefore, when the temperature is constant, the path of water diffusion in the sludge is reduced and the mass transfer resistance is small. Therefore, the thinner the sludge thickness is, the larger the $D_{eff}$ of the effective water diffusion coefficient in the sludge drying process is.

Taking the test values of the municipal sludge with a thickness of 2 mm at different temperature drying conditions as an example, the effective water diffusion coefficient

$D_{eff}$ at three stages at different temperatures can be calculated. The results are shown in Table 5 below.

**Table 5.** Effective water diffusivity of a 2 mm thick thin sludge layer at different temperatures.

| Temperature | Drying Stage | $D_{eff}$ | $R^2$ |
|---|---|---|---|
| 45 °C | Acc phase | $1.535 \times 10^{-10}$ | 0.99975 |
| | Constant phase | $2.3442 \times 10^{-10}$ | 0.98333 |
| | Slow downstage | $9.926 \times 10^{-10}$ | 0.93221 |
| 55 °C | Acc phase | $3.664 \times 10^{-10}$ | 0.99757 |
| | Constant phase | $6.391 \times 10^{-10}$ | 0.99136 |
| | Slow down stage | $1.669 \times 10^{-09}$ | 0.97264 |
| 75 °C | Acc phase | $9.157 \times 10^{-10}$ | 0.99445 |
| | Constant phase | $5.642 \times 10^{-9}$ | 0.94621 |
| | Slow downstage | $3.842 \times 10^{-9}$ | 0.98297 |
| 105 °C | Acc phase | $1.413 \times 10^{-9}$ | 0.974597 |
| | Constant phase | $3.015 \times 10^{-9}$ | 0.99546 |
| | Slow downstage | $1.271 \times 10^{-8}$ | 0.94625 |
| 135 °C | Acc phase | $2.351 \times 10^{-9}$ | 0.94549 |
| | Constant phase | $5.755 \times 10^{-9}$ | 0.99057 |
| | Slow downstage | $2.479 \times 10^{-8}$ | 0.95809 |

As can be seen from Table 5 above, in the sludge drying process of sludge with a thickness of 2 mm, the effective diffusion coefficient of municipal sludge moisture ranges from $1.535 \times 10^{-10}$ to $2.351 \times 10^{-9}$ $m^2 \cdot s^{-1}$ at 45~135 °C in the speed-increasing stage. At the constant velocity stage, the effective diffusion coefficient of municipal sludge moisture ranges from $2.344 \times 10^{-10}$ $m^2 \cdot s^{-1}$ to $5.755 \times 10^{-9}$ $m^2 \cdot s^{-1}$ at 45~135 °C. In the deceleration stage, the effective diffusion coefficient of water in municipal sludge is $9.926 \times 10^{-10}$~$2.479 \times 10^{-8}$ $m^2 \cdot s^{-1}$ at 45~135 °C. Therefore, when the thickness is constant, the higher the temperature is, the larger the temperature difference between the heat transfer of water inside and outside the sludge is, and the stronger the driving force of water diffusion outward. Therefore, the diffusion coefficient in the process of sludge drying is greater.

### *5.2. Apparent Activation Energy Calculation*

The functional relation between the effective diffusion coefficient and the reciprocal of the thermodynamic temperature T can be established by the Arrhenius equation. The Expression (4) is as follows:

$$D_{eff} = D_0 \exp\left(-\frac{E_a}{RT}\right) \quad (4)$$

where $D_0$ is the diffusion factor of the equation, and its unit is $m^2 \cdot s^{-1}$; $E_a$ is the apparent activation energy in $kJ \cdot mol^{-1}$; $T$ is the thermodynamic temperature of drying in K; $R$ is the gas constant, and its value is generally $8.314 \times 10^{-3}$ $kJ \cdot (mol \cdot k)^{-1}$. By taking logarithmic deformation of both sides of Equation (4), the following Expression (5) can be obtained:

$$\ln D_{eff} = \ln D_0 - \frac{E_a}{R}\frac{1}{T} \quad (5)$$

According to the Expression (5), we can know where the apparent activation energy $E_a$ was put forward by Arrhenius when he explained his empirical formula. The logarithm of the effective water diffusion coefficient is in a linear relationship with the reciprocal thermodynamic temperature corresponding to the test drying, so the apparent activation energy can be obtained by fitting the slope of the line according to $\ln D_{eff} - 1/T$. The diffusion factor $D_0$ of the equation can be obtained by fitting the intercept of the line according to $\ln D_{eff} - 1/T$. When the sludge thickness is 2 mm, the slope, intercept and correlation coefficient of the linear fitting of the three drying stages $\ln D_{eff} - 1/T$ are shown in the table below.

According to the slope and intercept in Table 6 above, there is a linear relationship between the logarithm of the effective diffusion coefficient of sludge and the inverse of the thermodynamic temperature of drying. Through the calculation of Expression (5), it is easy to calculate that the diffusion factor of municipal sludge in the rating-up drying stage is $1.789 \times 10^{-5}$ $m^2 \cdot s^{-1}$, and the apparent activation energy is 29.772 $kJ \cdot mol^{-1}$. In the same way, the diffusion factors in the constant drying stage and the reduced drying stage were $3.487 \times 10^{-4}$ $m^2 \cdot s^{-1}$ and $2.876 \times 10^{-4}$ $m^2 \cdot s^{-1}$, respectively, and the apparent activation energies were 39.202 $kJ \cdot mol^{-1}$ and 37.129 $kJ \cdot mol^{-1}$, respectively. When the thickness is constant, the apparent activation energy of sludge from large to small is the falling rate drying process, the constant speed drying process, and the higher speed drying; the reason is that the acc drying stage is mainly performed to remove sludge adsorption on the surface of the water, the constant-speed drying stage is mainly performed to remove sludge clearance of free water. Free water needs a certain amount of heat as a driving force to spread to the surface and evaporation, so the constant-speed drying stage of the activation energy is higher than acc drying stage. In the down-rate drying stage, to remove part of the water bound to the sludge, more energy is needed to destroy the molecular structure, so the activation energy in the down-rate drying stage is higher than that in the constant rate drying stage.

**Table 6.** The slope and intercept of three stages of linear fitting of 2 mm sludge $\ln D_{eff} - 1/T$.

| Drying Stage | Slope | Intercept | $R^2$ |
|---|---|---|---|
| Acc phase | −0.358 | −10.931 | 0.94348 |
| Constant phase | −0.447 | −7.961 | 0.97157 |
| Slow down stage | −0.472 | −5.851 | 0.99516 |

As evident from the results, the diffusion coefficient increases with the temperature but decreases with thickness. The apparent activation energy is related to the extent of moisture diffusion in each drying stage. The apparent activation energy for urban sludge increases sequentially during the up-rate drying stage, constant rate drying stage, and down-rate drying stage, with values of 29.772 $kJ \cdot mol^{-1}$, 37.129 $kJ \cdot mol^{-1}$, and 39.202 $kJ \cdot mol^{-1}$, respectively. This relationship is closely associated with the moisture migration, diffusion, and mass transfer resistances in the sludge, indicating that when treating urban sludge, spreading the sludge to reduce its thickness would be beneficial in shortening the drying time.

## 6. Conclusions

Based on experimental data from drying municipal sludge samples of different thicknesses and temperatures using simulated residual metallurgical heat, improvements were made to traditional drying models. The new model provides better descriptions of the moisture content, effective diffusion coefficient, and apparent water diffusion coefficient during the sludge drying process. The activation energy analysis of urban sludge drying under different conditions and scenarios led to the following conclusions:

1. In the urban sludge treatment process, methods such as timed stirring can be employed when the drying rate is low to enhance the sludge drying rate.
2. The modified Midilli model accurately describes the moisture content and drying equilibrium time within the range of 45–135 °C while also allowing for energy consumption calculations. This is of practical significance for sludge treatment and disposal.
3. The diffusion coefficient increases with temperature but decreases with thickness. The apparent activation energy is related to water diffusion in each stage, increasing successively in the ascending rate, constant rate, and falling rate stages. This is attributed to moisture migration, diffusion, and mass transfer resistances.

**Author Contributions:** Conceptualization, A.J., L.L., H.L. and Y.L.; Data curation, Z.T.; Formal analysis, H.G.; Funding acquisition, A.J., H.L. and Y.L.; Investigation, Z.T.; Methodology, H.G.; Project administration, A.J. and L.L.; Resources, H.G.; Software, Z.T.; Supervision, Y.L.; Validation, H.G. and Z.T.; Visualization, H.L.; Writing—original draft, H.G.; Writing—review and editing, H.G., Z.T., A.J. and L.L. All authors have read and agreed to the published version of the manuscript.

**Funding:** This research is funded by the Soft Science Research Program (Project No.: 22554005D), project name: Carbon Emission Status and Carbon Neutral Technology Innovation Roadmap of Hebei Iron and Steel Industry. And the Natural Science Foundation of Hebei Province, project number E2022209138.

**Institutional Review Board Statement:** Not applicable.

**Informed Consent Statement:** Not applicable.

**Data Availability Statement:** All the data generated or analyzed in this study are included in this published article.

**Conflicts of Interest:** The authors declare that they have no conflict of interests.

## References

1. Lei, W. Investigation on the Sludge Yield of Municipal Wastewater Treatment Plants in Key Watershed of China. *China Water Wastewater* **2018**, *34*, 23–27. [CrossRef]
2. Barati, M.; Jahanshahi, S. Granulation and Heat Recovery from Metallurgical Slags. *J. Sustain. Metall.* **2020**, *6*, 191–206. [CrossRef]
3. Zhang, W.; Qi, D.; Wu, S.; Dong, J.; Zhao, W. Effect of Different Pretreatment Methods on Sewage Sludge Anaerobic Digestion. *Trans. Chin. Soc. Agric. Mach.* **2014**, *45*, 187–198. [CrossRef]
4. Oladejo, J.; Shi, K.; Luo, X.; Yang, G.; Wu, T. A Review of Sludge-to-Energy Recovery Methods. *Energies* **2019**, *12*, 60. [CrossRef]
5. Werther, J.; Ogada, T. Sewage sludge combustion. *Prog. Energy Combust.* **1999**, *25*, 55–116. [CrossRef]
6. Fourti, O.; Jedidi, N.; Hassen, A. Humic substances change during the co-composting process of municipal solid wastes and sewage sludge. *World J. Microbiol. Biotechnol.* **2010**, *26*, 2117–2122. [CrossRef]
7. Schnell, M.; Horst, T.; Quicker, P. Thermal treatment of sewage sludge in Germany: A review. *J. Environ. Manag.* **2020**, *263*, 110367. [CrossRef]
8. Wang, W.; Luo, Y.; Qiao, W. Possible solutions for sludge dewatering in China. *Front. Environ. Sci. Eng. China* **2010**, *4*, 102–107. [CrossRef]
9. Werle, S.; Wilk, R.K. A review of methods for the thermal utilization of sewage sludge: The Polish perspective. *Renew. Energy* **2010**, *35*, 1914–1919. [CrossRef]
10. Praspaliauskas, M.; Pedišius, N. A review of sludge characteristics in Lithuania's wastewater treatment plants and perspectives of its usage in thermal processes. *Renew. Sustain. Energy Rev.* **2017**, *67*, 899–907. [CrossRef]
11. Rodrigo, D.F.B. Composition, Production, and Treatment of Sewage Sludge. In *Sustainable Sewage Sludge Management and Resource Efficiency*; Başak, K.T., Ed.; IntechOpen: Rijeka, Croatia, 2020; p. 3. ISBN 978-1-83962-707-1.
12. Elbaz, A.A.; Aboulfotoh, A.; ElGohary, E.H.; Reham, M. Review Classification of sludge drying beds SDB (conventional sand drying beds CSDB, Wedge-wire, Solar, and Vacuum assisted and paved drying beds PDB). *J. Mater. Environ. Sci.* **2020**, *11*, 593–608.
13. Tic, W.J.; Guziałowska-Tic, J.; Pawlak-Kruczek, H.; Woźnikowski, E.; Zadorożny, A.; Niedźwiecki, Ł. Novel Concept of an Installation for Sustainable Thermal Utilization of Sewage Sludge. *Energies* **2018**, *11*, 748. [CrossRef]
14. Bennamoun, L.; Arlabosse, P.; Léonard, A. Review on fundamental aspect of application of drying process to wastewater sludge. *Renew. Sustain. Energy Rev.* **2013**, *28*, 29–43. [CrossRef]
15. Dai, X.T.; Qi, Y.H.; Zhang, C.X.; Xu, H.C.; Yan, D.L.; Hong, Y.C. Analysis of Blast Furnace Slag Quenching Dry Granulation Process. *J. Iron Steel Res.* **2007**, *19*, 14–19. [CrossRef]
16. Chen, X.D.; Putranto, A. *Modelling Drying Processes: A Reaction Engineering Approach*; Cambridge University Press: Cambridge, UK, 2013; ISBN 9781107012103.
17. Ambawat, S.; Sharma, A.; Saini, R.K. Mathematical Modeling of Thin Layer Drying Kinetics and Moisture Diffusivity Study of Pretreated Moringa oleifera Leaves Using Fluidized Bed Dryer. *Processes* **2022**, *10*, 2464. [CrossRef]
18. Ramachandran, R.P.; Akbarzadeh, M.; Paliwal, J.; Cenkowski, S. Computational Fluid Dynamics in Drying Process Modelling—A Technical Review. *Food Bioprocess Technol.* **2018**, *11*, 271–292. [CrossRef]
19. Jayasuriya, H.; Pathare, P.B.; Al-Attabi, Z.; Al-Hamdani, A. Drying Kinetics and Quality Analysis of Coriander Leaves Dried in an Indirect, Stand-Alone Solar Dryer. *Processes* **2023**, *11*, 1596. [CrossRef]
20. Bennamoun, L.; Ndukwu, M.C. Modeling and simulation of drying kinetics/curves: Application to building materials. *J. Build. Pathol. Rehabil.* **2021**, *7*, 8. [CrossRef]
21. Zia, S.; Khan, M.R.; Aadil, R.M. Kinetic modeling of different drying techniques and their influence on color, bioactive compounds, antioxidant indices and phenolic profile of watermelon rind. *J. Food Meas. Charact.* **2023**, *17*, 1068–1081. [CrossRef]
22. Doymaz, I. Thin-Layer Drying of Spinach Leaves in a Convective Dryer. *J. Food Process. Eng.* **2009**, *32*, 112–125. [CrossRef]

23. Lewis, W.K. The Rate of Drying of Solid Materials. *J. Ind. Eng. Chem.* **1921**, *13*, 427–432. [CrossRef]
24. Ruixun, J.; Aimin, L.; Wang, W. Thin layer drying characteristics and kinetics model of dewatered sludge. *China Environ. Sci.* **2009**, *29*, 22–25. [CrossRef]
25. Page, G.E. *Factors Influencing the Maximum Rates of Air Drying Shelled Corn in Thin Layers*; ProQuest Dissertations Publishing: Ann Arbor, MI, USA, 1949.
26. Yin, L.B.; Xu, Q.S.; Peng, X.W.; Yu, Z.S.; Liao, Y.F.; Ma, X.Q. Dynamic analysis on thin layer drying of dyeing sludge. *Therm. Power Gener.* **2015**, *4*, 62–66. [CrossRef]
27. Henderson, S.M. Grain drying theory, I. Temperature effect on drying coefficient. *J. Agric. Eng. Res.* **1961**, *6*, 169–173.
28. Sharaf-Eldeen, Y.; Blaisdell, J.; Hamdy, M. A Model for Ear Corn Drying. *Trans. Asae* **1980**, *23*, 1261–1265. [CrossRef]
29. Midilli, A.; Kucuk, H.; Zhao, J. A New Model for Single-Layer Drying. *Dry Technol.* **2002**, *20*, 1503–1513. [CrossRef]
30. Kong, L.; Yang, X.; Dong, J.; Jingyi, Z. Research Progress of Thin-layer Drying Model for Paper Sludge. *China Pulp Pap.* **2019**, *38*, 70–75. [CrossRef]
31. Akgun, N.A.; Doymaz, I. Modelling of olive cake thin-layer drying process. *J. Food Eng.* **2005**, *68*, 455–461. [CrossRef]
32. Koua, K.B.; Fassinou, W.F.; Gbaha, P.; Toure, S. Mathematical modelling of the thin layer solar drying of banana, mango and cassava. *Energy* **2009**, *34*, 1594–1602. [CrossRef]

processes

*Article*

# Thermodynamic Analysis of Hydrogen Production from Bio-Oil Steam Reforming Utilizing Waste Heat of Steel Slag

**Zhijun Ding [1], Yang Liu [2], Xin Yao [1,*], Yuekai Xue [1,*], Chenxiao Li [1,*], Zhihui Li [1], Shuhuan Wang [1] and Jianwei Wu [3]**

[1] College of Metallurgy and Energy, North China University of Science and Technology, Tangshan 063210, China; sgdingzhijun@hbisco.com (Z.D.); lzh@ncst.edu.cn (Z.L.); wshh88@ncst.edu.cn (S.W.)
[2] Health Center, North China University of Science and Technology, Tangshan 063210, China; liuyangineur@outlook.com
[3] School of Metallurgy, Northeastern University, Shenyang 110819, China; wujianweiup@163.com
* Correspondence: yaoxin_0129@163.com (X.Y.); xueyuekai965@163.com (Y.X.); lichenxiao34@163.com (C.L.)

**Abstract:** (1) Background: The discharged temperature of steel slag is up to 1450 °C, representing heat having a high calorific value. (2) Motivation: A novel technology, integrating bio-oil steam reforming with waste heat recovery from steel slag for hydrogen production, is proposed, and it is demonstrated to be an outstanding method via thermodynamic calculation. (3) Methods: The equilibrium productions of bio-oil steam reforming in steel slag under different temperatures and S/C ratios (the mole ratio of steam to carbon) are obtained by the method of minimizing the Gibbs free energy using HSC 6.0. (4) Conclusions: The hydrogen yield increases first and then decreases with the increasing temperature, but it increases with the increasing S/C. Considering equilibrium calculation and actual application, the optimal temperature and S/C are 706 °C and 6, respectively. The hydrogen yield and hydrogen component are 109.13 mol/kg and 70.21%, respectively, and the carbon yield is only 0.08 mol/kg under optimal conditions. Compared with CaO in steel slag, iron oxides have less effect on hydrogen production from bio-oil steam reforming in steel slag. The higher the basicity of steel slag, the higher the obtained hydrogen yield and hydrogen component of bio-oil steam reforming in steel slag. It is demonstrated that appropriately decreasing iron oxides and increasing basicity could promote the hydrogen yield and hydrogen component of bio-oil steam reforming that utilizes steel slag as a heat carrier during the industrial application.

**Keywords:** thermodynamic analysis; heat recovery; steam reforming; bio-oil; steel slag

**Citation:** Ding, Z.; Liu, Y.; Yao, X.; Xue, Y.; Li, C.; Li, Z.; Wang, S.; Wu, J. Thermodynamic Analysis of Hydrogen Production from Bio-Oil Steam Reforming Utilizing Waste Heat of Steel Slag. *Processes* **2023**, *11*, 2342. https://doi.org/10.3390/pr11082342

Academic Editor: Shetian Liu

Received: 30 June 2023
Revised: 21 July 2023
Accepted: 27 July 2023
Published: 3 August 2023



## 1. Introduction

The goal of a dual-carbon strategy was proposed via China in 2020, which is that the $CO_2$ emissions reach a peak in 2030 and are neutral in 2060. Steel enterprises are a national pillar of life and production, but have high energy consumption and high carbon dioxide emissions [1,2]. The development of green metallurgy technology could provide a solid foundation for the dual-carbon strategy. Statistics indicates that China is the largest crude steel producer in the world, with production of 1032.8 million tons [3]. The energy consumption of steel enterprises in China is about 23% of the total consumption [4]. However, 70% of the energy used in steel enterprises is discharged as waste heat and waste energy [5,6]. Reasonable recovery of waste heat and waste energy is an effective method for the green metallurgy process. Steelmaking is the primary process of steel enterprises and consumes a huge amount of energy. Steel slag is a by-product of the steelmaking process, and has a discharge temperature of up to 1450 °C, representing heat having a high calorific value [7–9]. However, steel slag is disposed of via the water quenching method in the actual production process. The high calorific value of steel slag is not recovered reasonably. Thus, the recovery of waste heat from steel slag could become a key technology to recover waste heat and waste energy during the steel production process.

The methods of recovering waste heat from slag include physical methods and chemical methods. The physical methods use the waste heat of steel slag to heat water or cool College of Metallurgy and Energy, air. The chemical methods use chemical reactions to replace the waste heat of steel slag. Li investigated the thermodynamic analysis of the physical and chemical methods used to recover waste heat from slag via the enthalpy–exergy compass [10]. The results implied that the heat efficiency and exergy efficiency of physical methods were 76.9% and 34.2%, respectively, but the heat efficiency and exergy efficiency of chemical methods increased to 92.2% and 60%, respectively. Compared to physical methods, chemical methods can recover waste heat from slag with a high reaction rate and high efficiency [11–13]. Using chemical reactions to recover waste heat from slag was proposed and has been investigated from the perspectives of thermodynamics, dynamics, product characterization, etc. Kasai first proposed using steam reforming of methane recovery of waste heat from blast furnace slag [14]. The results implied that steam reforming of methane could efficiently recover waste heat from blast furnace slag. The feasibilities of coal gasification [15,16], biomass gasification [8], and bio-oil steam reforming [17] to recover waste heat from blast furnace slag were demonstrated via thermodynamic calculations. These chemical reactions could efficiently replace the waste heat of blast furnace slag. The dynamic characterizations of coal gasification [18,19], biomass gasification [13,20], and sludge gasification [21,22] in blast furnace slag were investigated via dynamic calculations. The results implied that blast furnace slag could decrease the reaction activation energy, and was thus regarded as an outstanding heat carrier. The product characterizations of biogas [23], municipal solid waste gasification [24], coal gasification [25,26], and sludge pyrolysis [27,28] using blast furnace slag as a heat carrier were investigated. The results implied that blast furnace slag could weaken the C-C bond, which catalyzed the process of using chemical reactions to recover waste heat from blast furnace slag. The waste heat of blast furnace slag could be efficiently replaced via chemical reactions and these chemical reactions could provide energy for the steel production process. This was a foundation for green metallurgy development.

Meanwhile, with the depletion of fossil fuels, renewable energy has gradually been explored by researchers. Hydrogen is regarded as a green energy resource for metallurgy industrial applications [29,30], and it can be obtained via petroleum cracking, coal pyrolysis, natural gas reforming, oil reforming, biomass pyrolysis, biomass gasification, etc. [31]. Biomass pyrolysis and gasification applications are outstanding methods used to obtain hydrogen. However, during the process of biomass pyrolysis and gasification, bio-oil can be obtained, decreasing the utilization efficiency of biomass. Utilization of bio-oil reforming has been proven to be a feasible and efficient method to increase the utilization efficiency of biomass and obtain hydrogen [32–34]. The reaction rate and heat source of bio-oil steam reforming are the main factors that limit the technological progress. In order to obtain hydrogen efficiently, studies have investigated the characterizations of the bio-oil steam reforming process over catalysts and obtained the catalyst property of the reforming process [32,33]. In order to identify a proper heat source for bio-oil reforming, our team proposed using reforming of bio-oil recovery of waste heat from blast furnace slag [17,35,36]. The results implied that CaO in blast furnace slag was the basic component for catalyzing the steam reforming process.

In a review of recovering waste heat from slag, researchers have primarily focused on recovering waste heat from blast furnace slag. Results showed that the higher the ratio of CaO to $SiO_2$, the higher the rate of the obtained methane–steam reaction [37]. The ratio of CaO to $SiO_2$ in steel slag is higher than that in blast furnace slag [8]. However, the characterizations of steam reforming of bio-oil in steel slag are ambiguous, and are first proposed in this paper. Temperature and steam are the main factors affecting the process of steam reforming of bio-oil. In addition, the components of steel slag vary enormously during the production processes. The effects of temperature, S/C, and type of steel slag on the equilibrium productions of steam reforming of bio-oil are explored via thermodynamic calculations in this paper, and the optimal condition is obtained. This can provide a theoretical foundation for the recovery of waste heat from steel slag during the process of metallurgy industrial application in the future.

## 2. Methodology

### 2.1. Materials

To obtain the thermodynamic characterizations of bio-oil steam reforming using steel slag as a heat carrier, the chemical compositions of bio-oil and steel slag should be determined. The bio-oil is complicated, but it mainly contains alcohols, acids, ketones, and phenols. In addition, the equal mass ratio of acetic acid, ethanol, acetone, and phenol can be used as the bio-oil model compound during the process of bio-oil reforming, as demonstrated in our previous studies [17,35]. The equal mass ratio of acetic acid, ethanol, acetone, and phenol is also used to replace bio-oil in the paper. In addition, the mass ratio of steel slag to bio-oil is 1:1 and the bio-oil is also defined as 1 kg during the process of thermodynamic calculations The contents of FeO and $Fe_2O_3$ in steel slag and the basicity of steel slag may affect the equilibrium production of bio-oil steam reforming using steel slag as a heat carrier. Steel slag with different quantities of FeO and $Fe_2O_3$ and levels of basicity is used in the thermodynamic calculation and their characterizations are shown in Table 1.

**Table 1.** The compositions of steel slag.

| Type of Steel Slag | CaO | $SiO_2$ | $Al_2O_3$ | MgO | $Fe_2O_3$ | FeO | $Fe_2O_3$ + FeO | R |
|---|---|---|---|---|---|---|---|---|
| 1 | 54.00 | 18.00 | 8.00 | 5.00 | 4.75 | 10.25 | 15.00 | 3 |
| 2 | 50.25 | 16.75 | 8.00 | 5.00 | 6.33 | 13.67 | 20.00 | 3 |
| 3 | 46.50 | 15.50 | 8.00 | 5.00 | 7.92 | 17.08 | 25.00 | 3 |
| 4 | 42.75 | 14.25 | 8.00 | 5.00 | 9.50 | 20.50 | 30.00 | 3 |
| 5 | 39.00 | 13.00 | 8.00 | 5.00 | 11.08 | 23.92 | 35.00 | 3 |
| 6 | 41.33 | 20.67 | 8.00 | 5.00 | 7.92 | 17.08 | 25.00 | 2 |
| 7 | 44.29 | 17.71 | 8.00 | 5.00 | 7.92 | 17.08 | 25.00 | 2.5 |
| 8 | 46.50 | 15.50 | 8.00 | 5.00 | 7.92 | 17.08 | 25.00 | 3 |
| 9 | 48.22 | 13.78 | 8.00 | 5.00 | 7.92 | 17.08 | 25.00 | 3.5 |
| 10 | 49.60 | 12.40 | 8.00 | 5.00 | 7.92 | 17.08 | 25.00 | 4 |

R is basicity, which is the mass ratio of CaO to $SiO_2$.

### 2.2. Thermodynamic Method

The chemical reaction of steam reforming of the bio-oil recovery of waste heat from steel slag contains complex mechanisms. When using the minimization of Gibbs free energy method, mastering the possible reactant production can only be used to obtain equilibrium productions. The minimization of Gibbs free energy has been proven to efficiently yield the equilibrium production and was applied in our previous studies [17,38]. The HSC 6.0 software with the function of the minimization of Gibbs free energy is also used in this paper. The thermodynamic characterizations of bio-oil steam reforming in blast furnace slag were investigated in our previous study [17]. The oxide species in steel slag are consistent with those of blast furnace slag. Thus, the possible reactant productions of bio-oil steam reforming in steel slag are consistent with those in blast furnace slag. The principle of minimization of Gibbs free energy in HSC 6.0 software, and the possible reactant production of bio-oil steam reforming in steel slag, are described in our previous study [17]. The primary reactions in the process of the steam reforming of the bio-oil recovery of waste heat from steel slag are shown as follows:

Thermal cracking reaction (TCR):

$$C_nH_mO_k \rightarrow C_xH_yO_z + oH_2 + pCO + qCO_2 + rCH_4 + sC, \Delta H^{\theta}_{800^{\circ}C} > 0 \quad (1)$$

Steam reforming reactions (SRRs):

$$C_2H_6O + 3H_2O \rightarrow 2CO_2 + 6H_2, \Delta H^{\theta}_{800^{\circ}C} = 210.26 \text{ kJ mol}^{-1} \quad (2)$$

$$C_2H_4O_2 + 2H_2O \rightarrow 2CO_2 + 4H_2, \Delta H^{\theta}_{800^\circ C} = 153.33 \text{ kJ mol}^{-1} \quad (3)$$

$$C_3H_6O + 5H_2O \rightarrow 3CO_2 + 8H_2, \Delta H^{\theta}_{800^\circ C} = 295.90 \text{ kJ mol}^{-1} \quad (4)$$

$$C_6H_6O + 11H_2O \rightarrow 6CO_2 + 14H_2, \Delta H^{\theta}_{800^\circ C} = 473.68 \text{ kJ mol}^{-1} \quad (5)$$

$$C_xH_yO_z + (2x - z)H_2O \rightarrow xCO_2 + (2x + y/2 - z)H_2,\ \Delta H^{\theta}_{800^\circ C} > 0 \quad (6)$$

$$CH_4 + 2H_2O \rightarrow CO_2 + 4H_2, \Delta H^{\theta}_{800^\circ C} = 191.10 \text{ kJ mol}^{-1} \quad (7)$$

Methanation reactions (MRs):

$$CO + 3H_2 \rightarrow CH_4 + H_2O, \Delta H^{\theta}_{800^\circ C} = -225.22 \text{ kJ mol}^{-1} \quad (8)$$

$$C + 2H_2 \rightarrow CH_4, \Delta H^{\theta}_{800^\circ C} = -89.45 \text{ kJ mol}^{-1} \quad (9)$$

Water gas reaction (WGR):

$$C + H_2O \rightarrow CO + H_2, \Delta H^{\theta}_{800^\circ C} = 135.77 \text{ kJ mol}^{-1} \quad (10)$$

Water gas shift reaction (WGSR):

$$CO + H_2O \rightarrow H_2 + CO_2, \Delta H^{\theta}_{800^\circ C} = -34.12 \text{ kJ mol}^{-1} \quad (11)$$

Bell reaction (BR):

$$2CO \rightarrow C + CO_2, \Delta H^{\theta}_{800^\circ C} = -169.88 \text{ kJ mol}^{-1} \quad (12)$$

Reactions of mineral oxides with the generated gases (RMGs):

$$CaO + CO_2 \rightarrow CaCO_3, \Delta H^{\theta}_{800^\circ C} = -167.62 \text{ kJ mol}^{-1} \quad (13)$$

$$CaO + H_2O \rightarrow Ca(OH)_2, \Delta H^{\theta}_{800^\circ C} = -61.88 \text{ kJ mol}^{-1} \quad (14)$$

$$Fe_2O_3 + H_2 \rightarrow 2FeO + H_2O, \Delta H^{\theta}_{800^\circ C} = 28.89 \text{ kJ mol}^{-1} \quad (15)$$

$$Fe_2O_3 + CO \rightarrow 2FeO + CO_2, \Delta H^{\theta}_{800^\circ C} = -5.23 \text{ kJ mol}^{-1} \quad (16)$$

$$4Fe_2O_3 + CH_4 \rightarrow 8FeO + CO_2 + 2H_2O, \Delta H^{\theta}_{800^\circ C} = 306.64 \text{ kJ mol}^{-1} \quad (17)$$

$$1.132Fe_2O_3 + H_2 \rightarrow 2.395Fe_{0.945}O + H_2O, \Delta H^{\theta}_{800^\circ C} = 40.95 \text{ kJ mol}^{-1} \quad (18)$$

$$1.132Fe_2O_3 + CO \rightarrow 2.395Fe_{0.945}O + CO_2, \Delta H^{\theta}_{800^\circ C} = 6.84 \text{ kJ mol}^{-1} \quad (19)$$

$$4.527Fe_2O_3 + CH_4 \rightarrow 9.581Fe_{0.945}O + CO_2 + 2H_2O, \Delta H^{\theta}_{800^\circ C} = 353.98 \text{ kJ mol}^{-1} \quad (20)$$

$$Fe_2O_3 + 3H_2 \rightarrow 2Fe + 3H_2O, \Delta H^{\theta}_{800^\circ C} = 63.356 \text{ kJ mol}^{-1} \quad (21)$$

$$Fe_2O_3 + 3CO \rightarrow 2Fe + 3CO_2, \Delta H^{\theta}_{800^\circ C} = -38.99 \text{ kJ mol}^{-1} \quad (22)$$

$$1.333Fe_2O_3 + CH_4 \rightarrow 2.667Fe + CO_2 + 2H_2O, \Delta H^{\theta}_{800^\circ C} = 275.35 \text{ kJ mol}^{-1} \quad (23)$$

$$FeO + H_2 \rightarrow Fe + H_2O, \Delta H^{\theta}_{800^\circ C} = 17.24 \text{ kJ mol}^{-1} \quad (24)$$

$$FeO + CO \rightarrow Fe + CO_2, \Delta H^{\theta}_{800^\circ C} = -16.88 \text{ kJ mol}^{-1} \quad (25)$$

$$4FeO + CH_4 \rightarrow 4Fe + CO_2 + 2H_2O, \Delta H^{\theta}_{800^\circ C} = 260.04 \text{ kJ mol}^{-1} \quad (26)$$

### 2.3. Performance Evaluation

Gas yield and dry gas concentration are used to evaluate the performance of the process of steam reforming of the bio-oil recovery of waste heat from steel slag. The gas yield encompasses the $H_2$ yield, CO yield, and $CH_4$ yield. The dry gas concentration includes the $H_2$ concentration, CO concentration, and $CH_4$ concentration. The explanations of gas yield and dry gas concentration are shown as follows:

$$\text{Gas yield} = \frac{\text{The mole of production of bio} - \text{oil steam reforming}}{\text{The quality of bio} - \text{oil}} (\text{mol/kg}) \quad (27)$$

$$\text{Dry gas concentration} = \frac{\text{The mole of production in dry reforming gas}}{\text{The totalmole of dry reforming gas}} (\%) \quad (28)$$

The effects of conditions (temperature, S/C (the mole ratio of steam to the carbon in bio-oil), basicity, and iron oxide) on the process of steam reforming of the bio-oil recovery of waste heat from steel slag were obtained via PY and DGC. Subsequently, the mechanism of steam reforming of the bio-oil recovery of waste heat from steel slag was obtained.

## 3. Results and Discussion

### 3.1. Effect of Temperature

The process of steam reforming of the bio-oil recovery of waste heat from steel slag contains endothermic reactions and exothermic reactions. The temperature can affect the coupling endothermic reactions and exothermic reactions, further affecting the equilibrium product distribution [39,40]. The effects of temperature on the PY and DGC of steam reforming of the bio-oil recovery of waste heat from steel slag are illustrated in this section.

Taking the characterizations of steel slag-3 as an example, the hydrogen yields under different temperatures and S/C ratios are shown in Figure 1. Figure 1 shows that hydrogen production from bio-oil steam reforming in steel slag is feasible, and the hydrogen yield first increases then decreases with increasing temperature. Taking S/C of 6 as an example, the hydrogen yield increases from 0.90 mol/kg to 90.42 mol/kg as the temperature increases from 200 °C to 739 °C. Then, hydrogen yield decreases from 90.42 mol/kg to 84.94 mol/kg as the temperature increases from 739 °C to 1000 °C. At low temperatures, the endothermic reactions of TCR (Equation (1)), SRR (Equations (2)–(7)), and WGR (Equation (10)), and the exothermal reactions of MR (Equations (8) and (9)), are the main factors that affect the hydrogen yield. With the increase in temperature, TCR (Equation (1)), SRR (Equations (2)–(7)), and WGR (Equation (10)) shift to the right, and MR (Equation (8)) shifts to the left, and the hydrogen yield increases. However, at high temperatures, the exothermal reaction of WGSR (Equation (11)) is the main factor that affects the hydrogen yield. With the increase in temperature, WGSR (Equation (11)) shifts to the left, decreasing the hydrogen yield. Xie [40] and Yao [17] also investigated the characterizations of hydrogen production from the process of bio-oil steam reforming in blast furnace slag, and a similar variation in hydrogen yield was obtained. They provide further evidence that the results of this paper are accurate.

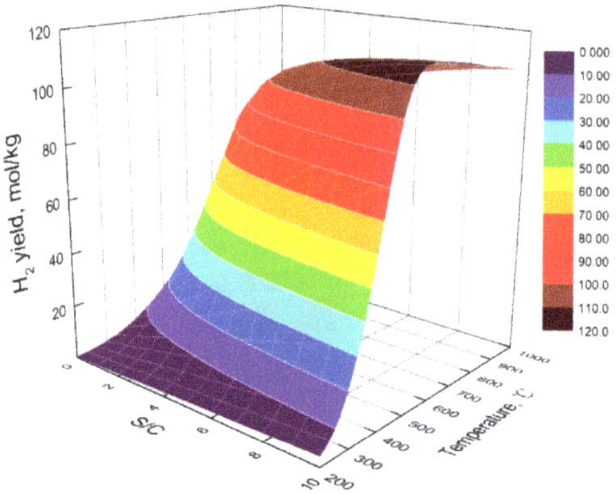


Figure 1. Hydrogen yields under different temperatures and S/C ratios.

Taking the characterizations of steel slag-3 as an example, the carbon monoxide yields under different temperatures and S/C ratios are illustrated in Figure 2. In Figure 2, the carbon monoxide yields increase with the increase in temperature. Taking S/C of 5 as an example, the carbon monoxide yield increases from 0.00 mol/kg to 22.49 mol/kg when the temperature increases from 200 °C to 1000 °C. The endothermic reactions of TCR (Equation (1)) and WGR (Equation (10)), and the exothermal reactions of MR (Equation (8)), WGSR (Equation (11)), BR (Equation (12)), and RMG (Equations (16), (22) and (25)), are the main factors that affect the carbon monoxide yield. With the increasing temperature, the endothermic reactions of TCR (Equation (1)) and WGR (Equation (10)) shift to the right, and the exothermal reactions of MR (Equation (8)), WGSR (Equation (11)), BR (Equation (12)), and RMG (Equations (16), (22) and (25)) shift to the right, decreasing the carbon monoxide yield. The results of increasing the carbon monoxide yield as the temperature increases are also demonstrated via fixed-bed experiments of bio-oil reforming in blast furnace slag [35].

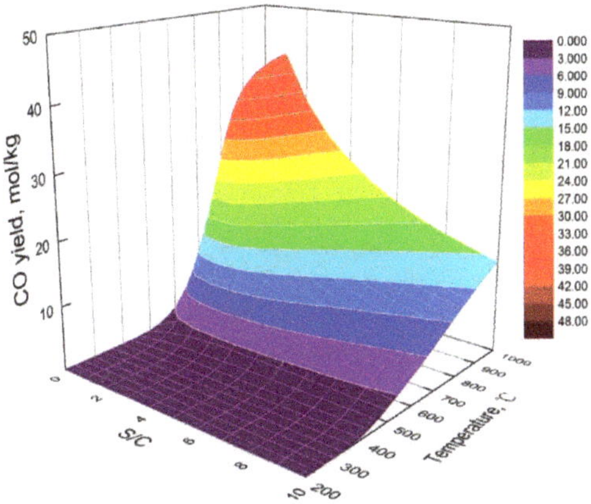


**Figure 2.** Carbon monoxide yields under different temperatures and S/C ratios.

Taking the characterizations of steel slag-3 as an example, the methane yields under different temperatures and S/C ratios are illustrated in Figure 3. The methane yield decreases with the increase in temperature, as shown in Figure 3. Taking S/C of 6 as an example, methane yield decreases from 24.40 mol/kg to 0.00 mol/kg as the temperature increases from 200 °C to 1000 °C. The endothermic reactions of SRR (Equation (7)) and RMG (Equations (17), (20), (23) and (26)), and the exothermal reactions of MR (Equations (8) and (9)) are the main factors that affect the methane yield. With the increasing temperature, the endothermic reactions of SRR (Equation (7)), RMG (Equations (17), (20), (23) and (26)) shift to the right, and the exothermal reactions of MR (Equations (8) and (9)) shift to the left, decreasing the methane yield. The characterizations of hydrogen production from the steam reforming of bio-oil over a Ce-Ni/Co catalyst were investigated [41]. The results showed that the higher temperature, the lower the methane yield. These results are consistent with this study.

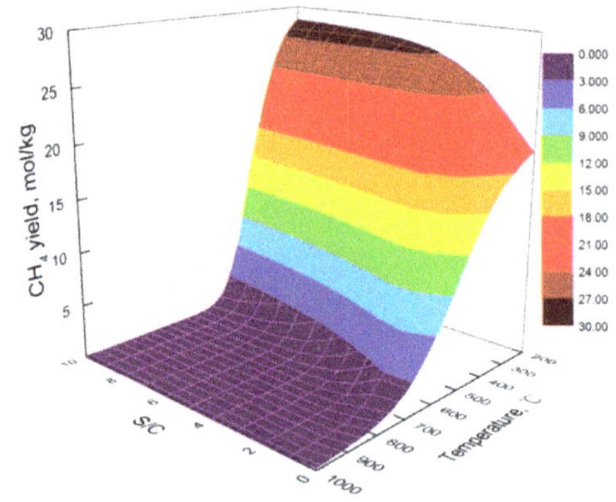


**Figure 3.** Methane yields under different temperatures and S/C ratios.

The process of bio-oil steam reforming using steel slag as a heat carrier generates carbon, which might decrease the reactive activation and have no benefit for the steam reaction. The carbon yield is regarded as a significant factor in the research on the steam reforming process. Taking the characterizations of steel slag-3 as an example, the carbon yields under different temperatures and S/C ratios are shown in Figure 4. As shown in Figure 4, the carbon yield decreases with the increasing temperature. Taking S/C of 5 as an example, the carbon yield decreases from 11.58 mol/kg to 0.03 mol/kg with the increase in temperature from 200 °C to 1000 °C. The endothermic reaction of WGR (Equation (10)) and the exothermal reaction of BR (Equation (12)) are the main factors that affect the carbon yield. With the increase in temperature, the endothermic reaction of WGR (Equation (10)) shifts to the right and the exothermal reaction of BR (Equation (12)) shifts to the left, decreasing the carbon yield.

Taking the characterizations of steel slag-3 as an example, the hydrogen component, carbon monoxide component, and methane component under different temperatures and S/C ratios are illustrated in Figure 5, Figure 6, and Figure 7, respectively. In Figure 5, the hydrogen component increases at the beginning and then decreases as the temperature increases. Taking S/C of 5 as an example, the hydrogen component increases from 4.95% to 70.93% with the increase in temperature from 200 °C to 608 °C, but the hydrogen component decreases from 70.93% to 67.04% with the increase in temperature from 608 °C to 1000 °C. In Figure 6, the carbon monoxide component increases with the increase in temperature. Taking S/C of 5 as an example, the carbon monoxide component increases from 0.00% to 15.42% with the increase in temperature from 200 °C to 1000 °C. In Figure 7, the methane component decreases with the increase in temperature. Taking S/C of 5 as an example, the methane component decreases from 72.24% to 0.00% with the increase in temperature from 200 °C to 1000 °C. The variations in the hydrogen component, carbon monoxide component, and methane component are attributed to their yields. Thus, the variation trends of the hydrogen component, carbon monoxide component, and methane component are similar to those of their yields.

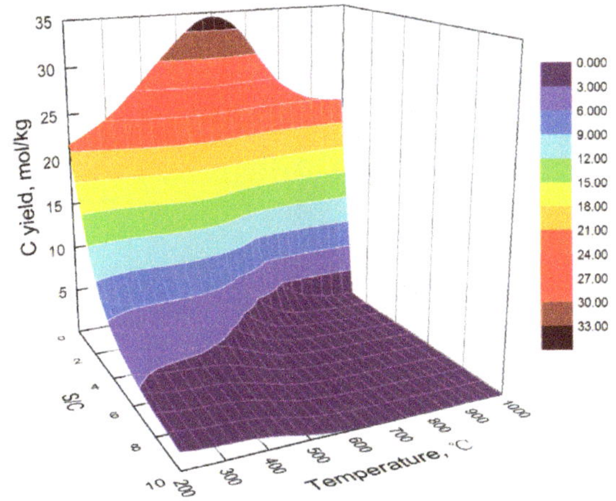


**Figure 4.** Carbon yields under different temperatures and S/C ratios.

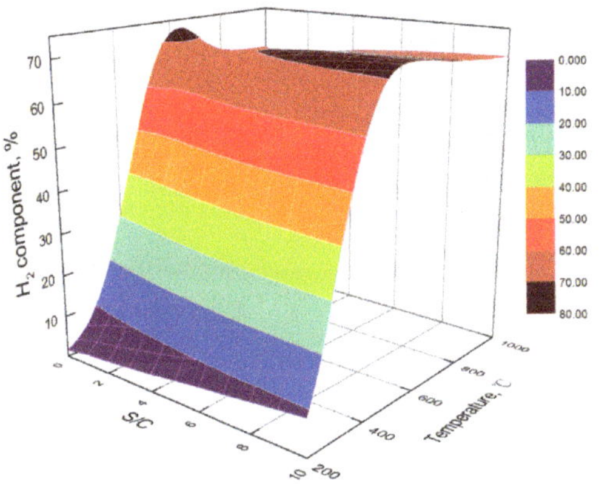


**Figure 5.** Hydrogen components under different temperatures and S/C ratios.

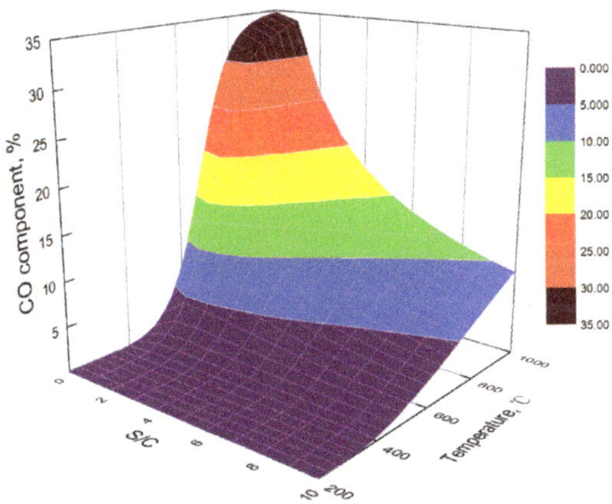


**Figure 6.** Carbon monoxide components under different temperatures and S/C ratios.

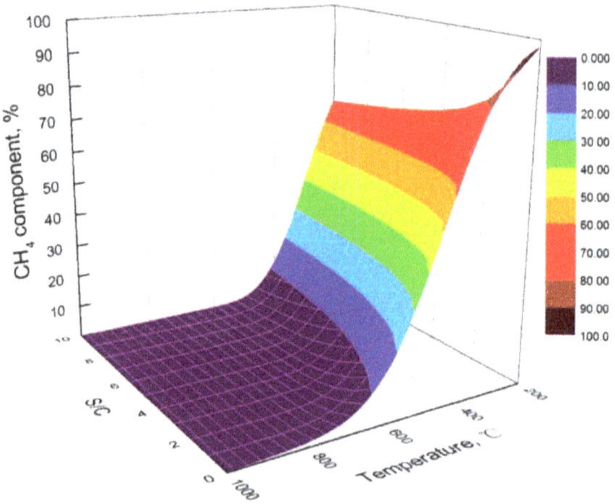


**Figure 7.** Methane components under different temperatures and S/C ratios.

### 3.2. *Effect of S/C*

Steam functions as the reactant during the process of steam reforming of the bio-oil recovery of waste heat from steel slag. The S/C impacts the partial pressure of steam, further affecting the equilibrium product distribution. The effects of S/C on the PY and DGC of the steam reforming of the bio-oil of recovery waste heat from steel slag are illustrated in this section.

The effects of S/C on the hydrogen yield are also shown in Figure 1. Figure 1 shows that the hydrogen yield increases with the increase in S/C. Taking the temperature of 706 °C as an example, the hydrogen yield increases from 36.08 mol/kg to 113.67 mol/kg as the S/C increases from 0 to 10. With the increase in S/C, SRR (Equations (2)–(7)), WGR (Equation (10)), and WGSR (Equation (11)) shift to the right, and MR (Equation (8)) and RMG (Equations (15), (18), (21) and (24)) shift to the left, increasing the hydrogen yield. The characterizations of bio-oil steam reforming for hydrogen production have been investigated via fixed-bed experiments [38]. The results implied that the hydrogen yield increased with the increase in S/C. These results are consistent with the findings in this study.

The effects of S/C on the carbon monoxide yield are shown in the Figure 2. Figure 2 shows that, at low temperatures (lower than 412 °C), the carbon monoxide yield increases with the increase in S/C. At low temperatures, MR (Equation (8)) and WGR (Equation (10)) are the main factors that affect the carbon monoxide yield and they shift to the left, increasing the carbon monoxide yield. However, at high temperatures (higher than 412 °C), the carbon monoxide yield increases first and then decreases with the increase in S/C. At high temperatures and low S/C, MR (Equation (8)) and WGR (Equation (10)) are the main factors that affect the carbon monoxide yield and they shift to the left, increasing the carbon monoxide yield. However, at high temperatures and high S/C, WGSR (Equation (11)) is the main factor affecting the carbon monoxide yield and it shifts to the right, decreasing the carbon monoxide yield.

The effects of S/C on the methane yield are shown in Figure 3. As shown in Figure 3, at low temperatures (lower than 216 °C), the methane yield increases with the increase in S/C. At low temperatures, RMG (Equations (17), (20), (23) and (26)) is the main factor that affects the methane yield and shifts to the left, increasing the methane yield. At high temperatures (higher than 216 °C), the methane yield increases first and then decreases with the increase in S/C. At high temperatures and low S/C, RMG (Equations (17), (20), (23) and (26)) is the main factor that affects the methane yield and shifts to the left, increasing the methane yield. However, at high temperatures and high S/C, SRR (Equation (7)) and MR (Equation (8)) are the main factors that affect the methane yield. With the increase in S/C, SRR (Equation (7)) shifts to the right and MR (Equation (8)) shifts to the left, decreasing the methane yield.

The effects of S/C on the carbon yield are also shown in Figure 4. As shown in Figure 4, the carbon yield decreases with the increase in S/C. Taking the temperature of 690 °C as an example, the hydrogen yield increases from 32.41 mol/kg to 0.02 mol/kg with the increase in S/C from 0 to 10. WGR (Equation (7)) is the main factor that affects the carbon yield and shifts to the right with the increase in S/C, decreasing the carbon yield.

Combining the effects of temperature and S/C, when S/C is 6, the maximal hydrogen yield is 109.13 mol/kg at 706 °C. In addition, the hydrogen component and carbon yield are 70.21% and 0.08 mol/kg at 706 °C, respectively. When S/C increases to 10, the maximal hydrogen yield and hydrogen component are 115.68 mol/kg and 72.34%, respectively. The purpose of the novel technology is to utilize hydrogen to recover the waste heat of steel slag. In addition, carbon may go against the process of steam reforming of bio-oil [38,42]. The hydrogen yield and carbon yield via thermodynamic calculations are regarded as the primary indexes for evaluating the process. When S/C increases from 6 to 10, the hydrogen yield increases, although not obviously. However, the energy consumption of the process increases on account of the increasing steam [5]. The optimal temperature and S/C are 706 °C and 6, respectively, where the hydrogen yield and hydrogen component are 109.13 mol/kg and 70.21%, respectively.

### 3.3. *Effect of Type of Steel Slag*

The compositions of steel slag mainly depend on the different kinds of steelmaking processes and steel types. The iron oxides and basicity of steel slag can affect the equilibrium production of the process of the bio-oil steam reforming recovery of waste heat from steel slag. Using bio-oil steam reforming recovery of waste heat from steel slag aims to obtain hydrogen efficiently. The effects of the type of steel slag on the hydrogen yield, carbon yield, and hydrogen component are illustrated in this section. Taking S/C of 2 as an example, the hydrogen yields and hydrogen components under different types of steel slag are shown in Figures 8 and 9, respectively. As shown in Figures 8 and 9, the hydrogen yield and hydrogen component decrease with the increase in iron oxides in steel slag. Taking the temperature of 592 °C as an example, the hydrogen yield and hydrogen component decrease from 72.36 mol/kg and 65.96% to 71.79 mol/kg and 65.32%, respectively, with the iron oxides increasing from 15% to 35%. However, the hydrogen yield and hydrogen component increase with the increasing basicity of steel slag. Taking the temperature of 706 °C as an example, the hydrogen yield and hydrogen component increase from 89.85 mol/kg and 66.12% to 90.36 mol/kg and 67.53%, respectively, with the basicity increasing from 2 to 4. The basicity of steel slag is calculated via the mole ratio of CaO to $SiO_2$. With the increase in the basicity of steel slag, CaO in the steel slag increases accordingly. CaO absorbs carbon dioxide via RMG (Equation (11)), decreasing the carbon dioxide yield. The decreasing $CO_2$ yield shifts TCR (Equation (1)), SRR (Equations (2)–(7)), and WGSR (Equation (11)) to the right, increasing the hydrogen yield and hydrogen component. However, with the increase in iron oxides, CaO in the steel slag decreases accordingly, which has adverse effects on hydrogen production. In addition, with the increase in iron oxides, RMG (Equations (15), (18), (21) and (24)) is the main factor that affects the methane yield and shifts to the right, decreasing the hydrogen yield and hydrogen component. Compared with CaO in steel slag, iron oxides have less effect on hydrogen production from bio-oil steam reforming in steel slag. The appropriate decrease in iron oxides and increase in basicity can be beneficial for the process of bio-oil steam reforming recovery of waste heat from steel slag in the actual application.

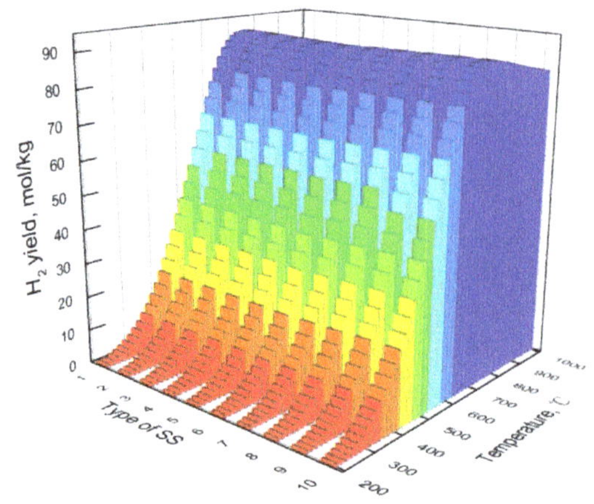


**Figure 8.** Hydrogen yields under different types of steel slag.

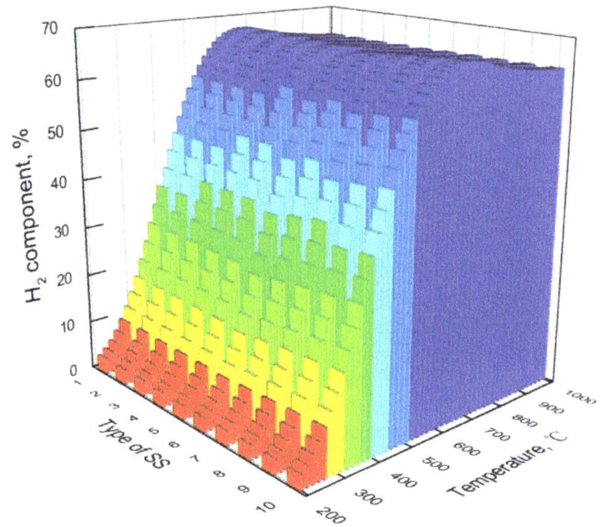


**Figure 9.** Hydrogen components under different types of steel slag.

*3.4. The Mechanism of Recovering Waste Heat from Steel Slag*

Based on the equilibrium productions and our previous studies of the chemical reactions in the recovery of waste heat from slag [2,35,36], the mechanism of bio-oil steam reforming in steel slag was obtained, as shown in Figure 10. In the practical process, steel slag is discharged via the process of steelmaking and it is shown in the slag storage pot. Bio-oil and steam in a certain proportion are fed into the slag storage pot via pipelines. The reactions of bio-oil steam reforming in steel slag take place. The syngas (containing $H_2$, CO, $CO_2$, and $CH_4$) is obtained and discharged from the slag storage pot. The transfer of heat due to the chemical reaction of bio-oil steam reforming and steel slag is accomplished. As shown in Figure 10, the ions of $Ca^{2+}$, $Fe^{2+}$, and $Fe^{3+}$ in steel slag weaken the bonds of C–C and H–O [2,35,43,44], catalyzing the process of bio-oil steam reforming recovery of waste heat from steel slag. Utilization of bio-oil steam reforming efficiently recovers waste heat from steel slag.

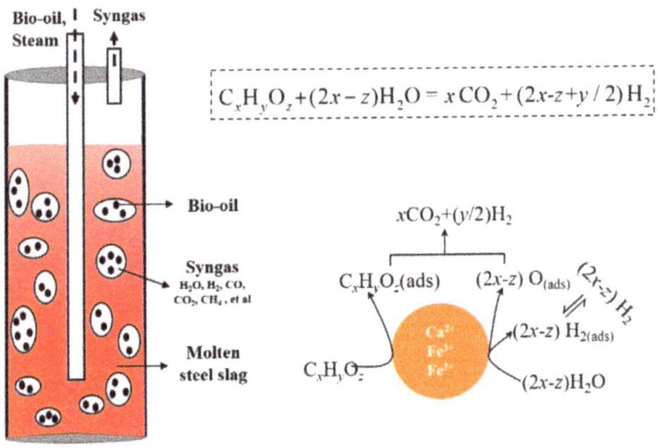


**Figure 10.** The mechanism of bio-oil steam reforming recovery of waste heat from steel slag.

## 4. Conclusions

The novel technology of the bio-oil steam reforming recovery of waste heat from steel slag was investigated via thermodynamic calculations. The primary remarks are summarized as follows:

1. The novel method of utilizing bio-oil steam reforming to recover waste heat from steel slag is proposed in this study. It is proven that bio-oil steam reforming can be used to obtain hydrogen and utilize the waste heat of steel slag, and can provide an opportunity for green metallurgy technology.
2. At low temperatures, the hydrogen yield of bio-oil steam reforming in steel slag increases with the increase in temperature. However, at high temperatures, the water gas shift reaction is the main reaction affecting the hydrogen yield, resulting in the decrease in the hydrogen yield with the increase in temperature. The carbon yield decreases with the increase in temperature. The hydrogen component increases first and then decreases with the increase in temperature.
3. The higher the S/C, the higher the obtained hydrogen yield and the lower the obtained carbon yield. By combining the hydrogen yield and energy consumption, the optimal S/C is 6 and the optimal temperature is 706 °C. At the optimal condition, the hydrogen yield, hydrogen component, and carbon yield are 109.13 mol/kg, 70.21%, and 0.08 mol/kg, respectively.
4. Compared with CaO in steel slag, iron oxides have less effect on hydrogen production from bio-oil steam reforming in steel slag. The basicity of steel slag promotes hydrogen production. During industrial application, the hydrogen yield and hydrogen component of bio-oil steam reforming in steel slag can be promoted via an appropriate decrease in iron oxides and increase in basicity.

**Author Contributions:** Methodology, Z.L., S.W. and J.W.; Writing—original draft, Z.D. and Y.L.; Writing—review & editing, Z.D., X.Y., Y.X. and C.L. All authors have read and agreed to the published version of the manuscript.

**Funding:** This research was funded by the National Natural Science Foundation of China (52274334), the Natural Science Foundation of Hebei (E2021209106), the Youth Scholars Promotion Plan of North China University of Science and Technology (QNTJ202205), the Talent Foundation of Tangshan (A202110039), the Science and Technology Planning Project of Tangshan (22130233H), and the Funds of Basic Scientific Research of Hebei (JQN2020006). And the APC was funded by the Natural Science Foundation of Hebei (E2021209106).

**Data Availability Statement:** No new data were created or analyzed in this study. Data sharing is not applicable to this article.

**Conflicts of Interest:** The authors declare no conflict of interest.

## References


1. Yao, X.; Yu, Q.; Wang, K.; Xie, H.; Qin, Q. Kinetic study on recovery heat of granulated blast-furnace slag through biomass gasification using $CO_2$ as gasification agent. *J. Therm. Anal. Calorim.* **2018**, *131*, 1313–1321. [CrossRef]
2. Yao, X.; Yu, Q.; Han, Z.; Xie, H.; Duan, W.; Qin, Q. Kinetic and experimental characterizations of biomass pyrolysis in granulated blast furnace slag. *Int. J. Hydrogen Energy* **2018**, *43*, 9246–9253. [CrossRef]
3. Available online: https://worldsteel.org/steel-topics/statistics/world-steel-in-figures-2022/ (accessed on 2 April 2023).
4. Lin, B.; Wu, R. Designing energy policy based on dynamic change in energy and carbon dioxide emission performance of China's iron and steel industry. *J. Clean. Prod.* **2020**, *256*, 120412. [CrossRef]
5. Yao, X.; Liu, Y.; Yu, Q.; Wang, S. Energy consumption of two-stage system of biomass pyrolysis and bio-oil reforming to recover waste heat from granulated BF slag. *Energy* **2023**, *273*, 127204. [CrossRef]
6. Wang, R.Q.; Jiang, L.; Wang, Y.D.; Roskilly, A.P. Energy saving technologies and mass-thermal network optimization for decarbonized iron and steel industry: A review. *J. Clean. Prod.* **2020**, *274*, 122997. [CrossRef]
7. Sun, Y.; Seetharaman, S.; Zhang, Z. Integrating biomass pyrolysis with waste heat recovery from hot slags via extending the C-loops: Product yields and roles of slags. *Energy* **2018**, *149*, 792–803. [CrossRef]
8. Sun, Y.; Zhang, Z.; Liu, L.; Wang, X. Integration of biomass/steam gasification with heat recovery from hot slags: Thermodynamic characteristics. *Int. J. Hydrogen Energy* **2016**, *41*, 5916–5926. [CrossRef]
9. Sun, Y.; Zhang, Z. Disposal of High-Temperature Slags: A Review of Integration of Heat Recovery and Material Recycling. *Metall. Mater. Trans. E* **2016**, *3*, 114–122. [CrossRef]
10. Li, P. Thermodynamic analysis of waste heat recovery of molten blast furnace slag. *Int. J. Hydrogen Energy* **2017**, *42*, 9688–9695. [CrossRef]
11. Sun, Y.; Zhang, Z.; Liu, L.; Wang, X. Heat Recovery from High Temperature Slags: A Review of Chemical Methods. *Energies* **2015**, *8*, 1917–1935. [CrossRef]
12. Yao, X.; Yu, Q.; Xie, H.; Duan, W.; Han, Z.; Liu, S.; Qin, Q. Syngas production through biomass/$CO_2$ gasification using granulated blast furnace slag as heat carrier. *J. Renew. Sustain. Energy* **2017**, *9*, 053101. [CrossRef]
13. Yao, X.; Yu, Q.; Wang, K.; Xie, H.; Qin, Q. Kinetic characterizations of biomass char $CO_2$-gasification reaction within granulated blast furnace slag. *Int. J. Hydrogen Energy* **2017**, *42*, 20520–20528. [CrossRef]
14. Kasai, E.; Kitajima, T.; Akiyama, T.; Yagi, J.; Saito, F. Rate of methane-steam reforming reaction on the surface of molten BF slag—For heat recovery from molten slag by using a chemical reaction. *ISIJ Int.* **1997**, *37*, 1031–1036. [CrossRef]
15. Duan, W.; Yu, Q.; Xie, H.; Liu, J.; Wang, K.; Qin, Q.; Han, Z. Thermodynamic analysis of synergistic coal gasification using blast furnace slag as heat carrier. *Int. J. Hydrogen Energy* **2016**, *41*, 1502–1512.
16. Duan, W.; Yu, Q.; Xie, H.; Qin, Q.; Zuo, Z. Thermodynamic analysis of hydrogen-rich gas generation from coal/steam gasification using blast furnace slag as heat carrier. *Int. J. Hydrogen Energy* **2014**, *39*, 11611–11619. [CrossRef]
17. Yao, X.; Yu, Q.; Xie, H.; Duan, W.; Han, Z.; Liu, S.; Qin, Q. The production of hydrogen through steam reforming of bio-oil model compounds recovering waste heat from blast furnace slag. *J. Therm. Anal. Calorim.* **2018**, *131*, 2951–2962. [CrossRef]
18. Li, P.; Yu, Q.; Xie, H.; Qin, Q.; Wang, K. $CO_2$ gasification rate analysis of Datong coal using slag granules as heat carrier for heat recovery from blast furnace slag by using a chemical reaction. *Energy Fuels* **2013**, *27*, 4810–4817. [CrossRef]
19. Li, P.; Yu, Q.; Qin, Q.; Lei, W. Kinetics of $CO_2$/coal gasification in molten blast furnace slag. *Ind. Eng. Chem. Res.* **2012**, *51*, 15872–15883. [CrossRef]
20. Yao, X.; Yu, Q.; Han, Z.; Xie, H.; Duan, W.; Qin, Q. Kinetics of $CO_2$ gasification of biomass char in granulated blast furnace slag. *Int. J. Hydrogen Energy* **2018**, *43*, 12002–12012. [CrossRef]
21. Sun, Y.; Zhang, Z.; Liu, L.; Wang, X. Two-stage high temperature sludge gasification using the waste heat from hot blast furnace slags. *Bioresour. Technol.* **2015**, *198*, 364–371. [CrossRef]
22. Sun, Y.; Zhang, Z.; Liu, L.; Wang, X. Integrated carbon dioxide/sludge gasification using waste heat from hot slags: Syngas production and sulfur dioxide fixation. *Bioresour. Technol.* **2015**, *181*, 174–182. [CrossRef] [PubMed]
23. Purwanto, H.; Akiyama, T. Hydrogen production from biogas using hot slag. *Int. J. Hydrogen Energy* **2006**, *31*, 491–495. [CrossRef]
24. Zhao, L.; Wang, H.; Qing, S.; Liu, H. Characteristics of gaseous product from municipal solid waste gasification with hot blast furnace slag. *J. Nat. Gas Chem.* **2010**, *19*, 403–408. [CrossRef]
25. Duan, W.; Yu, Q.; Liu, J.; Wu, T.; Yang, F.; Qin, Q. Experimental and kinetic study of steam gasification of low-rank coal in molten blast furnace slag. *Energy* **2016**, *111*, 859–868. [CrossRef]
26. Duan, W.; Yu, Q.; Wu, T.; Yang, F.; Qin, Q. Experimental study on steam gasification of coal using molten blast furnace slag as heat carrier for producing hydrogen-enriched syngas. *Energy Convers. Manag.* **2016**, *117*, 513–519. [CrossRef]
27. Luo, S.; Guo, J.; Feng, Y. Hydrogen-rich gas production from pyrolysis of wet sludge in situ steam agent. *Int. J. Hydrogen Energy* **2017**, *42*, 18309–18314. [CrossRef]
28. Luo, S.; Fu, J.; Zhou, Y.; Yi, C. The production of hydrogen-rich gas by catalytic pyrolysis of biomass using waste heat from blast-furnace slag. *Renew. Energy* **2017**, *101*, 1030–1036. [CrossRef]
29. Wang, P.; Xie, H.; Zhang, J.; Jia, L.; Yu, Z.; Li, R. Optimization of two bio-oil steam reforming processes for hydrogen production based on thermodynamic analysis. *Int. J. Hydrogen Energy* **2022**, *47*, 9853–9863. [CrossRef]

30. Jin, K.; Ji, D.; Xie, Q.; Nie, Y.; Yu, F.; Ji, J. Hydrogen production from steam gasification of tableted biomass in molten eutectic carbonates. *Int. J. Hydrogen Energy* **2019**, *44*, 22919–22925. [CrossRef]
31. Krzywanski, J.; Fan, H.; Feng, Y.; Shaikh, A.; Fang, M.; Wang, Q. Genetic algorithms and neural networks in optimization of sorbent enhanced $H_2$ production in FB and CFB gasifiers. *Energy Convers. Manag.* **2018**, *171*, 1651–1661. [CrossRef]
32. Bizkarra, K.; Bermudez, J.; Arcelus-Arrillaga, P.; Barrio, V.L.; Cambra, J.; Millan, M. Nickel based monometallic and bimetallic catalysts for synthetic and real bio-oil steam reforming. *Int. J. Hydrogen Energy* **2018**, *46*, 11706–11718. [CrossRef]
33. Valle, B.; Aramburu, B.; Benito, P.; Bilbao, J.; Gayubo, A.G. Biomass to hydrogen-rich gas via steam reforming of raw bio-oil over Ni/$La_2O_3$-$\alpha Al_2O_3$ catalyst: Effect of space-time and steam-to-carbon ratio. *Fuel* **2018**, *216*, 445–455. [CrossRef]
34. Rodrigues, C.; Alonso, C.; Machado, G.; de Souza, T. Optimization of bio-oil steam reforming process by thermodynamic analysis. *Int. J. Hydrogen Energy* **2020**, *45*, 28350–28360. [CrossRef]
35. Yao, X.; Yu, Q.; Xu, G.; Han, Z.; Qin, Q. Production of syngas from dry reforming of bio-oil model compound in granulated blast furnace slag. *Korean J. Chem. Eng.* **2019**, *36*, 722–728. [CrossRef]
36. Yao, X.; Yu, Q.; Xu, G.; Han, Z.; Xie, H.; Duan, W.; Qin, Q. The characteristics of syngas production from bio-oil dry reforming utilizing the waste heat of granulated blast furnace slag. *Int. J. Hydrogen Energy* **2018**, *43*, 22108–22115. [CrossRef]
37. Shimada, T.; Kochura, V.; Akiyama, T.; Kasai, E.; Yagi, J. Effects of slag compositions on the rate of methane–steam reaction. *ISIJ Int.* **2000**, *41*, 111–115. [CrossRef]
38. Xie, H.; Yu, Q.; Zuo, Z.; Han, Z.; Yao, X.; Qin, Q. Hydrogen production via sorption-enhanced catalytic steam reforming of bio-oil. *Int. J. Hydrogen Energy* **2016**, *41*, 2345–2353. [CrossRef]
39. Zuo, Z.; Yu, Q.; Xie, H.; Yang, F.; Qin, Q. Thermodynamic analysis of reduction in copper slag by biomass molding compound based on phase equilibrium calculating model. *J. Therm. Anal. Calorim.* **2018**, *132*, 1277–1289. [CrossRef]
40. Xie, H.; Yu, Q.; Zhang, Y.; Zhang, J.; Liu, J.; Qin, Q. New process for hydrogen production from raw coke oven gas via sorption-enhanced steam reforming: Thermodynamic analysis. *Int. J. Hydrogen Energy* **2017**, *42*, 2914–2923. [CrossRef]
41. Xie, H.; Yu, Q.; Wei, M.; Duan, W.; Yao, X.; Qin, Q.; Zuo, Z. Hydrogen production from steam reforming of simulated bio-oil over Ce–Ni/Co catalyst with in continuous $CO_2$ capture. *Int. J. Hydrogen Energy* **2015**, *40*, 1420–1428. [CrossRef]
42. Xie, H.; Zhang, J.; Yu, Q.; Zuo, Z.; Liu, J.; Qin, Q. Study on Steam Reforming of Tar in Hot Coke Oven Gas for Hydrogen Production. *Energy Fuels* **2016**, *30*, 2336–2344. [CrossRef]
43. Luo, S.; Feng, Y. The production of fuel oil and combustible gas by catalytic pyrolysis of waste tire using waste heat of blast-furnace slag. *Energy Convers. Manag.* **2017**, *136*, 27–35. [CrossRef]
44. Huang, B.-S.; Chen, H.-Y.; Chuang, K.-H.; Yang, R.-X.; Wey, M.-Y. Hydrogen production by biomass gasification in a fluidized-bed reactor promoted by an Fe/CaO catalyst. *Int. J. Hydrogen Energy* **2012**, *37*, 6511–6518. [CrossRef]

processes

MDPI

Article

# A Comprehensive Model for Evaluating Titanium Industry Security in China

Xiangwei Hui, Ming Qi, Wenhao Wang *, Sen Yang and Chengqi Zhang

School of Materials & Metallurgy, Guizhou University, Guiyang 550025, China; gs.xwhui21@gzu.edu.cn (X.H.); gs.mqi21@gzu.edu.cn (M.Q.); gs.yangshen22@gzu.edu.cn (S.Y.); gs.cqzhang20@gzu.edu.cn (C.Z.)
* Correspondence: whwang@gzu.edu.cn

**Abstract:** Currently, China is the largest consumer of titanium (Ti), yet the development of its Ti industry is limited by numerous factors, such as industrial structure imbalance. This study aimed to evaluate the security of China's Ti industry from 2010 to 2020, seeking to identify relevant issues and propose policy strategies. Firstly, a comprehensive evaluation system for Ti industry security was established, encompassing aspects of availability, economics, and sustainability. Secondly, the entropy weight technique for order preference by similarity to an ideal solution (TOPSIS) combination method and gray correlation method were employed to assess the safety level of China's Ti industry chain in each year from 2010 to 2020. Additionally, the coupling degree and sensitivity were used to analyze the dimension layers and index system to determine those that negatively impact the safety level of the Ti industry chain. The analysis results reveal that the economic level exerts a significant influence on the development of the Ti industry. Accordingly, under the same level of change, while considering availability, equal attention should be provided to economic considerations for a well-rounded evaluation of the industry's safety level.

**Keywords:** titanium industry chain; security system; sustainability; grey correlation analysis; entropy weight-TOPSIS

**Citation:** Hui, X.; Qi, M.; Wang, W.; Yang, S.; Zhang, C. A Comprehensive Model for Evaluating Titanium Industry Security in China. *Processes* **2023**, *11*, 2286. https://doi.org/10.3390/pr11082286

Academic Editors: Xin Yao, Huaqing Xie and Carlos Sierra Fernández

Received: 24 June 2023
Revised: 22 July 2023
Accepted: 24 July 2023
Published: 30 July 2023



## 1. Introduction

As China enters a new stage of development, the security and stability of industry and supply chains play a critical role in aiding its progress. Titanium (Ti) metal and its alloys hold strategic importance as high-performance non-ferrous metals and serve as crucial military resources. Ti finds extensive applications not only in traditional industries, such as chemical, electric power, metallurgy, and salt [1], but also in emerging fields, including aerospace [2,3], national defense [4], automobile industry [5–7], marine engineering, medicine, leisure, and sports. Its role in the development of weapons and equipment is irreplaceable. Acknowledging its strategic value, the United States has listed Ti in the Key Mineral Catalogue [8], while Japan has included it in the Rare Metal Protection Strategy. Furthermore, China has recognized the significance of the Ti industry in Made in China 2025 and Guidelines for the Development of New Materials Industry [9]. Compared with the United States, Russia, and Japan [10], which have employed Ti in advanced technologies, with aerospace as the main downstream application, China exhibits a unique consumption of Ti metal. The proportion of industrial consumption is relatively high, while that of aerospace and civil products is relatively low.

Despite the abundance of Ti resources in China, the overall grade of these resources tends to be low. The ores are more commonly of average quality, and high-grade ores are relatively scarce. More than 90% of Ti ore is rock-type vanadium–Ti–magnetite, characterized by a dense structure and high gangue content, which makes beneficiation and sorting challenging [11]. Figure 1a,c illustrate the distributions of Ti ore and ilmenite, respectively. Additionally, domestic beneficiation technology is still in its nascent stages, leading to insufficient production of high-quality Ti concentrates and a heavy reliance on imports. Ti materials, especially those of high quality, have stringent requirements. While global Ti resources are abundant, high-quality rutile accounts for less than 10% of the total rutile, making it a scarce resource. Figure 1b illustrates the distribution of rutile in China, the majority of which is imported [12]. The rapid growth in demand for aviation-grade Ti sponge, coupled with the military industry's heightened requirements for stable supply and material quality, poses challenges. Most Ti sponge production enterprises in China do not possess Ti-ore resources, which impacts the long-term stable supply, product quality, and cost of high-end Ti products, resulting in difficulties in meeting long-term demand. Despite an increase in the annual production of Ti materials, the low quality of Ti in China hampers the ability to meet the requirements for high-end Ti materials in aerospace and marine engineering applications. In the current Chinese Ti sponge market, export prices are lower than import prices, with high-end imported Ti sponges having a higher proportion than high-end exported ones, highlighting the insufficiency in China's production capacity for high-quality Ti sponges [13].

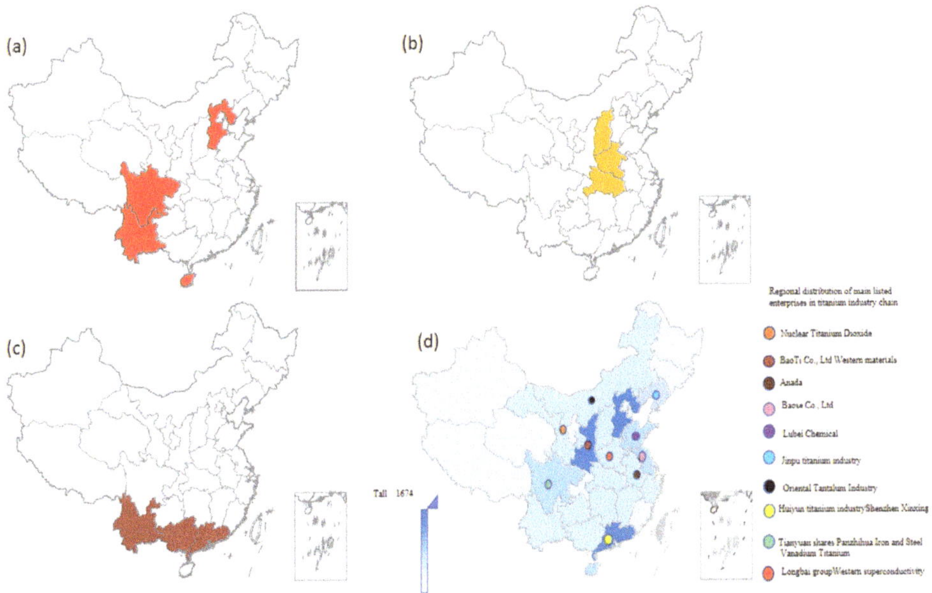


**Figure 1.** Distribution of (**a**) titanium ore, (**b**) ilmenite illustrates, and (**c**) rutile in China. The regional heat map of (**d**) the number of enterprises in the titanium industry chain and the distribution of major listed companies in the titanium industry chain.

Furthermore, due to the relatively late development of the Ti industry in China, most high-end products heavily rely on imports, resulting in high technical barriers. As a result, there are a limited number of enterprises in China's Ti industry, and their production scale is generally small. This constraint leads to limitations in the tonnage and capacity of existing equipment for processing Ti materials, preventing the processing of large Ti materials for aerospace and marine engineering and highlighting the need for improved industry concentration [14]. In addition, at the upstream end of the Ti industry lies the resource sector, including mining. Most enterprises are concentrated in resource-rich areas, resulting in a relatively scattered distribution. Figure 1d illustrates the distribution of Ti enterprises across various provinces in China [15].

With the rapid growth in demand for high-end Ti materials in China, challenges such as high production costs, unbalanced product structure, and low industry concentration have emerged, making it difficult to meet the requirements of high-end industries reliant on Ti metal and its alloys. Consequently, there is a requirement to update and enhance the Ti industrial chain in China.

Therefore, the assessment of the safety of China's Ti industry and subsequent updates are critical for its sustainable development. However, comprehensive safety assessment of the entire Ti industry chain is limited. Therefore, this study proposes an evaluation of the entire Ti industry chain by constructing indicators related to availability, economics, and sustainability, which are crucial for addressing the aforementioned challenges.

The main contributions of this study can be summarized as follows:

(1) Presentation of comprehensive evaluation criteria for assessing the security of the Ti industry chain. These criteria comprise 34 indicators that cover various aspects of the industry's safety.
(2) Development of a systematic method for assessing the safety of China's Ti industry chain from 2010 to 2020. This method considers the entire industrial chain, dimension layers, and individual indicators.
(3) Analyses of the changes observed in different dimensional layers and indicators of China's Ti industry, along with the formulation of policy suggestions based on the findings.

The remainder of this article is organized as follows:

Section 2 presents a literature review of the industrial chain. Section 3 introduces the evaluation standard system and methodologies employed for evaluating the Ti industry chain. Section 4 presents the results of the safety evaluation, including dimensional layer coupling analysis, sensitivity analysis, gray correlation, and sensitivity analysis of the index system. Finally, Section 5 presents policy recommendations and concludes the paper.

## 2. Literature Review

The term "Industrial chain" is unique to China and has not been extensively studied as an independent concept by foreign scholars. However, similar concepts, such as commodity, supply, and value chains, have been proposed by Grey and Miguel [16], Priem and Swimk [17], Reuver and Bouwman [18], and Stabell and Fjeldstad [19]. There is no consensus among academics regarding the specific connotations of industrial chain security. In the context of specific industries, researchers have conducted studies on the analysis of industrial chains and their characteristics. For example, Liu et al. [20] used the material flow analysis method to analyze the cobalt (Co) industrial chain and calculated the resource dependence between the industrial chain links in each country/region. Li et al. [21] investigated the current situation of the Co trade and the impact of national risks on Co trade patterns from the perspective of the entire industrial chain. Similarly, Li et al. [22] analyzed the global copper industry chain and trade characteristics of important countries by building a trade network of different links. Huang et al. [23] used complex network methods and extended the gravity model to analyze the evolution of international tungsten industry competition and its influencing factors. Liu et al. [24] quantified the flow of indium (In) in China from 2000 to 2019 using material flow analysis, focusing on the problems faced by China's In industry.

Compared to these metals, the Ti industry has been less researched because it is a new industry with a relatively short history. Most Ti-related research has focused on catalytic processes, such as optimization with other ions. In addition, the synergistic physicochemical effects of the forming compounds [25,26] have few studies devoted to Ti flow, partially describing the flow of metallic Ti or TD [27,28]. Li et al. [29] used a dynamic material flow analysis method to reveal the evolutionary trend of China's Ti cycle from 2005 to 2020. However, only limited studies have examined the impacts of changes in the availability, economy, and sustainability of Ti resources and products on the development of the entire industry.

Under these circumstances, by drawing on the assessment methods of energy security and the development status of the Ti industry, this study aimed to construct an index system based on the three dimensions of availability, economy, and sustainability. To comprehensively evaluate the safety of China's Ti industry, 34 indicators were selected according to Ti resources and the availability, economy, and sustainability of Ti products. To objectively and comprehensively evaluate the safety level of the Ti industry, the entropy weight technique for order preference by similarity to an ideal solution (TOPSIS) method was used in combination with gray correlation, coupling analysis, and sensitivity analysis. The evaluation process is from whole to part, which is conducive to identifying and analyzing problems. The main research ideas and innovation points are presented below.

To address the research gaps, we quantitatively measured and evaluated the system safety level of the Ti industry chain from 2010 to 2020 based on the entropy weight-TOPSIS distance function model. Subsequently, we analyzed the coordination relationship between the Ti industry and each dimension layer through coupling coordination using gray correlation and sensitivity analysis. The main innovation points are: (1) focusing on industrial chain security as a multidimensional concept, which helps eliminate the limitation of unilaterally describing industrial chain security based solely on the evaluation of trade or supply security; (2) proposing a systematic method to measure the safety performance of an industrial chain by analyzing its weak links and presenting corresponding solutions; and (3) developing an industrial chain security level index to help determine the security levels of different dimensions within the industrial chain and identify trade-offs in areas that require improvement. This allows a comprehensive evaluation and establishes a reliable risk assessment system for industrial chain security. This facilitates the formulation of a complete mechanism concerning organization, monitoring, and management when detecting early warnings. (4) Valuable insights are offered, and policies are suggested, such as better management of Ti resources and optimization of the Ti industry structure.

## 3. Materials and Methods

### 3.1. System Definition

Based on the research methods used in other metal industry chains and the existing literature on the development of the Ti industry, this study conducted a safety evaluation of the Ti industry chain. The evaluation process considered the entire industry as the evaluation object and took into account the development status and industrial composition of China's Ti industry. It analyzed the main factors affecting the safety of the Ti industry, constructed a quantifiable evaluation index system, and used certain theoretical methods to evaluate the Ti industry. The evaluation aimed to provide a comprehensive assessment of the historical and current state of the Ti industry. Due to data limitations, it was not feasible to cover every aspect of the industry. Therefore, this study focused on the quantitative analysis of important indicators upstream (raw material supply), midstream (Ti product manufacturing), and downstream (market application) in the Ti industry chain. The Ti industry chain is illustrated in Figure 2.

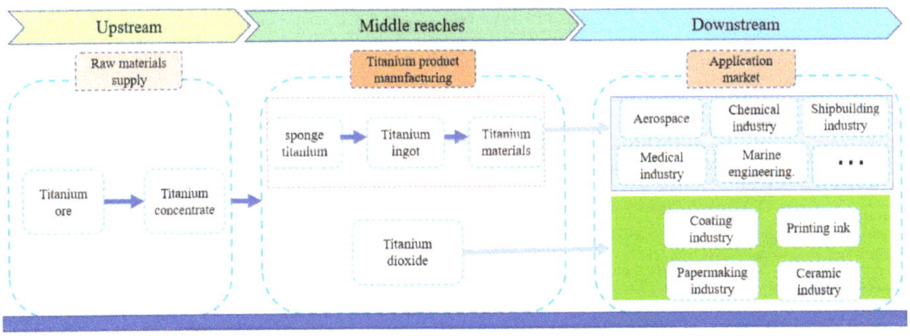


**Figure 2.** Titanium industry industrial chain structure.

**Table 1.** *Cont.*

| Dimension Layer | | Basic Indicators | | Indicator Attribute | Meaning and Interpretation of the Indicators |
|---|---|---|---|---|---|
| | | C4 | Sponge titanium capacity | + | The size of the production capacity of titanium products in China reflects the supply of titanium product production equipment in China. |
| | | C5 | Titanium dioxide production capacity | + | |
| | | C6 | Titanium ingot production capacity | + | |
| | | C7 | CR3 | + | The output of titanium materials in my country ranks among the top three (CR3), and the ratio of the output of titanium materials in the top five (CR5) enterprises to the total output reflect the concentration of my country's titanium materials industry. |
| | | C8 | CR5 | + | |
| | | C9 | The gross profit margin of titanium products | + | Gross margin = (gross margin / operating income)× 100%, reflecting the competitiveness of the enterprise |
| | Security capability | C10 | National Defense Support Capability | + | Defense spending as a percentage of GDP. In the long run, the security of the industrial chain cannot be separated from the guarantee of strong national defense capabilities. |

### 3.3. Assessment Model

In this study, the entropy weight-TOPSIS [31] distance function model was used to quantitatively model the safety evaluation of the Ti industry in China. The model consists of three main parts, as illustrated in Figure 3, to ensure a robust and accurate assessment process.

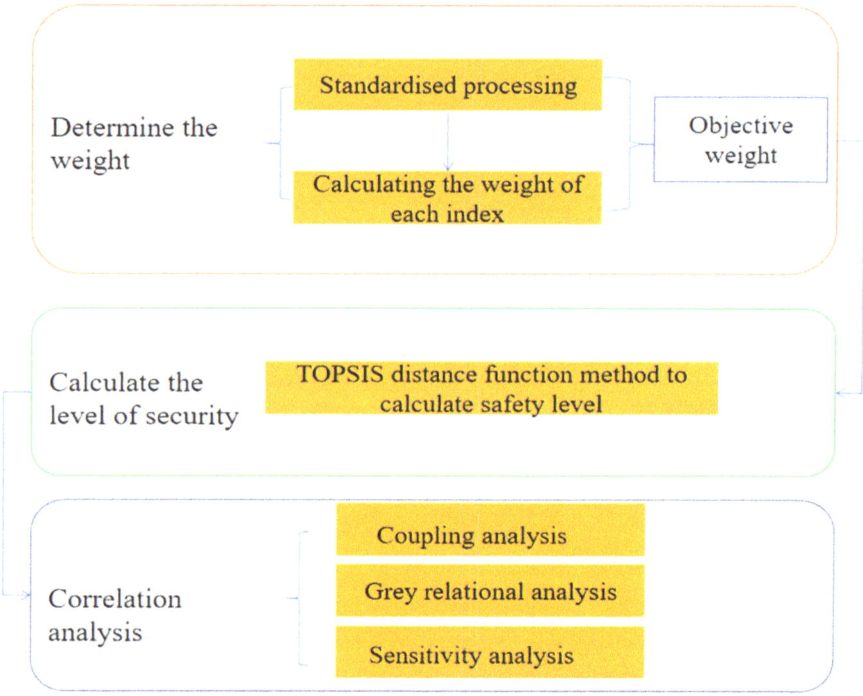


**Figure 3.** The model mainly has three parts.

3.3.1. Normalisation of the Raw Data

Because the measurement units of various indicators are not uniform, it is necessary to perform standardization before using and converting the absolute value of the indicator into a relative value to address the problem of homogeneous comparison of different indicators [32]. The specific standardization method is as follows: $m$ objects and $n$ indicators are selected, where $x_{ij}$ is the value of the $j$-th indicator of the $i$-th object ($i$ = 1, 2, ..., $m$; $j$ = 1, 2, ..., $n$). The positive and negative indices in the index system are standardized in the following ways, and standardized non-dimensionalized evaluation index data are obtained.

(1) Positive indicators:

$$x_{ij} = \frac{x_{ij} - \min\{x_{1j}, \ldots, x_{mj}\}}{\max\{x_{1j}, \ldots, x_{mj}\} - \min\{x_{1j}, \ldots, x_{mj}\}} \quad (1)$$

(2) Negative indicators:

$$x_{ij} = \frac{\max\{x_{1j}, \ldots, x_{mj}\} - x_{ij}}{\max\{x_{1j}, \ldots, x_{mj}\} - \min\{x_{1j}, \ldots, x_{mj}\}} \quad (2)$$

where $x_{ij}$ is the value of the $j$-th index of the $i$-th dimensional layer ($i$ = 1, 2, ... $n$; $j$ = 1, 2, ... $m$), and min $\{x_{1j}, \ldots, x_{mj}\}$ and max $\{x_{1j}, \ldots, x_{mj}\}$ represent the minimum and maximum values of the $j$-th column of the matrix, respectively. For convenience, the normalized data were recorded as $x_{ij}$.

3.3.2. Obtaining the Weights of Resource Security Evaluation Indicators

The index weight is determined using the direct right method [33]. The entropy weight method is an objective weight determination method that uses index data to determine the index weight and has good operability and practicability. Compared with the expert scoring method, the entropy weight method enhances the difference and identification of the evaluation indicators and can more objectively reflect implicit information in the data [34]. The specific process is as follows:

(1) Calculate the feature proportion of the evaluation object under the evaluation index ($P_{ij}$):

$$P_{ij} = \frac{x_{ij}}{\sum_{j=1}^{n} x_{ij}} \quad (3)$$

(2) Calculate the entropy of the index ($e_j$):

$$\begin{aligned} &e_j = \frac{(-1)\sum_{i=1}^{n} p_{ij} \cdot Ln(p_{ij})}{Ln(m)} \\ &In, like p_{ij} = 0, but \frac{\lim}{p_{ij} \to 0} p_{ij} \cdot \frac{Ln(p_{ij})}{1/p_{ij}} = 0, So e_{ij} > 0. \end{aligned} \quad (4)$$

(3) Calculate the direct redundancy. The greater the direct redundancy, the more important the index ($d_j$):

$$d_j = 1 - e_j \quad (5)$$

(4) Calculate the weight of each indicator ($w_j$):

$$w_j = \frac{d_j}{\sum_{j=1}^{n} d_j} \quad (6)$$

3.3.3. Measurement of the Safety Level of China's Titanium Industry Chain

The safety level of the Ti industry in China was quantitatively evaluated using the TOPSIS distance metric method [35]. This method makes complete use of the original data, is not affected by the selection of reference sequences, and has an intuitive geometric meaning. It is primarily used for risk decision analyses, environmental benefit evaluations, cost-effectiveness analyses, and multi-attribute optimal scheme selection. The idea behind the TOPSIS model is to construct positive and negative ideal solutions for a multi-attribute problem [36,37]. The most satisfactory solution is selected using two benchmarks: solutions that are close to the positive ideal solution and far from the negative ideal solution. These criteria are used for ranking the solutions [38]. When a solution is closest to the positive ideal solution and far from the negative ideal solution, it is considered the optimal scheme for the evaluation object and is ranked as the most advanced; otherwise, it is ranked as the worst scheme, as it is the most backward. The specific application process is as follows:

(1) Form a weighted judgment weight matrix ($Y$):

$$Y = \sum_{j=1}^{n} w_j \cdot p_{ij} = (y_{ij})_{m \times n} \tag{7}$$

(2) Determine the ideal solution

The ideal solution is divided into a positive ideal solution and a negative ideal solution. The positive ideal solution is represented by $y_j^+$, and the negative ideal solution is represented by $y_j^-$. The specific formula is as follows:

$$y_j^+ = \max\{y_{1j}, \ldots, y_{mj}\} \; y_j^- = \min\{y_{1j}, \ldots, y_{mj}\} \tag{8}$$

(3) The distance between each index and the ideal solution is determined:

The distance from each index to the positive ideal solution is represented by $d_i^+$, and that to the negative ideal solution is represented by $d_i^-$, according to the Euclidean distance formula:

$$d_i^+ = \sqrt{\sum_{j=1}^{n} (y_{ij} - y_j^+)^2} \; d_i^- = \sqrt{\sum_{j=1}^{n} (y_{ij} - y_j^-)^2} \tag{9}$$

(4) The closeness of the evaluation object to the ideal solution is calculated:

The higher the degree of closeness, the better the object, where the value range of the degree of closeness is [0, 1]. For the evaluation of industrial chain security, the closer the degree of closeness is to 1, the higher the degree of industrial chain security in the year; conversely, when the degree of closeness is 0, the level of industrial chain security is the lowest. The closeness is calculated as follows:

$$c_i = \frac{d_i^-}{d_i^+ + d_i^-}, (i = 1, 2, 3 \ldots m) \tag{10}$$

3.3.4. Correlation Analysis of Factors Affecting the Safety Level of the Ti Industry Chain

During the system development process, if the trends of the two factors are consistent, that is, if the degree of synchronous change is high, the degree of correlation between the two is high, and vice versa [39]. The gray relational analysis method measures the degree of correlation between each index and the level of energy security according to the similarity and dissimilarity of development trends among the factors. This is a multi-factor statistical analysis method that is mainly used to compare the strength of similar associations between items in the system and other factors. Finally, these factors are sorted according to their degree of influence, and an analysis result that is suitable for dynamic history is obtained. The specific application process is as follows:

(1) Analyze the target sequences.

The safety levels of the Ti industry in China were used as the target sequence. Safety evaluation indicators of the Ti industry chain were used as comparison sequences. The degree of correlation between the safety level of the country's Ti industry chain and each evaluation index was analyzed.

(2) Calculate the difference sequence matrix.

To find the difference sequence matrix between the target sequence and the comparison sequence, the calculation formula is (Z):

$$Z = \left|x_{ij} - c_i\right| = \left(\lambda_{ij}\right)_{m\times n} \tag{11}$$

(3) Two-level maximum and minimum values were calculated.

$$\begin{gathered}\lambda_{\max} = \max\{\lambda_{1j}, \lambda_{2j}, \ldots, \lambda_{ij}\},\ \lambda_{\min} = \min\{\lambda_{1j}, \lambda_{2j}, \ldots, \lambda_{mj}\} \\ m = \max(\lambda_{\max}),\ n = \max(\lambda_{\min})\end{gathered} \tag{12}$$

(4) Correlation coefficients are calculated ($\xi_j$):

$$\xi_j = \frac{n + \rho m}{\left|x_{ij} - c_i\right| + \rho m} \tag{13}$$

Among them, $\rho \in [0, 1]$ is the resolution coefficient, which is inversely proportional to the resolution. When $\rho \leq 0.5463$, the resolution is the most ideal, usually $\rho = 0.5$.

(5) Determine the degree of correlation:

Average the formula for calculating correlation coefficients for each. The indicators are as follows:

$$\eta_j = \frac{\sum_{i=1}^{m} \xi_{ij}}{m} \tag{14}$$

## 4. Results and Discussions

### *4.1. Model Results*

#### 4.1.1. The Weight of the Safety Level Indicator

Based on Equations (1)–(6), the weight of each basic index is obtained. The weight of the index reflects the proportion of the index in the system and plays a significant role in the calculation of the safety level. Table 2 lists the weights and changes of the safety evaluation index of the Ti industry.

**Table 2.** The weight and index change in the safety evaluation index of the titanium industry chain.

| Index | $W_j$ | Years | | | | | | | | | | |
|---|---|---|---|---|---|---|---|---|---|---|---|---|
| | | 2010 | 2011 | 2012 | 2013 | 2014 | 2015 | 2016 | 2017 | 2018 | 2019 | 2020 |
| A1 | 0.029 | 0.662 | 0.6240 | 0.5484 | 0.6011 | 0.6543 | 0.6947 | 0.6061 | 0.5531 | 0.5738 | 0.6463 | 0.6777 |
| A2 | 0.032 | 1.045 | 1.1506 | 1.1098 | 1.0464 | 1.0895 | 1.0592 | 0.9794 | 0.9740 | 0.9533 | 0.9330 | 0.9672 |
| A3 | 0.027 | 0.993 | 1.1173 | 1.0556 | 1.1309 | 1.0676 | 1.1669 | 1.2586 | 1.2738 | 1.3149 | 1.3569 | 1.4232 |
| A4 | 0.057 | 1.495 | 1.1445 | 1.1986 | 1.1544 | 1.1622 | 1.3001 | 1.3597 | 1.1852 | 1.3476 | 1.3428 | 1.1484 |
| A5 | 0.017 | 0.648 | 0.5891 | 0.5478 | 0.5321 | 0.4264 | 0.2234 | 0.3145 | 0.4530 | 0.5474 | 0.5890 | 0.6281 |
| A6 | 0.018 | 0.553 | 0.4135 | 0.9553 | 0.9344 | 1.0000 | 0.9915 | 0.9216 | 0.9325 | 0.8588 | 0.6000 | 0.8125 |
| A7 | 0.029 | 0.338 | 0.3760 | 0.4516 | 0.3989 | 0.3457 | 0.3054 | 0.3941 | 0.4631 | 0.4239 | 0.3523 | 0.3204 |
| A8 | 0.017 | −0.003 | −0.1506 | −0.0599 | −0.0464 | −0.0895 | −0.0592 | 0.0260 | 0.0208 | 0.0467 | 0.0670 | 0.0328 |
| A9 | 0.024 | 0.066 | −0.1033 | −0.1320 | −0.1073 | −0.1770 | −0.0430 | −0.2568 | −0.2732 | −0.3168 | −0.3568 | −0.4243 |
| A10 | 0.036 | −0.038 | −0.1056 | −0.1654 | −0.1544 | −0.1622 | −0.1622 | −0.1488 | −0.1860 | −0.1962 | −0.2049 | −0.1036 |
| B1 | 0.018 | 0.877 | 1.0829 | 1.0829 | 0.7370 | 1.3305 | 1.0161 | 1.2327 | 1.2773 | 1.0567 | 0.9743 | 1.1749 |
| B2 | 0.036 | 1.273 | 1.0612 | 1.1351 | 1.1621 | 1.3312 | 1.7625 | 1.4908 | 1.1228 | 1.2071 | 1.3583 | 1.4715 |
| B3 | 0.029 | 1.976 | 1.7236 | 1.7236 | 3.0951 | 3.1576 | 3.3439 | 3.1915 | 3.0015 | 2.8495 | 3.0036 | 2.6922 |
| B4 | 0.011 | −415.918 | −8705.5954 | −3204.7844 | −33,418.7158 | −3563.7000 | −2108.1418 | 950.6452 | 1627.0000 | 2717.0000 | 4723.0000 | 3267.0000 |
| B5 | 0.031 | 44,940 | −45,097 | −58,774 | −41,743 | −54,313 | −45,000 | −72,900 | −140,900 | −168,600 | −170,700 | −186,900 |
| B6 | 0.031 | 10,588.328 | −658.1663 | −989.8339 | 35,091.6428 | 10,222.800 | 16,358.5376 | 11,068.7651 | 12,128.000 | 11,777.000 | 8298.000 | 3518.000 |
| B7 | 0.039 | 2.000 | 2.000 | 0.500 | 2.000 | 2.200 | 0.500 | 4.600 | 0.800 | 1.500 | 1.900 | 2.000 |
| B8 | 0.035 | 1.100 | 1.4000 | 1.3000 | 1.5000 | 0.0000 | 0.8000 | 4.6000 | 0.1000 | 1.4000 | 1.4000 | 1.5000 |
| B9 | 0.027 | 0.558 | 0.5055 | 0.5485 | 0.5411 | 0.4522 | 0.7049 | 0.7619 | 0.7841 | 0.7005 | 0.5372 | 0.6947 |
| B10 | 0.023 | 0.613 | 0.6285 | 0.7308 | 0.7714 | 0.7320 | 0.8345 | 0.8497 | 0.8969 | 0.8676 | 0.8368 | 0.9129 |
| B11 | 0.042 | 0.519 | 0.6422 | 0.4372 | 0.5708 | 0.4600 | 0.4425 | 0.4924 | 0.4841 | 0.4729 | 0.4997 | 0.6027 |
| B12 | 0.027 | 412,119 | 487,940.2 | 538,580.0 | 592,963.2 | 643,563.1 | 676,708.0 | 746,395.1 | 832,035.9 | 919,281.1 | 986,515.2 | 1,015,986.2 |
| B13 | 0.030 | 5321.0 | 6011.0 | 6702.7 | 7202.0 | 8082.0 | 8869.0 | 9543.5 | 10,211.0 | 11,069.0 | 12,680.0 | 13,795.4 |
| B14 | 0.013 | 4.500 | 4.2000 | 0.3000 | 0.5000 | 2.1000 | −10.0000 | 3.3000 | 2.3400 | 7.6200 | 8.8300 | 9.2500 |
| C1 | 0.020 | 1309.091 | 1090.9091 | 750.0000 | 745.0980 | 793.7500 | 898.8235 | 935.7143 | 975.0000 | 395.2381 | 404.7619 | 159.6330 |
| C2 | 0.051 | 0.381 | 0.3942 | 0.3525 | 0.3552 | 0.3745 | 0.4007 | 0.4334 | 0.5125 | 0.5796 | 0.5783 | 0.5753 |
| C3 | 0.043 | 13.500 | 12.5000 | 12.2000 | 14.3000 | 15.0000 | 19.8000 | 25.6000 | 24.7000 | 25.0000 | 26.0000 | 29.0000 |
| C4 | 0.043 | 103,500 | 128,500 | 148,500 | 150,000 | 150,000 | 88,000 | 88,000 | 93,000 | 107,000 | 158,000 | 177,000 |
| C5 | 0.026 | 2,400,000 | 2,880,000 | 2,600,000 | 2,800,000 | 3,060,000 | 2,780,000 | 3,060,000 | 3,200,000 | 3,400,000 | 3,800,000 | 3,845,000 |
| C6 | 0.027 | 89,200 | 96,200 | 148,500 | 109,000 | 124,000 | 135,000 | 135,000 | 146,700 | 158,700 | 177,500 | 199,000 |
| C7 | 0.023 | 46.000 | 45.0000 | 35.9000 | 32.0000 | 35.0000 | 40.1000 | 39.2000 | 36.0000 | 39.8000 | 44.0000 | 46.0000 |
| C8 | 0.032 | 50.100 | 52.0000 | 45.0000 | 48.0000 | 45.8000 | 51.5000 | 50.5000 | 62.6000 | 53.0000 | 58.6000 | 59.0000 |
| C9 | 0.022 | 16.60 | 17.30 | 18.00 | 14.20 | 19.30 | 18.82 | 20.26 | 22.75 | 24.74 | 21.34 | 25.90 |
| C10 | 0.032 | 0.0129 | 0.0123 | 0.0124 | 0.0121 | 0.0126 | 0.0131 | 0.0128 | 0.0123 | 0.0120 | 0.0129 | 0.0136 |

It can be observed from Table 2 that during the evaluation period, the largest average weight is that of the titanium self-sufficiency rate (A4) indicator. The smallest indicator is the net profit margin of titanium products (B14), whose weight value is 0.013–0.057.

According to the development status of the Ti industry during the evaluation period, the weight of each index is essentially in line with the development status.

4.1.2. Safety Level of the Titanium Industry

According to the calculation of the comprehensive safety level of the Ti industry using Equations (7)–(10), Table 3 lists the safety levels in the Ti industry from 2010 to 2020.

**Table 3.** Titanium Industry Chain Safety Level from 2010 to 2020.

| Years | 2010 | 2011 | 2012 | 2013 | 2014 | 2015 | 2016 | 2017 | 2018 | 2019 | 2020 |
|---|---|---|---|---|---|---|---|---|---|---|---|
| $d_i^+$ | 0.386 | 0.422 | 0.455 | 0.402 | 0.413 | 0.377 | 0.301 | 0.399 | 0.363 | 0.332 | 0.319 |
| $d_i^-$ | 0.317 | 0.261 | 0.210 | 0.261 | 0.232 | 0.291 | 0.349 | 0.290 | 0.303 | 0.341 | 0.407 |
| Safety level | 0.451 | 0.382 | 0.316 | 0.394 | 0.360 | 0.436 | 0.537 | 0.421 | 0.455 | 0.507 | 0.561 |
| Security level [40–42] | General | Early warning | Early warning | Early warning | Early warning | General | General | General | General | General | General |

As shown in Table 3, from 2010 to 2020, the security level of China's Ti industry first declined and then increased. The years 2011–2014 demonstrated early warning levels, while the other exhibited only general levels. In the early warning years, from the perspective of the dimension level, the levels of sustainability and economy were lower than in other years. From the perspective of basic indicators, the Ti self-sufficiency rate, GDP, and $TiO_2$ trade deficit were relatively low. Indicators such as the import concentration of sponge Ti, external dependence on Ti concentrate, external dependence on Ti materials, and external dependence on $TiO_2$ were higher than in other years. From the perspective of industrial development, this is primarily because of market demand. China emerged as a frontrunner in mitigating the repercussions of the international financial crisis from 2010 to 2012. With the rapid development of the national economy, the demand for Ti metal products was strong; however, this favorable situation only lasted until the second half of 2011, when the country began to vigorously control inflation and intensify the regulation of real estate; thus, the demand for Ti products began to decline. Since April 2012, the price and output of Ti products have declined. In 2013, the domestic demand for Ti products declined with the development of China's macro economy. Since 2013, a year of difficult economic recovery in Europe and the United States, the Japanese economy has had even more difficulty recovering, and the international demand for Ti materials has dropped sharply. Although China has excess structural capacity, the Ti industry continues to struggle. In 2014, due to the sluggish growth of the world economy and the persistent downturn in the global Ti industry, the Ti industry of China also entered a cold winter period marked by high production capacity, low profit, and low demand. This situation resulted in Ti-processing enterprises operating at a low profit.

*4.2. Security Dimension Layer Analysis of the Ti-Industry Chain*

4.2.1. Analysis of Coupling Coordination Degree

The degree of coupling and coordination [43] for the safety level of the Ti industry chain and economic availability sustainability level from 2010 to 2020 were calculated, as shown in Tables 4–6. The C coupling degree reflects the strength of the interaction between the systems, while the D coupling coordination model can further reflect the coordination degree of system development. T is a comprehensive coordination index, which mainly reflects the coordination level of various dimensions and industrial safety development.

**Table 4.** Coupling coordination values of the titanium industry chain and availability level.

| Years | The Safety Level of Titanium Industry Chain | Availability Level | C | T | D | Coupling Coordination Type |
|---|---|---|---|---|---|---|
| 2010 | 0.451 | 0.651 | 0.641 | 0.451 | 0.538 | Grudging coordination |
| 2011 | 0.382 | 0.435 | 0.392 | 0.382 | 0.387 | Mild maladjustment |
| 2012 | 0.316 | 0.419 | 0.618 | 0.316 | 0.442 | On the verge of maladjustment |
| 2013 | 0.394 | 0.380 | 0.802 | 0.394 | 0.562 | Grudging coordination |
| 2014 | 0.360 | 0.372 | 0.780 | 0.360 | 0.530 | Grudging coordination |
| 2015 | 0.436 | 0.465 | 0.508 | 0.436 | 0.470 | On the verge of maladjustment |
| 2016 | 0.537 | 0.480 | 0.901 | 0.537 | 0.696 | Primary coordination |
| 2017 | 0.421 | 0.374 | 0.676 | 0.421 | 0.534 | Grudging coordination |
| 2018 | 0.455 | 0.442 | 0.723 | 0.455 | 0.573 | Grudging coordination |
| 2019 | 0.507 | 0.425 | 0.484 | 0.507 | 0.496 | On the verge of maladjustment |
| 2020 | 0.561 | 0.425 | 0.480 | 0.561 | 0.519 | Grudging coordination |

**Table 5.** Coupling coordination values of the titanium industry chain and economic level.

| Years | The Safety Level of Titanium Industry Chain | Economical Level | C | T | D | Coupling Coordination Type |
|---|---|---|---|---|---|---|
| 2010 | 0.451 | 0.361 | 0.507 | 0.451 | 0.478 | On the verge of maladjustment |
| 2011 | 0.382 | 0.369 | 0.473 | 0.382 | 0.425 | On the verge of maladjustment |
| 2012 | 0.316 | 0.233 | 0.443 | 0.316 | 0.374 | Mild maladjustment |
| 2013 | 0.394 | 0.473 | 0.491 | 0.394 | 0.440 | On the verge of maladjustment |
| 2014 | 0.360 | 0.362 | 0.595 | 0.360 | 0.463 | On the verge of maladjustment |
| 2015 | 0.436 | 0.441 | 0.651 | 0.436 | 0.533 | Grudging coordination |
| 2016 | 0.537 | 0.628 | 0.932 | 0.537 | 0.707 | Intermediate coordination |
| 2017 | 0.421 | 0.384 | 0.679 | 0.421 | 0.535 | Grudging coordination |
| 2018 | 0.455 | 0.409 | 0.816 | 0.455 | 0.609 | Primary coordination |
| 2019 | 0.507 | 0.435 | 0.831 | 0.507 | 0.649 | Primary coordination |
| 2020 | 0.561 | 0.516 | 0.708 | 0.561 | 0.630 | Primary coordination |

**Table 6.** Coupling coordination values of the titanium industry chain and sustainability level.

| Years | The Safety Level of Titanium Industry Chain | Sustainability Level | C | T | D | Coupling Coordination Type |
|---|---|---|---|---|---|---|
| 2010 | 0.451 | 0.344 | 0.433 | 0.451 | 0.442 | On the verge of maladjustment |
| 2011 | 0.382 | 0.343 | 0.663 | 0.382 | 0.504 | Grudging coordination |
| 2012 | 0.316 | 0.310 | 0.425 | 0.316 | 0.367 | Mild maladjustment |
| 2013 | 0.394 | 0.271 | 0.548 | 0.394 | 0.465 | On the verge of maladjustment |
| 2014 | 0.360 | 0.346 | 0.781 | 0.360 | 0.530 | Grudging coordination |
| 2015 | 0.436 | 0.393 | 0.548 | 0.436 | 0.489 | On the verge of maladjustment |
| 2016 | 0.537 | 0.437 | 0.719 | 0.537 | 0.621 | Primary coordination |
| 2017 | 0.421 | 0.524 | 0.795 | 0.421 | 0.579 | Grudging coordination |
| 2018 | 0.455 | 0.525 | 0.660 | 0.455 | 0.548 | Grudging coordination |
| 2019 | 0.507 | 0.720 | 0.932 | 0.507 | 0.687 | Primary coordination |
| 2020 | 0.561 | 0.797 | 0.701 | 0.561 | 0.627 | Primary coordination |

It can be observed from the table that in the years when the safety level of the Ti industry chain was at the early warning level, the degree of coupling between the economic level and safety level of the Ti industry chain was between imminent and mild imbalance. In other words, the development of the economic level provides a great guarantee for the safety level of the Ti industry chain. In 2012, the year with the lowest safety level in the Ti industry chain within the scope of this study, the safety, economic, and sustainability levels of the Ti industry chain were in a state of mild imbalance, and the availability level was on the verge of imbalance. Thus, it is necessary to further promote the synchronous coupling development of the Ti industry and economic level to achieve benign interaction and healthy and sustainable development of the Ti industry.

#### 4.2.2. Security Level Analysis of Dimension Layer

Availability level. From 2010 to 2020, the level of safety availability in the Ti industry in China fluctuated significantly, as illustrated in Figure 4. In 2010, the level of safety in the Ti industry was relatively high, including external dependence on Ti dioxide, external dependence on Ti materials, $TiO_2$ self-sufficiency rate, and external dependence on Ti sponges, which had a positive impact on the Ti industry's availability level. The self-sufficiency rate of Ti sponges, import concentration of Ti ore, import concentration of Ti sponges, self-sufficiency rate of Ti concentrate, and external dependence on Ti concentrate have negative impacts on the availability of the Ti industry. Among these, the self-sufficiency rate of Ti materials and its external dependence on Ti materials has a significant impact on the availability level.

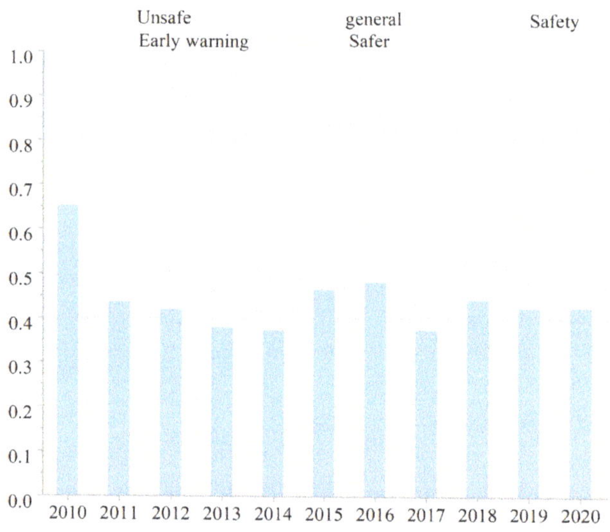


**Figure 4.** Safety availability level of China's titanium industry chain security from 2010 to 2020.

Economic level. From 2010 to 2020, the safety and economic levels of the Ti industry in China fluctuated significantly, as illustrated in Figure 5. The economy in 2012 was significantly lower than that in other years. In 2012, the economic level of the Ti industry was primarily affected by the operating rates. The state began to vigorously control inflation and intensify its efforts to regulate real estate, and demand for Ti products began to decline. Since April and May 2012, the price of Ti products declined, directly leading to a decline in Ti production in the second half of 2012. The economic status of the Ti industry fluctuated significantly from 2016 to 2017. In 2017, with the improvement of the TiO2 market, the pressure for environmental protection in the production of raw Ti materials in various places increased. The shortage of Ti concentrate resources led to a continuous rise in domestic raw Ti material prices, which subsequently impacted the prices of sponge Ti and Ti materials, causing significant fluctuations and reaching a bottoming point.

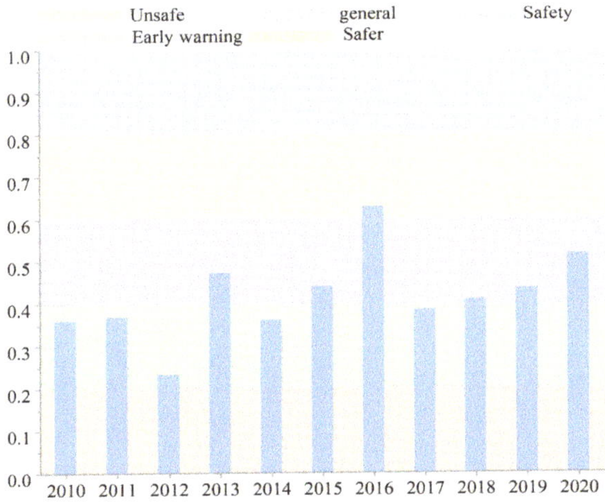


**Figure 5.** Safety and economic level of China's titanium industry chain security from 2010 to 2020.

Sustainability level. From 2010 to 2020, the safety sustainability level of the Ti industry in China showed an initial declining trend, followed by a rise. This is mainly due to the development of high-end industries, as illustrated in Figure 6. In 2014, under the guidance of the country's policy of stabilizing growth, adjusting structure, transforming growth modes, scientific and technological innovation and development, harmonious development with the environment, and promoting healthy and sustainable development of the industry, the main themes of economic development were as follows: capacity expansion, industrial structure optimization, and gradual improvement in the sustainable development of the Ti industry. In 2020, China's Ti processing industry, through structural adjustment, transformation, and upgrading in recent years, formed a company with Baoti Group Co., Ltd. (Baoji, China) and Western Superconducting Materials Technology Co., Ltd. Co., Ltd. (Xi'an, China) Hunan Xiangtou Jintian Titanium Metal Co., Ltd. (Changsha, China) Western Metal Materials Technology Co., Ltd. (Xi'an, China) and other leading enterprises, such as large state-owned enterprises. As such, sustainable development has reached a relatively safe level.

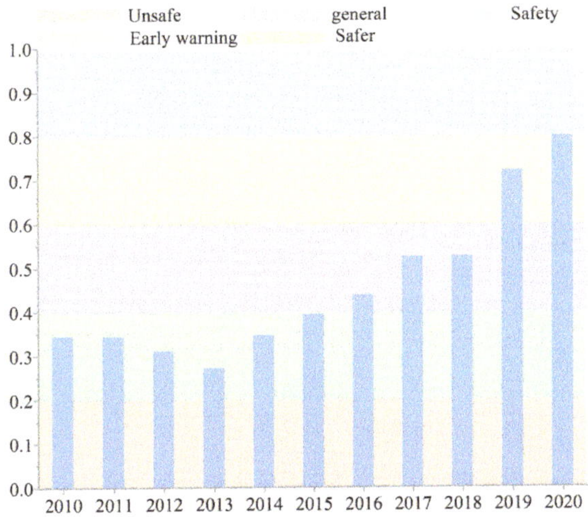


**Figure 6.** Safety sustainability level of China's titanium industry chain security from 2010 to 2020.

Overall, from 2010 to 2020, the safety of the availability level in the Ti industry has shown a downward trend; the economic level fluctuated less, and the sustainable safety level fluctuated significantly. Since 2013, the sustainability level has improved annually. Overall, the area of the quadrilateral enclosed by each dimension layer continued to increase, particularly the sustainable safety level, as illustrated in Figure 7.

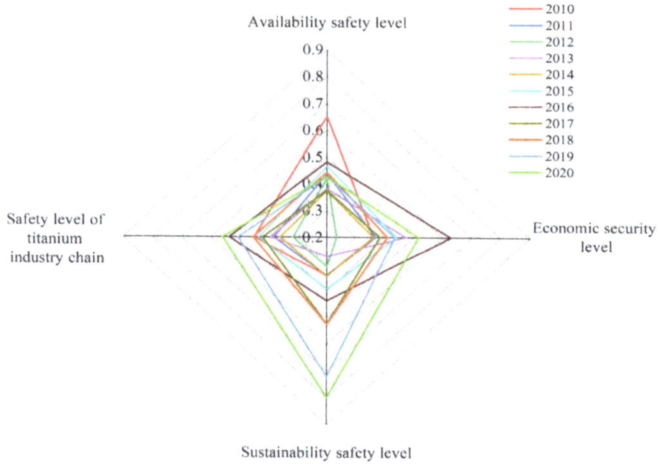


**Figure 7.** Radar chart of China's titanium industry chain security from 2010 to 2020.

4.2.3. Sensitivity Analysis

Sensitivity analysis is an analytical method that quantitatively describes the degree of influence of the model input variables on the output variables. A sensitivity analysis of the safety evaluation index system of the Ti industry chain helps identify the direction and degree of influence of different combinations of factors on the safety level of the Ti industry chain. Based on the security level of the Ti industry chain in China and the weight of each indicator from 2010 to 2020, assuming that the weight of each indicator did not change, it is classified according to the security dimension of the Ti industry chain.

Based on 2020, the combination of influencing factors was adjusted at the rate of change of −15, −10, −5, 5, 10, and 15%, and sensitivity analysis of the security dimension of the Ti industry chain was performed.

In general, the sensitivity of the Ti industry chain to changes in security availability is higher than that of economics and sustainability, as shown in Figure 8. For usability indicators, the sensitivity of positive changes in the security of the Ti industry chain is significantly higher than that of negative changes. In other words, compared with the negative impact of reducing the security and availability of the Ti industry chain, more attention should be paid to positive reinforcement. At a change level of 15%, the positive sensitivity was approximately nine times the negative sensitivity, and a small increase in the safety and availability in the Ti industry chain would have a greater positive impact on the safety level of the entire Ti industry chain.

For sustainability indicators, the sensitivity to negative changes in the safety level of the Ti industrial chain was significantly higher than that to positive changes. That is, compared to the positive strengthening of the safety and sustainability of the Ti industry chain, it is necessary to pay more attention to the negative factors of sustainability.

For the economic indicators, the positive and negative sensitivity changes in the safety level of the Ti industrial chain were small, and the sensitivity level was relatively stable.

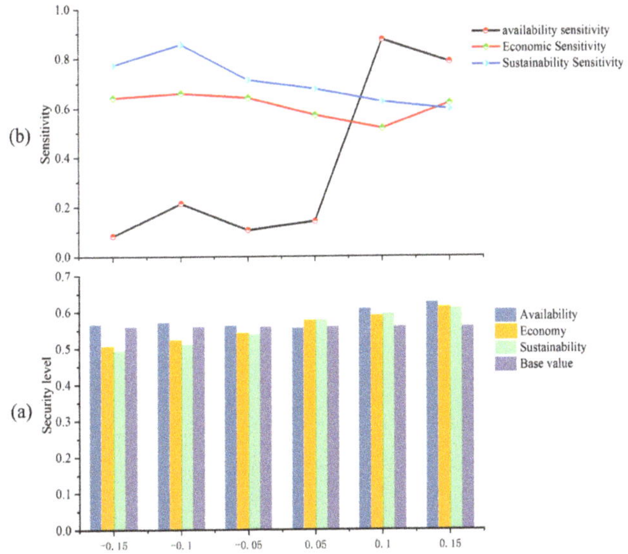


**Figure 8.** Changes in the level of security (**a**) and sensitivity analysis of dimensions (**b**).

*4.3. Correlation Analysis of the Ti Industry Chain and Sensitivity Analysis*

According to the grey correlation analysis, the indicators related to the changing trend of the safety level of China's Ti industry chain greater than 0.8 include availability indicators (titanium dioxide ($TiO_2$) self-sufficiency rate, $TiO_2$ external dependence), economic indicators (defense and military expenditure, GDP, $TiO_2$ trade deficit, Ti product net profit margin), and sustainability indicators (CR3, CR5, $TiO_2$ production capacity, Ti ingot production capacity, high-end industry share, Ti product gross margin, $TiO_2$ production capacity utilization rate, sponge Ti capacity utilization rate, China Innovation Index), as shown in Figure 9. Based on the 2020 data, assuming that the weight of each indicator remains unchanged, sensitivity analysis was performed for indicators with a correlation greater than 0.8 at −15, −10, −5, 10, and 15%, as shown in Figure 10.

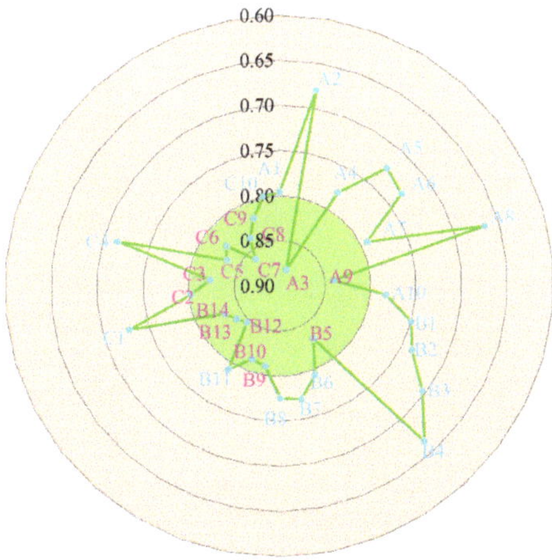

**Figure 9.** Correlation diagram between the safety level of the titanium industry chain and various influencing factors.

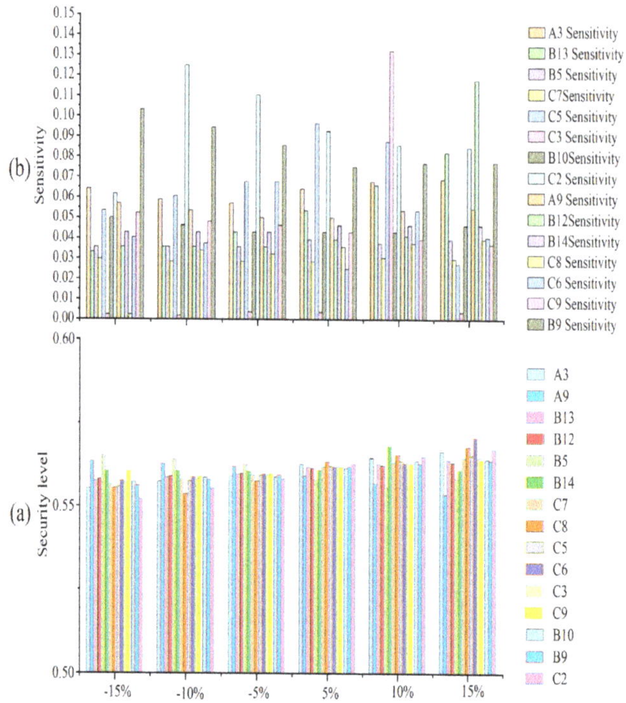


**Figure 10.** The changes in indicators arouse changes in the level of security of the titanium industry chain (**a**) and sensitivity analysis of indicators (**b**).

Overall, under the positive and negative 15, 10, and 5% variables, the change in the degree of correlation (greater than 0.8) made the change in the safety index of the Ti industry chain small. Among them, the indicators with high safety sensitivity in the Ti industry chain are CR5 and the China Innovation Index.

First, the self-sufficiency rate of $TiO_2$ in the $\pm$15, 10, and 5% variables was caused by a change in the range of the Ti industry chain safety index from 0.0054 to 0.0058. The positive and negative sensitivity changes in the safety level of the Ti industry chain are small, and the sensitivity to positive changes is higher than that to negative changes; that is, instead of considering the negative impact of the decline in the self-sufficiency rate of Ti dioxide, the self-sufficiency rate of $TiO_2$ should be appropriately improved.

The change in the external dependence of $TiO_2$ under the variables of plus or minus 15, 10, and 5% caused changes in the safety index of the Ti industry chain between $-0.0069$ and 0.0028.

The positive and negative sense of the safety level of the Ti industry chain changes significantly, and under the 15% change, the positive sensitivity is approximately 2.5 times the negative sensitivity. In other words, while considering the positive impact of reducing the external dependence on $TiO_2$, the negative impact of excessive dependence on imports of $TiO_2$ should be strengthened.

Second, the changes in the safety index of the Ti industry chain caused by investment in national defense and GDP in the positive and negative 15, 10, and 5% variables were between $-0.003$ and 0.0033 and between $-0.0025$ and 0.0021, respectively. Relative to the GDP and defense expenditures, the positive and negative sensitivity changes in the safety level of the Ti industry chain are small, and the sensitivity level is relatively stable. The trade deficit in $TiO_2$ under the variables of plus or minus 15, 10, and 5% caused a change in the safety index of the Ti industry chain between $-0.0049$ and 0.0045. The positive and negative sense of the safety level of the Ti industry chain changes significantly when the amount of change is 5%, and the positive and negative sense of 10% is higher than the positive and negative sense of the change rate of 15%. In other words, it is crucial to carefully consider the effects of a 15% change in the trade deficit of $TiO_2$. Additionally, when examining the positive sense, a 5% increase should be regarded as approximately 1.5 times more impactful than a 10% decrease in the trade deficit. In essence, when considering the positive impact of the trade deficit of Ti dioxide, the negative impact of the increase in the trade deficit of $TiO_2$ should be considered.

The net profit margin of Ti products in the positive and negative 15, 10, and 5% variables caused the Ti industry chain safety index to change between −0.0001 and 0.0074. The safety level of the Ti industry chain changes significantly in the positive sensitivity under the 10% variable, and the positive sensitivity is higher than the negative sensitivity; in other words, the small increase in the net profit margin of Ti products under this variable has a relatively positive impact on the safety of the Ti industry chain.

Third, the industry concentrations CR3 and CR5 caused changes in the safety index of the Ti industry chain under positive and negative 15, 10, and 5% variables, between −0.0042 and 0.0039 and between −0.0077 and 0.0067, respectively. The change in CR3 in the industry concentration causes small positive and negative sensitivity changes in the safety level of the Ti industry chain, and the sensitivity level is relatively stable. The change in CR5 causes a large change in the positive and negative sense of the safety level of the Ti industry chain, while the sensitivity of the negative change in CR5 is higher than that of the positive change under the variables of 5 and 10%. In other words, compared with the improvement of CR5, more attention should be paid to the negative impact of CR5 decline.

Fourth, the production capacity of $TiO_2$ and Ti ingots caused by changes in the safety index of the Ti industry chain under the variables plus and minus 15, 10, and 5% are between −0.0048 and 0.0046, and −0.003 and 0.0099, respectively.

Relative to $TiO_2$ production capacity, the safety level of the Ti industry chain is equal in the positive and negative directions under the variables of 5 and 10%, that is, when the Ti ingot production capacity is reduced or increased by 10 or 5%, the impact of positive and negative changes equally considered. Under the 15% variable, the positive sensitivity was less than the negative sensitivity, and more consideration should be given to the impact of fewer negative changes on the safety of the Ti industry chain under this variable.

Relative to the Ti ingot capacity, the sensitivity under negative variables remained unchanged. The positive change under the 15% variable is approximately 3.3 times the negative sensitivity; that is, the small increase in Ti ingot production capacity at this time will have a significant positive impact on the safety of the Ti industry chain.

The proportion of high-end industries under the 15, 10, and 5% variables caused the Ti industry chain safety index to vary between −0.0036 and 0.0039.

The positive and negative sensitivity changes in the safety level of the Ti industry chain are small, and the sensitivity to positive changes is higher than negative changes and remains unchanged. In other words, within the scope of this study, the proportion of high-end industries should be increased.

The gross profit margin of Ti products changes in the safety index of the Ti industry chain under the variables of plus or minus 15, 10, and 5% and is between −0.0019 and 0.0033. For the gross profit margin of Ti products, the positive and negative sense of the safety level of the Ti industry chain changes greatly, and under the 15% variable, the positive sensitivity is approximately 17 times the negative sensitivity; that is, at the level of this change, a small increase in the gross profit margin of Ti products has a higher positive impact on the safety level of the Ti industry chain.

The capacity utilization rate of $TiO_2$ and the capacity utilization rate of sponge Ti under the variables of plus or minus 15, 10, and 5% caused the changes in the safety index of the Ti industry chain between −0.0034 and 0.0036 and between −0.0044 and 0.0031, respectively. Relative to the capacity utilization rate of Ti dioxide, the positive and negative sense of the safety level of the Ti industry chain changes greatly. Under the 5% variable, the negative sense of the safety level of the Ti industry chain is approximately 2.7 times the positive sensitivity; that is, compared with the improvement in the capacity utilization rate of Ti dioxide, more attention should be paid to the negative impact of the decline in the capacity utilization rate of Ti dioxide.

Relative to the capacity utilization rate of sponge Ti, the negative sensitivity of the safety level of the Ti industry chain is higher than the positive sensitivity; that is, at the same level of change, more attention should be paid to the negative impact of the decline in the capacity utilization rate of sponge Ti.

The China Innovation Index causes a change in the safety index of the Ti industry chain between −0.0087 and 0.0065 under positive and negative 15, 10, and 5% variables, respectively. The negative sense of the safety level of the Ti industry chain was higher than its positive sensitivity; that is, more attention should be paid to improving the innovation index in the Ti industry under the same level of change.

## 5. Conclusions and Policy Suggestions

### 5.1. Conclusions

By constructing a comprehensive evaluation system for the Ti industry, security was established, including availability, economics, and sustainability. The entropy weight-TOPSIS method was used to analyze the safety level of the Ti industry chain from 2010 to 2020, and the coupling degree and sensitivity of the Ti industry and each dimension layer were analyzed by coupling degree analysis and sensitivity. The grey correlation analysis method was used to analyze the factors with the highest correlation with the Ti industry chain, and the sensitivity analysis of the factors with high correlation in the Ti industry chain was analyzed.

The results show that during the evaluation period, the overall safety level of China's Ti industry chain first declines and then rises, which is highly similar to the development status of China's Ti industry, showing that the model we constructed has certain feasibility and can help China's Ti industry structure and other aspects of policy formulation. According to the dimension layer, the safety availability level of the Ti industry chain was relatively weak and showed a downward trend. The coupling of the economic and safety levels is low in the years when the safety level of the Ti industry was low. According to the sensitivity analysis, the sensitivity of the Ti industry chain to changes in safety availability is higher than that of economy and sustainability. At a change level of 15%, the positive sensitivity is approximately nine times the negative sensitivity. That is, the safety availability of the Ti industry chain under the 15% variable is slightly improved, which has a greater positive impact on the safety level of the entire Ti industry chain.

According to the results of the gray correlation analysis, indicators with a correlation greater than 0.8 comprise those of availability ($TiO_2$ self-sufficiency rate, $TiO_2$ external dependence), economic (defense and military expenditure, GDP, $TiO_2$ trade deficit, TiO2 products), net profit margin, and sustainability (CR3, CR5, $TiO_2$ production capacity, Ti ingot production capacity, high-end industry ratio, $TiO_2$ product gross margin, $TiO_2$ capacity utilization rate, Ti sponge capacity utilization rate, and China Economic Innovation Index). Overall, under the variables of plus or minus 15, 10, and 5%, the change in a single indicator with a correlation degree greater than 0.8 causes a small change in the safety index of the Ti industry chain, and the indicators with higher sensitivity to the entire Ti industry chain are CR5 and the China Innovation Index.

Among them, under the change in CR5 at 5% and 10%, the sensitivity to negative change is greater than that to positive change, which means that it is necessary to pay more attention to the negative rather than the positive impact of the decline in industry concentration. Compared with the China Innovation Index, the negative sensitivity of the safety level of the Ti industry chain is higher than the positive sensitivity; that is, more attention should be paid to improving the innovation index in the Ti industry under the same level of change.

The advantage of this study is that the 34 indicators can reflect the safety level of the Ti industry, and a set of evaluation systems from the whole to the dimension of the indicators was constructed; however, the disadvantage is that the influence of non-directional factors at different stages is not considered. It is suggested that readers can evaluate the security of the Ti industry supply chain based on the security of the supply of high-end Ti products and Ti resources and different forms of changes to provide more opinions on the supply of Ti resources in China. The framework of this study is suitable for evaluating the safety level of the Ti industry, the impact of certain factors on the development of the Ti industry, and providing reference opinions on the formulation of Ti industry policies.

### 5.2. Polic Suggestions

As the safety availability level of the Ti industry chain was relatively weak and showed a downward trend, the economic level exerted a greater influence on the development of the Ti industry. The indicators with higher sensitivity to the entire Ti industry chain are CR5, the China Innovation Index, and the developmental status of the Ti industry. As a result, several policy recommendations have been proposed, including promoting the recycling of Ti resources.

To further improve safety in the Ti industry chain. The following policy recommendations are proposed based on an analysis of the industry chain in China's academic circles and the current situation of the Ti industry in China. Upstream in the Ti industry, it is necessary to accelerate research, improve the utilization rate and utilization level of Ti resources, and enhance the quality of raw Ti materials. The concentration of enterprise production has improved in the middle reaches of the Ti industry. A complete consumption structure exists downstream of the Ti industry.

The details are as follows:

(1) First, we consider the existing patented technologies in the Ti industry chain. Some developed countries have monopoly advantages over the Ti industry and cannot be surpassed in a short period, such as the preparation of high-end Ti materials and recycling of Ti waste. In the processing field, the Ti industry in China should adhere to bottom-line thinking, such as the production of high-end Ti materials, the production technology of high-grade sponge Ti, and other fields to focus on tackling key problems to maintain adequate stability and safety in the face of any form.

(2) The availability of Ti resources is important for consolidating the security of the Ti industry. Based on the distribution characteristics of Ti resources in China and related policies, the first task is to promote market-oriented reform in the upstream areas of the Ti industry chain, stimulate the vitality of upstream exploration and mining enterprises of Ti resources, and increase the strength of exploration and mining development through the reform of upstream mining rights, and opening and exit mechanisms to improve the self-sufficiency guarantee level of Ti resources in China. Second, the import channels of Ti resources must be broadened, and the impact of Ti raw materials, Ti product import concentrations, and geopolitical risks on the security of the Ti industry in China must be reduced. Simultaneously, research and development of Ti waste recycling technology is needed to improve the recycling rate of Ti waste in China.

(3) To adhere to long- and short-term safety relationships of Ti resources, from the perspective of import and export trades, China mainly exports low-grade Ti at low prices. Most of the total trade balance has been reduced in quantity, thereby reducing the availability of Ti resources in China. Therefore, the exploitation of Ti resources in China should be reasonably conducted such that the lean ore is mined by force and the rich ore is mined by fines. According to the distribution characteristics of Ti resources in China, the cost of domestic Ti resources, international Ti resource prices, geopolitical reforms, and other factors are comprehensively considered to achieve the best balance between self-sufficiency and import supplementation of Ti raw materials and products. For example, when the price of international Ti resources is reason-

able, the import of Ti resources should increase, the mineable life of Ti resources in China should increase, and the future security level of Ti resources in China should be improved.

(4) By carrying forward alliance thinking, attaching importance to industry alliances, and strengthening industry concentration, currently, the evolution of the industry chain is progressing toward a decentralized high-end industry. Enterprises on the same information platform can form strategic alliances to achieve complementary advantages, such as process docking, resource sharing, and cultural integration.

**Author Contributions:** Conceptualization, X.H. and W.W.; methodology, X.H.; software, X.H.; validation, X.H., W.W. and M.Q.; formal analysis, X.H.; investigation, W.W.; data curation, X.H.; writing—original draft preparation, X.H.; writing—review and editing, W.W.; project administration, C.Z. and S.Y.; funding acquisition, W.W. All authors have read and agreed to the published version of the manuscript.

**Funding:** This research was funded by the Science and Technology Foundation of Guizhou Province [grant numbers QKHJC-ZK 2021-YB 261 and QKHJC 2019-1406], the National Natural Science Foundation of China [grant number 51874108], the Talent Projects of Guizhou University and the Education Department of Guizhou Province [grant number GDPY 2019-20, and QJHKYZ 2021-097].

**Data Availability Statement:** Data are contained within the article.

**Acknowledgments:** The author is grateful to senior engineers Qiang Liang and Jinze Li of Zunyi Titanium Co., Ltd. for their valuable discussions and advice.

**Conflicts of Interest:** The authors declare no conflict of interest.

## References

1. Zhang, L.; Wang, X.D.; Qu, X.H. Application Status and Market Analysis of Non-Aero Titanium in China. *Key Eng. Mater.* **2012**, *520*, 8–14. [CrossRef]
2. Santhosh, R.; Geetha, M.; Nageswara Rao, M. Recent Developments in Heat Treatment of Beta Titanium Alloys for Aerospace Applications. *Trans. Indian Inst. Met.* **2017**, *70*, 1681–1688. [CrossRef]
3. Ares, G.G.; Mandal, P.; Diego, G.; Nicola, Z.; Paul, B. Studies on Titanium Alloys for Aerospace Applications. *Defect Diffus. Forum* **2018**, *385*, 419–423. [CrossRef]
4. Akinribide, O.; Obadele, B.; Akinwamide, S.; Bilal, H.; Ajibola, O.; Ayeleru, O.; Ringer, S.; Olubambi, P. Sintering of binderless TiN and TiCN-based cermet for toughness applications: Processing techniques and mechanical properties: A review. *Ceram. Int.* **2019**, *45*, 21077–21090. [CrossRef]
5. Froes, F.; Qian, M.; Niinomi, M. *Titanium for Consumer Applications*; Elsevier: Amsterdam, The Netherlands, 2019; pp. 77–90.
6. Mekgwe, G.N.; Akinribide, O.J.; Akinwamide, S.O.; Olubambi, P.A. Fabrication of graphite reinforced TiCxNy by spark plasma sintering technique: A comparative assessment of microstructural integrity and nanoindentation properties. *Vacuum* **2021**, *187*, 110144. [CrossRef]
7. You, S.H.; Lee, J.H.; Oh, S.H. A Study on Cutting Characteristics in Turning Operations of Titanium Alloy used in Automobile. *Int. J. Precis. Eng. Man.* **2019**, *20*, 209–216. [CrossRef]
8. Schulz, K.J.; Deyoung, J.H.; Seal, R.R.; Bradley, D.C. Critical mineral resources of the United States—Economic and environmental geology and prospects for future supply. *Geol. Surv.* **2017**, *1802*, 797. [CrossRef]
9. Chong, Y.; Poschmann, M.; Zhang, R.; Zhao, S.; Hooshmand, M.S.; Rothchild, E.; Olmsted, D.L.; Morris, J.W.; Chrzan, D.C.; Asta, M.; et al. Mechanistic basis of oxygen sensitivity in titanium. *Sci. Adv.* **2020**, *6*, eabc4060. [CrossRef]
10. Zhang, T. Analyzing titanium patents at home and abroad. *Iron Steel Vanadium Titan.* **2017**, *38*, 158–164.
11. Wu, X.; Zhang, J. Geographical Distribution and Characteristics of Titanium Resources in China. *Titan. Ind. Prog.* **2006**, *6*, 8–12.
12. Li, Q.L.; Li, S.G.; Zheng, Y.F.; Li, H.M.; Massonne, H.J.; Wang, Q.C. A high precision U–Pb age of metamorphic rutile in coesite-bearing eclogite from the Dabie Mountains in central China: A new constraint on the cooling history. *Chem. Geol.* **2003**, *200*, 255–265. [CrossRef]
13. Wang, X.D.; Lu, F.S.; Jia, H.; Hao, B. Report on China Titanium Industry Progress in 2010. *Titan. Ind. Prog.* **2011**, *28*, 1–6. [CrossRef]
14. Qiu, G.Z.; Guo, Y.F. Current situation and development trend of the titanium metal industry in China. *Int. J. Miner. Metall. Mater.* **2022**, *29*, 599–610. [CrossRef]
15. Wang, J.; Huang, Y. Exploration of Cooperation Mode of Vanadium and Titanium Enterprises in Panzhihua and Xichang and Applied Undergraduate Colleges. *J. Chengdu Text. Coll.* **2016**, *2*, 242–244.
16. Mcmichael, P.; Gereffi, G.; Korzeniewicz, M. Commodity chains, and global capitalism. *Contemp. Sociol.* **1995**, *24*, 348. [CrossRef]
17. Constantin, B.; Tobias, S.; Daniel, R. Antecedents and enablers of supply chain agility and its effect on performance: A dynamic capabilities perspective. *Int. J. Prod. Res.* **2013**, *51*, 1295–1318. [CrossRef]

18. Priem, R.L.; Swink, M. A Demand-side Perspective on Supply Chain Management. *J. Supply Chain. Manag.* **2012**, *48*, 7–13. [CrossRef]
19. Stabell, C.B.; Fjeldstad, Ø.D. Configuring value for competitive advantage: On chains, shops, and networks. *Strateg. Manag. J.* **1998**, *19*, 413–437. [CrossRef]
20. Liu, M.; Li, H.; Zhou, J.; Feng, S.; Wang, Y.; Wang, X. Analysis of material flow among multiple phases of cobalt industrial chain based on a complex network. *Resour. Policy* **2022**, *77*, 102691. [CrossRef]
21. Li, Y.L.; Huang, J.B.; Zhang, H.W. The impact of country risks on cobalt trade patterns from the perspective of the industrial chain. *Resour. Policy* **2022**, *77*, 102641. [CrossRef]
22. Li, B.H.; Li, H.J.; Dong, Z.L.; Lu, Y.; Liu, N.R.; Hao, X.Q. The global copper material trade network and risk evaluation: A industry chain perspective. *Resour. Policy* **2021**, *74*, 102275. [CrossRef]
23. Huang, J.B.; Ding, Q.; Wang, Y.; Hong, H.J.; Zhang, H.W. The evolution and influencing factors of international tungsten competition from the industrial chain perspective. *Resour. Policy* **2021**, *73*, 102185. [CrossRef]
24. Lin, J.; Li, X.; Wang, M.X.; Liu, L.T.; Dai, T. How Can China's Indium Resources Have a Sustainable Future? Research Based on the Industry Chain Perspective. *Sustainability* **2021**, *13*, 12042. [CrossRef]
25. Svanidze, E.; Besara, T.; Ozaydin, M.F.; Tiwary, C.S.; Wang, J.K.; Radhakrishnan, S.; Mani, S.; Xin, Y.; Han, K.; Liang, H.; et al. High hardness in the biocompatible intermetallic compound $\beta$-$Ti_3Au$. *Sci. Adv.* **2016**, *2*, e1600319. [CrossRef]
26. Shon, H.K.; Phuntsho, S.; Vigneswaran, S.; Kandasamy, J.; Nghiem, L.D.; Kim, G.J.; Kim, J.B.; Kim, J. Preparation of titanium dioxide nanoparticles from electrocoagulation sludge using sacrificial titanium electrodes. *Environ. Sci. Technol.* **2010**, *44*, 5553–5557. [CrossRef] [PubMed]
27. Sun, J.F.; Li, A.S.; Su, F.H. The excellent lubricating ability of functionalization graphene dispersed in perfluoropolyether for titanium alloy. *ACS Appl. Nano Mater.* **2019**, *2*, 1391–1401. [CrossRef]
28. Li, X.; Liu, J.; Zhang, D.; Xiong, Z.H.; He, X.Q.; Yuan, M.; Wang, M.X. Material flow analysis of titanium dioxide and sustainable policy suggestion in China. *Resour. Policy* **2020**, *67*, 101685. [CrossRef]
29. Li, M.H.; Geng, P.Y.; Liu, G.; Gao, Z.Y.; Rui, X.; Xiao, S.J. Uncovering spatiotemporal evolution of titanium in China: A dynamic material flow analysis. *Resour. Conserv. Recycl.* **2022**, *180*, 106166. [CrossRef]
30. Song, M.; Tao, W. Research on the evaluation of China's regional energy security and influencing factors. *Energy Sources Part B Econ. Plan. Policy* **2022**, *17*, 1993383. [CrossRef]
31. Ma, L.; Liu, S.; Ai, L.; Sun, Y. Research on an Evaluation of the Work Suitability in the New First-Tier Cities. In *International Conference on Harmony Search Algorithm*; Springer: Cham, Switzerland, 2020; Volume 1063, pp. 175–185. [CrossRef]
32. Zhang, S.L.; Liu, L.L.; Ran, J.X. Evaluation and Analysis of Competitiveness of Pharmaceutical Industry in China—An Empirical Study Based on Provincial Data From the Mainland. *J. Hangzhou Dianzi Univ.* **2011**, *1*, 1–5. [CrossRef]
33. Hu, H.Q.; Ma, Y.; Wu, S.J. Fuzzy comprehensive evaluation on high-quality development of China's rural economy based on entropy weight. *J. Intell. Fuzzy Syst.* **2020**, *38*, 7531–7539. [CrossRef]
34. Ding, Q.Y.; Wang, Y.M. Intuitionistic Fuzzy TOPSIS Multi-attribute Decision Making Method Based on Revised Scoring Function and Entropy Weight Method. *J. Intell. Fuzzy Syst.* **2019**, *36*, 625–63510. [CrossRef]
35. Wang, L.-J.; Shao, X.-L. Study on the safety production evaluation of the coal mine based on entropy-TOPSIS. *J. Coal Sci. Eng.* **2010**, *16*, 284–287. [CrossRef]
36. Lu, J.P.; Wei, C.; Wu, J.; Wei, G.W. TOPSIS method for probabilistic linguistic MAGDM with entropy weight and its application to supplier selection of new agricultural machinery products. *Entropy Switz.* **2019**, *21*, 953. [CrossRef]
37. Naeem, M.; Khan, M.A.; Abdullah, S.; Qiyas, M.; Khan, S. Extended TOPSIS method based on the entropy measure and probabilistic hesitant fuzzy information and their application in decision support system. *J. Intell. Fuzzy Syst.* **2021**, *40*, 11479–11490. [CrossRef]
38. Niu, D.X.; Li, S.; Dai, S.Y. A comprehensive evaluation of the operating efficiency of electricity retail companies based on the improved TOPSIS method and LSSVM optimized by modified ant colony algorithm from the view of sustainable development. *Sustainability* **2018**, *10*, 860. [CrossRef]
39. Tang, J.; Zhu, H.L.; Liu, Z.; Jia, F.; Zheng, X.X. Urban Sustainability Evaluation under the Modified TOPSIS Based on Grey Relational Analysis. *Int. J. Env. Res. Pub. He.* **2019**, *16*, 256. [CrossRef]
40. Chang, J.X. Study on Evaluation System of Energy Security and Policy in China. Ph.D. Thesis, China University of Geosciences, Beijing, China, 2010.
41. Wu, C.G.; He, S.J.; Sheng, C.M.; Liu, Z.J.; Wan, H. Discussion on Comprehensive Evaluation Method of Energy Security. *J. Nat. Resour.* **2011**, *26*, 964–970.
42. Yang, B.; Ding, L.J.; Zhan, X.Y.; Tao, X.Z.; Peng, F. Evaluation and analysis of energy security in China based on the DPSIR model. *Energy Rep.* **2022**, *8*, 607–615. [CrossRef]
43. Dong, F.; Li, W. Research on the coupling coordination degree of "upstream-midstream-downstream" of China's wind power industry chain. *J. Clean. Prod.* **2021**, *283*, 124633. [CrossRef]

processes

Article

# Behavior of Carbothermal Dephosphorization of Phosphorus-Containing Converter Slag and Its Resource Utilization

**Shuai Tong, Chenxiao Li *, Liqun Ai, Shuhuan Wang and Shuai Zhang**

Metallurgy and Energy College, North China University of Science and Technology, Tangshan 063210, China; ts961109@163.com (S.T.)
* Correspondence: lichenxiao34@163.com; Tel.: +86-188-1159-0603

**Abstract:** Phosphorus-containing converter slag is a common waste in the iron and steel industry, and has the characteristics of high generation and low secondary-utilization values; however, the high-phosphorus content in converter slag limits its ability to be recycled during the steelmaking process. In this study, the dephosphorization behavior of converter slag by carbothermal reduction was studied through experiments and thermodynamic calculations. The results showed that the gas product of the converter slag produced by carbothermal reduction was mainly $P_2$, and that part of $P_2$ entered the iron phase to generate iron phosphate compounds. With the increase in Fe content, the amount of $P_2$ also increased, which may provide a suitable new direction for the production of ferrophosphorus. Based on the carbothermal reduction theory, a new "circulating steelmaking process of converter steel slag gasification" was proposed and applied to Chengde Iron and Steel Group Co., Ltd. (Chengde, China). The industrial production practice showed that the process did not affect the dephosphorization effect of the next furnace, and that the average iron, steel, and slagging-material consumption per furnace was reduced by 4.74 kg and 608 kg, respectively. $CO_2$ emissions were reduced by 4.86 kg, thus achieving the goals of energy saving, environmental protection, and efficient dephosphorization.

**Keywords:** converter slag; dephosphorization; carbothermal reduction; $CO_2$ emission; ferrophosphorus

**Citation:** Tong, S.; Li, C.; Ai, L.; Wang, S.; Zhang, S. Behavior of Carbothermal Dephosphorization of Phosphorus-Containing Converter Slag and Its Resource Utilization. *Processes* **2023**, *11*, 1943. https://doi.org/10.3390/pr11071943

Academic Editor: Blaž Likozar

Received: 8 June 2023
Revised: 20 June 2023
Accepted: 22 June 2023
Published: 27 June 2023



## 1. Introduction

The iron and steel industry is a high-consumption and high-emission industry, in which the main emissions are gas wastes (such as $CO_2$, $SO_2$, and $NO_x$) and solid wastes (such as converter slag, granulated slag, dust and sludge) [1]. At present, the output of solid waste in China's iron and steel industry accounts for about 15% of the national industrial solid waste output, and the accumulated storage of converter slag in China is nearly one billion tons [2]. The utilization rate of converter slag is less than 30% because of its high alkalinity, high free calcium oxide content and its many metastable phases. A large amount of converter slag is left idle, which not only consumes land resources, but also pollutes the environment and harms human health [3]. Therefore, how to realize the recycling of converter slag in the furnace has become an important research direction. Bao et al. [4] studied the leaching law of phosphorus in different acids from high-phosphorus converter slag using a selective leaching process. Wang et al. [5] studied the kinetics of microwave carbothermal dephosphorization of converter slag, indicating that the carbothermal reduction reaction is a second-order reaction, and that the limiting link may be an interfacial chemical reaction. Zhang et al. [6], based on a micro-sintering experiment, analyzed the influence of alkalinity, temperature, and carbon content on the dephosphorization rate of high-phosphorus steel-slag gasification. Ai et al. [7] studied the carbothermal reduction behavior of converter steel slag via microwave heating and found that the maximum reduction dephosphorization rate of steel slag can reach 91.5% at low temperatures. Zhao's research indicated that when reducing converter slag using a carbothermic method, coke powder was used as both a reducing agent and heating agent, and that the exothermic reaction between coke powder and oxygen could ensure the dephosphorization reaction at a higher temperature [8].

China's crude steel output is huge, and high-quality iron ore is constantly consumed, which has led to greater interest in high-phosphorus iron ore with large reserves. However, the phosphorus content in high-phosphorus iron ore ranges from 0.3% to 1.8%, and the ore structure is compact, which will inevitably increase the difficulty in treating high-phosphorus converter slag during smelting. If the phosphorus in high-phosphorus converter slag can be separated and utilized, and the remaining residue used for internal circulation in the steelmaking process, it will reduce environmental pressures and smelting costs [9,10].

At present, there are many basic research studies on the carbothermal reduction process and reaction thermodynamic conditions of converter slag, that have mainly been conducted in the laboratory. However, there are few reports on the dephosphorization behavior of converter slag by carbothermal reduction and the actual smelting effect of furnace recovery. This experiment mainly studied the dephosphorization behavior of converter slag by carbothermal reduction, which provided a new idea for the resource utilization of phosphorus-containing converter slag. The recovery and utilization of phosphorus resources after carbothermal reduction of phosphorus-containing converter slag were investigated.

Based on the carbothermal reduction theory, our team put forward the process of "circulating steelmaking by gasification dephosphorization of converter slag", which was applied by the Chengde Iron and Steel Group Co., Ltd. (Cheng steel) and achieved considerable economic indicators. Gasification dephosphorization is a novel method that could make it easier to reuse convert slag. In this process, carbon is added to the converter before slag splashing. The slag and carbon are then fully mixed and reacted under dynamic conditions during the slag-splashing process. The reduction products of $P_2O_5$ in converter slag are removed from the slag as gas, which makes it possible to reuse the treated slag in the next heat. This process provides a better direction for the comprehensive utilization of converter slag from carbothermal reduction.

## 2. Materials and Methods

### 2.1. Materials

The components of converter slag and coke are shown in Tables 1 and 2, respectively. The converter slag and coke were crushed to less than 0.063 mm by an electromagnetic crusher to obtain a larger specific surface area, and 100 g of converter slag was thoroughly mixed with coke powder with 2 carbon equivalent for each heat (100 g of converter slag for every 10.26 g of coke powder). The unit carbon equivalent was defined as the carbon content used for the total reduction of FeO, $P_2O_5$, and MnO in the slag.

**Table 1.** Chemical composition of converter slag (wt.%).

| FeO | CaO | $SiO_2$ | MgO | MnO | $P_2O_5$ | Basicity (R) |
|---|---|---|---|---|---|---|
| 15.82 | 42.85 | 12.55 | 7.15 | 3.35 | 3.08 | 3.41 |

**Table 2.** Chemical composition of coke (wt.%).

| Fixed Carbon | CaO | $SiO_2$ | MgO | $P_2O_5$ | S |
|---|---|---|---|---|---|
| 86.55 | 1.37 | 6.42 | 0.33 | 0.36 | 0.98 |

### 2.2. Methods

The converter-slag-dephosphorization experiment by carbothermal reduction was conducted in a high-temperature heating furnace (Figure 1) comprising a gas control system and temperature control system, and the temperature was measured with a dual platinum rhodium thermocouple. Before the experiment, the mixture was placed in an MgO crucible, covered with a graphite crucible, and placed in the heating furnace, which was heated to 1813 K and kept warm for 1 h. During the experiments, $N_2$ was circulated in the furnace with a flow rate of 0.7 L/min, and the gas flow rate was detected by a precise rotameter; cooling circulating water was continuously used. Finally, the slag was cooled to room temperature with the furnace temperature, and the chemical composition was determined by X-ray fluorescence (XRF). In addition, the component content of the slag was adjusted by analyzing pure reagents, and univariate adjustment was adopted. In this experiment, only the basicity and FeO content of the final slag were changed.

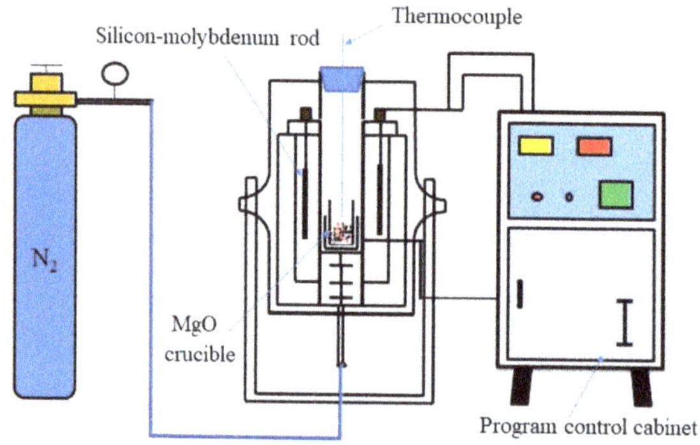


**Figure 1.** Carbon thermal reduction experimental apparatus.

The dephosphorization rate of the converter slag was calculated by comparing the mass fractions of $P_2O_5$ in the slag before and after the reaction, as shown in Equation (1). The chemical composition of the carbothermal reduction slag is listed in Table 3. It can be seen from Table 3 that the average dephosphorization rate under the experimental conditions was 32.57%.

$$\varphi = \frac{w0 - wt}{w0} \times 100\% \quad (1)$$

where $\varphi$ is the dephosphorization rate of converter slag, $w0$ is the initial $P_2O_5$ mass fraction of converter slag, and $wt$ is the $P_2O_5$ mass fraction of converter slag reduced by the coke.

**Table 3.** Chemical composition of carbothermal reduction slag (wt.%).

| Reduction Condition | TFe | CaO | $SiO_2$ | MgO | MnO | $P_2O_5$ | Dephosphorization Rate/% |
|---|---|---|---|---|---|---|---|
| Final slag | 19.69 | 44.22 | 16.32 | 8.30 | 3.09 | 2.22 | 27.92 |
| Final slag of R = 2.81 | 20.43 | 42.05 | 19.03 | 7.38 | 3.47 | 1.90 | 38.31 |
| Final slag of w(FeO) = 24% | 23.66 | 42.01 | 15.96 | 7.17 | 3.27 | 2.11 | 31.49 |

## 2.3. Characterization Methods

The compositions of the slag and coke were detected by X-ray fluorescence (XRF). The microscopic morphology of the slag was examined using a scanning electron microscope (SEM), and the element content was analyzed by energy-dispersive X-ray (EDS) spectroscopy attached to the SEM. The composition of molten iron was determined by inductively coupled plasma (ICP).

## 3. Experimental Results and Discussion

### 3.1. Thermodynamic Analysis

The P in converter slag mainly consists of of $P_2O_5$, which has been confirmed by most scholars. The gas products containing the P element formed from the reduction of pure $P_2O_5$ under standard conditions are given by Equations (2)–(5) [11–15].

$$\frac{1}{2}P_2O_{5(l)} + \frac{5}{2}C_{(s)} = P_{(g)} + \frac{5}{2}CO_{(g)},\ \Delta G^0 = 1099262 - 767.09T \tag{2}$$

$$P_2O_{5(l)} + 5C_{(s)} = P_{2(g)} + 5CO_{(g)},\ \Delta G^0 = 603609 - 648.94T \tag{3}$$

$$2P_2O_{5(l)} + 10C_{(s)} = P_{4(g)} + 10CO_{(g)},\ \Delta G^0 = 965073 - 1141.52T \tag{4}$$

$$\frac{1}{2}P_2O_{5(l)} + \frac{3}{2}C_{(s)} = PO_{(g)} + \frac{3}{2}CO_{(g)},\ \Delta G^0 = 613650 - 461.50T \tag{5}$$

Equations (2)–(5) show that four products, P, $P_2$, $P_4$, and PO are formed. According to the above Eqs, the relationships between the standard Gibbs free energy ($\Delta G^0$) and temperature were established, as shown in Figure 2a. It can be seen from Figure 2a that with the increase of temperature, the standard Gibbs free energy decreased. When the reaction was $\Delta G^0 = 0$, the corresponding temperatures of Equations (2)–(5) were 1433 K, 930 K, 845 K, and 1330 K, respectively. Thus, as long as the temperature was higher than 1433 K, the carbothermal reaction could occur, and also at 930 K, 845 K, and 1330 K, respectively, the $\Delta G^0 < 0$.

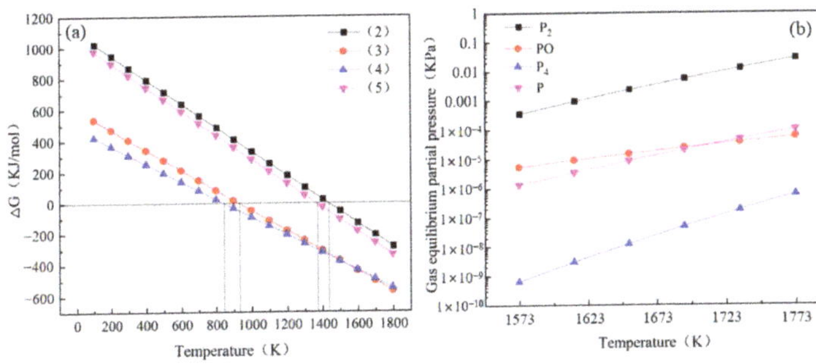


**Figure 2.** Thermodynamic analysis: (**a**) the relationship between Gibbs of carbothermal reduction and temperature; and (**b**) the relationship between temperature and partial pressure of gas products.

In order to further determine the dephosphorization products, the equilibrium partial pressure of the gas products of the carbothermal reaction was calculated by Factsage, as shown in Figure 2b. It can be seen from Figure 2b that, with the increase in temperature, the equilibrium partial pressure of the gas products increased, and it was found that the equilibrium partial pressure of $P_2$ gas was the largest. Therefore, it can be speculated that the main phosphorus gas product of the carbothermal reduction converter slag was $P_2$. This result was consistent with other scholars' research [16–18].

### 3.2. Dephosphorization Behavior

Because it was impossible to capture the location of P during carbothermal reduction at high temperatures, the Fe phase micro-area of the slag before and after the reduction was analyzed. Figure 3 shows the microstructure before and after the carbothermal reduction of the converter slag. The SEM showed that the Fe phase of the final slag consisted of small particles, while the Fe phase in the carbothermal reduction slag consisted of large-diameter particles or strips. The EDS indicated that the Fe phase in the carbothermal reduction slag contained P element, as shown in Figure 4. It can be seen from Figure 4 that the Fe content in the Fe phase micro-area of the final slag was between 10% and 50%, and the P content was between 0.4% and 0.8%, with few fluctuations. However, the content of P in the Fe phase micro-area of carbothermal reduction slag was between 0.15% and 7.5%, which was generally higher than that in the Fe phase micro-area of the final slag.

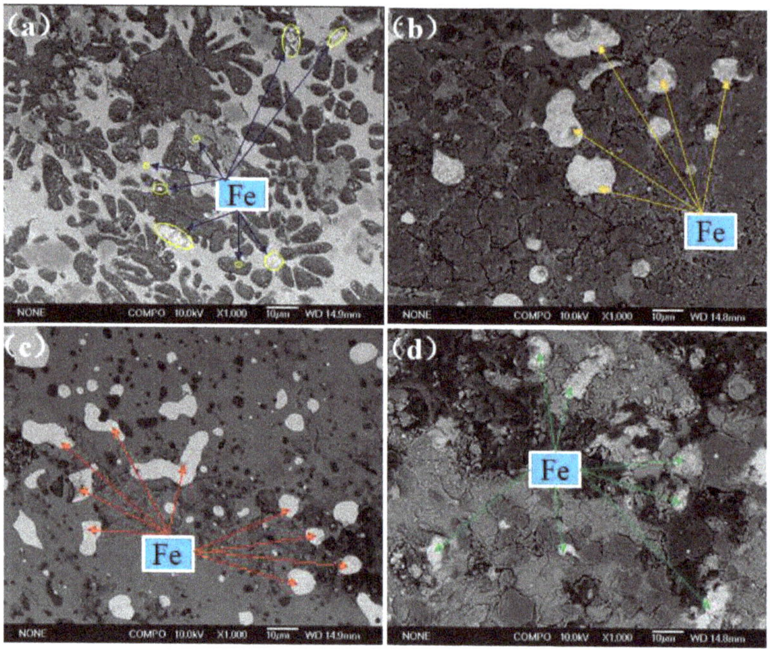


**Figure 3.** SEM images of carbothermally reduced slag in different conditions: (**a**) final slag; (**b**) the final slag was reduced; (**c**) final slag of w(FeO) = 24% was reduced; and (**d**) final slag of R = 2.81 was reduced.

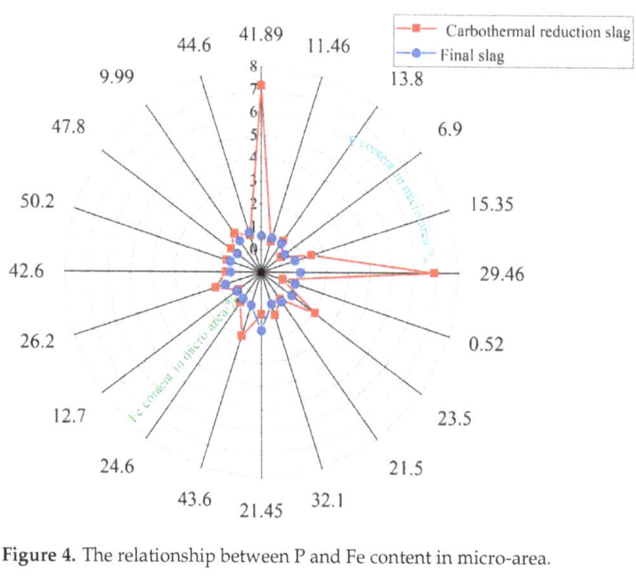


**Figure 4.** The relationship between P and Fe content in micro-area.

On the whole, there was no obvious linear relationship between the Fe content and P content in the carbothermic reduction slag, but when the Fe content was less than 30%, the P content increased with the increase in Fe content. When the Fe content was greater than 30%, with an increase in Fe content, the fluctuations of the P content were few, but there still were large P micro-regions. Therefore, in the reduction process, part of the gas product $P_2$ will persist in the metallic iron, and with the increase in Fe content generated by the reduction of FeO by coke, the content of P in metallic iron will also increase.

## 4. Resource Utilization of Converter Slag after Carbothermal Dephosphorization

### *4.1. Application of Carbon Thermal Reduction Slag Circulation Steelmaking*

Our team has conducted a considerable amount of research on the carbothermic reduction of slag [19–24], and, based on the efficient resource utilization of the carbothermic reduction of converter slag, a new process of "circulating steelmaking by gasification dephosphorization of converter slag" was put forward. Coke was added during the slag-splashing and furnace-protection stages of the converter to realize gasification dephosphorization, and the slag after gasification dephosphorization was reserved for recycling in subsequent furnaces, which could then be used as hot slag to realize low-temperature dephosphorization.

#### 4.1.1. Industrial Test Conditions

The industrial test of this process was carried out in Cheng steel's 100 t converter, and the smelting steel grade was HRB400E. The enterprise adopts a semi-steel process, and the composition of the molten steel is shown in Table 4. Taking the coke powder produced by Cheng steel as the reducing agent, part of the slag was poured out after tapping, and the reducing agent was added all at once from a high bunker before the slag-splashing operation. Then, the gun was lowered, and nitrogen was blown to carry out the slag-splashing operation.

**Table 4.** Composition of molten steel (wt.%).

| Type | C | P | S | V |
|---|---|---|---|---|
| Semi-molten steel | 3.5~4.0 | 0.17~0.20 | 0.025~0.06 | <0.04 |
| Terminal molten steel | 0.04~0.11 | 0.017~0.032 | 0.022~0.045 | vestige |

#### 4.1.2. Changes of Phosphorus Content in Slag

Figure 5 shows the change in P content of the slag before and after gasification dephosphorization. It can be seen in Figure 5 that the P content in the slag decreased after carbothermal reduction; the average P content in the slag before and after gasification dephosphorization was 1.78% and 1.12%, respectively, and the average gasification dephosphorization rate was about 37%. Because the P in the converter slag mainly resided in $C_2S$ in the form of $P_2O_5$, the partial $P_2O_5$ in converter slag could be effectively removed by gasification dephosphorization.

#### 4.1.3. Composition of Terminal Molten Steel

The phosphorus enrichment of converter slag limits its recycling in the furnace, and the high-phosphorus content of converter slag will affect the dephosphorization effect of the subsequent furnace. After gasification-dephosphorization treatment, the phosphorus capacity of converter slag is increased, and it can thus continue to participate in dephosphorization processes. In order to better study the influence of the recycling of gasification-dephosphorization slag on subsequent dephosphorization processes, the composition changes of the molten steel at the end of each smelting procedure in the test were analyzed.

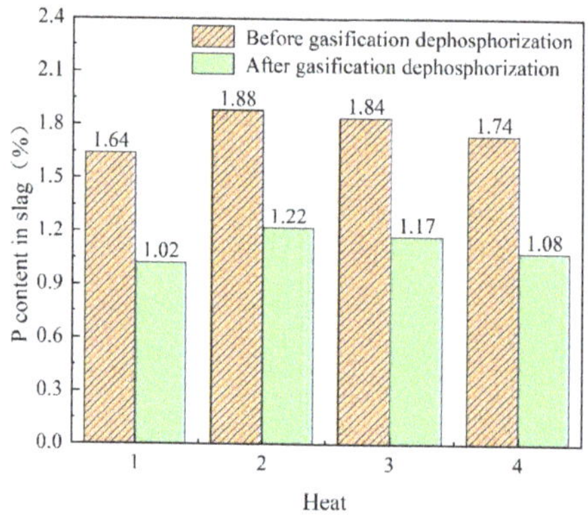


**Figure 5.** Content of P in slag before and after gasification dephosphorization.

The composition of the terminal molten steel in the test furnace is shown in Figure 6. It can be seen that the composition of the terminal molten steel was relatively stable. All the elements met the requirements of the steel grade; the P content was between 0.009% and 0.013%; the S content was between 0.020% and 0.026%; and the C content was between 0.053% and 0.082%. It was proven that the recycling of gasification-dephosphorization slag will not affect the dephosphorization effect of subsequent heats, and the gasification-dephosphorization slag could thus participate in the slagging of the next heat with the appropriate slag alkalinity and FeO content to ensure the dephosphorization rate.

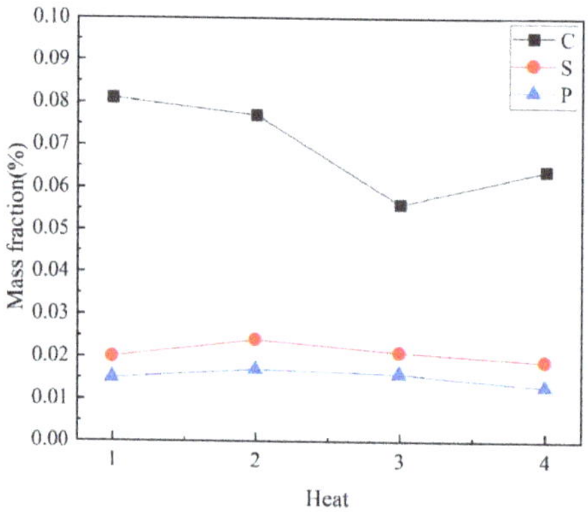


**Figure 6.** Composition of terminal molten steel after gasification dephosphorization.

4.1.4. Comparison of Economic Indicators

Figure 7a,b shows a comparison between iron and steel consumption and slagging material consumption at conventional and experimental heats, respectively. Figure 7a shows that the circulating heat of the gasification-dephosphorization slag could effectively reduce steel consumption. The average iron and steel consumption in the conventional heat was 1032.25 kg, and that in the test heat was 1027.5 kg, with an average reduction of 4.74 kg. The slag contained a lot of metallic iron and iron oxide. A carbon- reducing agent was added at the slag-splashing stage to reduce some of the FeO in the slag to Fe at a high temperature, which reduced the iron loss and improved the steelmaking level. The recycling of gasification-dephosphorization slag can reduce slagging-material consumption. The average slagging-material consumption for the conventional furnace was 1826.25 kg, and that for the experimental furnace was 1218.25 kg, which was 608 kg lower than that for the conventional furnace, as shown in Figure 7b. Because the gasification-dephosphorization slag had the characteristics of "high CaO and low FeO" [25], the amount of lime could be appropriately reduced when it stayed in the furnace to participate in subsequent dephosphorization procedures.

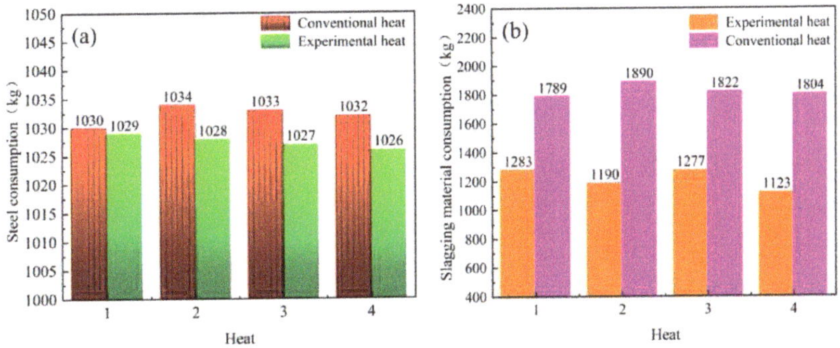


**Figure 7.** Comparison of economic indicators after gasification dephosphorization: (**a**) iron and steel consumption; and (**b**) slagging-material consumption.

The recycling of converter slag can not only reduce the generation of new slag, but can also reduce the consumption of lime because of the high content of CaO in the slag, and at the same time increase the steel output of the subsequent smelting process using the same amounts of iron and steel.

The production practice showed that the gasification-dephosphorization slag had a certain dephosphorization ability, and could directly participate in the early dephosphorization of the next heat if it stayed in the furnace. Gasification-dephosphorization slag itself belonged to the categories of "hot slag" and "cooked slag", which have a microphase suitable for phosphorus occurrence without undergoing the melting and slagging stages. It could be quickly dephosphorized by making full use of the low-temperature conditions during the early stages. However, the gasification-dephosphorization slag was still phosphorus-containing slag, and its dephosphorization ability was limited. Therefore, the gasification-dephosphorization slag cannot be used as the leading dephosphorization agent, and needs to be mixed with fluxes such as lime and dolomite to achieve a better dephosphorization effect.

### 4.2. Assessment and Analysis of Carbon Emission Reduction

Under the background of the "double carbon" target strategy, carbon emissions reductions in iron and steel metallurgical processes play a leading role, so putting controlled lime into use is conducive to reducing $CO_2$ emissions. The lifecycle assessment method is used to quantify carbon emissions and carbon footprint data [26,27]. Taking the boundary of the traditional lime-slagging-mode system as the research object, the starting point of limestone mining and the ending point of converter slag are presented in Figure 8.

According to the emission factor method provided by China Energy Conversion Standard IPCC [28,29], $CO_2$ emissions are represented by the calorific value of unit standard coal. The calorific value of 1 kg standard coal was thus 29.308 MJ, which was equivalent to 2.5 kg of $CO_2$ emissions, and its calculation formula is shown in Equation (6). The lime kiln parameters and $CO_2$ emission factors are listed in Table 5 [30].

$$E_{CO_2} = \sum_{i}^{n} E_i \times F_i \tag{6}$$

where $E_{CO_2}$ is $CO_2$ emitted per unit product, kg; $E_i$ is the emission factor of fuel, kg/kg; $F_i$ is the consumption of the $i$ fuel, kg.

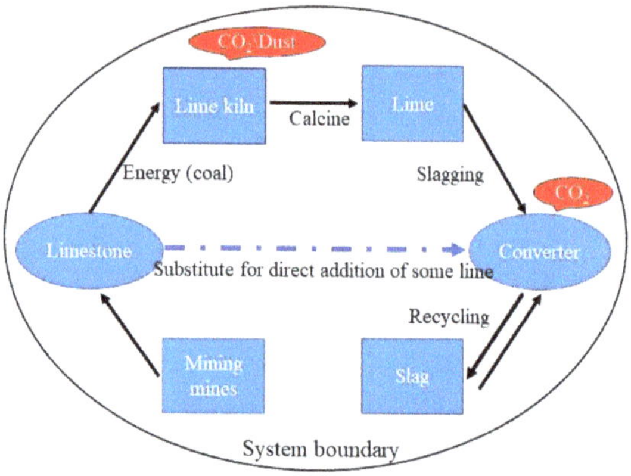


**Figure 8.** System boundary used in calculation.

**Table 5.** Parameters of lime kiln and $CO_2$ emission factor used in calculation.

| Energy Type | Ton of Lime Consumes Energy | $CO_2$ Emission Factor |
|---|---|---|
| Anthracite | 90 kg | 2.68 kg/kg |
| Converter gas | 770 $m^3$ | 0.79 kg/$m^3$ |
| Electric power | 60 kW·h | 1.03 (kg/kW·h) |

The lime kiln mainly used anthracite and converter gas as fuels, and the converter gas can be produced for self-use, so its energy consumption was not considered. The energy consumption and $CO_2$ emission of other links are shown in Table 6. It can be seen from Table 6 that $CO_2$ emissions were the highest during the process of limestone decomposition and lime calcification, followed by lime slagging. Therefore, from the start point to the end point of the system boundary, the estimated emissions per ton of lime was about 1034 kg $CO_2$. Taking Cheng steel's 100 t converter as an example, the average slagging material was reduced by 4.74 kg, and $CO_2$ emissions were reduced by about 4.86 kg per heat.

**Table 6.** Calculation results of carbon emissions in production links.

| Emission Link | Standard Coal Quantity Converted into Energy Consumption per Ton of Lime/Kg | $CO_2$ Emission/Kg |
|---|---|---|
| Electricity consumption (production and anthracite supply) | 24.88 | 62.2 |
| Limestone decomposition and lime calcination | 339.8 | 849.5 |
| Lime slagging | 48.75 | 121.88 |

### *4.3. Thoughts on Recovery and Utilization of Phosphorus Resources*

The above research demonstrates that it is practical to reduce slag using the carbothermic method in the furnace. Under the conditions of high-temperature carbothermal reduction, the $P_2O_5$ in slag can be effectively gasified. A part of the $P_2$ persists in the molten metal to generate $Fe_xP_y$, while the other part escapes in the form of gas. Figure 9 shows the schematic diagram of the reaction process of converter-slag gasification-dephosphorization steelmaking and the flow direction of the phosphorus.

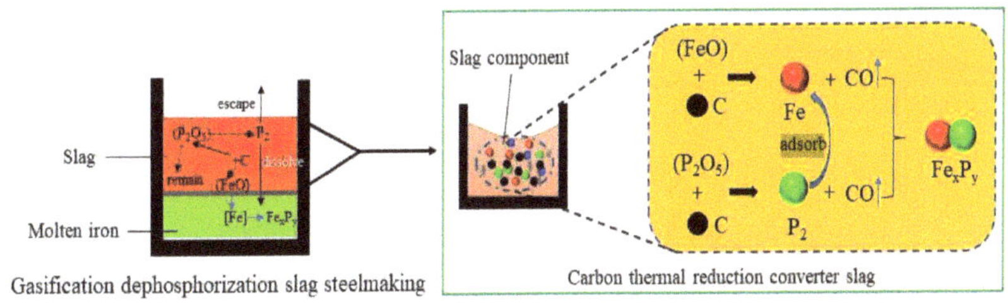


**Figure 9.** Schematic diagram of the reaction process of converter-slag gasification-dephosphorization steelmaking.

For the recovery and resource utilization of phosphorus resources after the carbothermal reduction of phosphorus-containing slag, the appropriate carbothermal reduction conditions and the best gasification-dephosphorization parameters should be considered first to ensure the full reduction of $P_2O_5$ in converter slag. Second, the efficient recovery and utilization of reduced phosphorus resources should be considered. We can consider the following two aspects [31] by:

1. Making more $P_2$ overflow and collecting it, and then carrying out an oxidation–hydration reaction to prepare the phosphoric acid for use as phosphate fertilizer, thereby making it less difficult for entry into the industry. This idea can provide a new direction for a steel chemical co-production project, which is an innovative development model of the iron and steel industry combined with the chemical industry. At present, there have been successful cases reported in China, such as the project of Shandong Shiheng Special Steel, in which formic acid was produced through the use of converter gas, and the production of sodium bicarbonate raw materials by Handan Xinxing Cast Pipe, and so on [32,33];
2. After phosphorus-containing slag is processed by carbothermal reduction gasification dephosphorization, favorable conditions for dissolving the reduced phosphorus vapor into metallic iron phase are controlled, so that more phosphorus vapor can enter the molten iron to generate ferrophosphorus, which can be used as electrode raw materials or directly used as ferrophosphorus in steelmaking, with high added value [31,34].

Because the FeO in converter slag is reduced prior to $P_2O_5$, part of the $P_2$ gas will be dissolved into molten iron to generate ferrophosphorus, which will reduce the overflow of $P_2$. At the same time, the process will seriously pollute the environment and the equipment investment is expensive for the preparation of phosphoric acid. Therefore, the recycling of phosphorus-containing slag is a suitable method for preparing ferrophosphorus.

## 5. Conclusions

The resource utilization of converter slag is of great significance to green and low-carbon production in the iron and steel industry. In this paper, the dephosphorization behavior of converter slag by carbothermal reduction and its industrial applications, were mainly studied, and ways of using the phosphorus in the dephosphorization process of converter slag by carbothermal reduction and gasification were analyzed. The main conclusions and prospects are as follows:

(1) On the basis of our experiments and thermodynamic calculations, the feasibility of phosphorus removal from converter slag by carbothermal reduction was demonstrated. When the temperature is higher than 1433 K, the carbothermal reduction reaction can take place, and the main reduction product is $P_2$. The average dephosphorization rate was 32.57% under the experimental conditions;

(2) The production practice of Cheng steel shows that the dephosphorization effect of the subsequent furnace procedure will not be affected by the carbothermic reduction of converter slag. The composition of the terminal molten steel was stable, and the average consumption of iron and steel per heat was reduced by 4.74 kg, the slagging

material was reduced by 608 kg, and $CO_2$ emissions were reduced by 4.86 kg, which will result in emissions reductions and energy savings;

(3) Part of the $P_2$ produced by the carbothermal reduction enters the Fe phase, and the other part is discharged with the furnace gas. With an increase in Fe content, the amount of $P_2$ also increases. In order to efficiently recycle phosphorus resources, appropriate carbothermal reduction conditions and optimal gasification-dephosphorization parameters should be fully considered so that more $P_2$ can enter the molten iron to generate ferrophosphorus.

However, the preparation of ferrophosphorus by adding $P_2$ into metallic iron should also be comprehensively studied while ensuring appropriate carbothermal reduction conditions and the optimal parameters for gasification dephosphorization, which will be studied through thermodynamic calculations and thermal experiments in our follow-up work.

**Author Contributions:** S.T. and C.L. conceived and designed the study; S.T. and S.Z. performed the calculation; S.T conducted the experiment; L.A. and S.W. analyzed the experimental data; and S.T. and C.L. wrote the paper. All authors have read and agreed to the published version of the manuscript.

**Funding:** This research was funded by the National Natural Science Foundation of China, grant number 52004097 and 52274334.

**Informed Consent Statement:** Written informed consent has been obtained from the patient(s) to publish this paper.

**Data Availability Statement:** Not applicable.

**Acknowledgments:** Thanks to S. Cui of Tangshan Normal University for checking and revising the manuscript.

**Conflicts of Interest:** The authors declare no conflict of interest.

## References

1. Zhou, C.G.; Yang, H.Z.; Ai, L.Q. Research status and prospect of recycling technology of converter slag containing phosphorus. *Iron Steel* **2021**, *56*, 22–39.
2. Zhang, J.; Yan, D.L.; Qi, Y.H. Difficulty analysis on treatment and utilization of iron and steel smelting slag. *Steel* **2020**, *55*, 1–5.
3. Wang, Z.L.; Bao, Y.P.; Wang, D.Z.; Wang, M. Effective removal of phosphorus from high phosphorus steel slag using carbonized rice husk. *J. Environ. Sci.* **2023**, *124*, 156–164. [CrossRef] [PubMed]
4. Wang, Z.L.; Bao, Y.P.; Wang, D.Z. Leaching rule of phosphorus element in high phosphorus converter slag with different acids. *Iron Steel* **2021**, *56*, 103–110.
5. Wang, Y.C.; Li, H.Y.; Luo, G.P. Microkinetic of gasification dephosphorization of converter slag by microwave carbon thermal reduction. *J. Iron Steel Res.* **2017**, *29*, 93–97.
6. Zhang, W.; Liu, W.X.; Li, J. Main factors of gasification dephosphorization of high phosphorus steel slag. *Iron Steel* **2015**, *50*, 11–14.
7. Ai, L.Q.; Zhang, Y.L.; Zhu, W.H. Research on carbothermic reduction for dephosphorization from converter slag by microwave heating. *Iron Steel Vanadium Titan.* **2015**, *36*, 63–67.
8. Zhao, C.L.; Zhang, N.; Kang, L. Experiment on dephosphorization of converter slag by carbothermal reduction. *Iron Steel* **2016**, *51*, 41–44.
9. Lin, L.L.; Zhi, Y.Q.; He, J.G. Influence of slag temperature on phosphorus enrichment in P-bearing steelmaking slag. *Ironmak. Steelmak.* **2021**, *48*, 334. [CrossRef]
10. Kubo, H.; Matsubae, Y.K.; Nagasaka, T. Magnetic separation of phosphorus enriched phase from multiphase dephosphorization slag. *Trans. Iron Steel Inst. Jpn.* **2010**, *50*, 59–64. [CrossRef]
11. Wu, M. Low carbon steel production with BOF slag-remaining practice. *Steelmaking* **2009**, *25*, 16–19.
12. Wu, X.R.; An, J.N.; Chen, R.H. Distribution and concentration of phosphorus in factory converter slags and growth of P-concentrating phase. *J. Anhui Univ. Technol. Nat. Sci.* **2010**, *3*, 233–237.
13. Wang, N.; Liang, Z.G.; Chen, M. Enrichment behavior of phosphorus in $CaO\text{-}SiO_2\text{-}Fe_tO\text{-}P_2O_5$ Slag. *J. Northeast. Univ. Nat. Sci.* **2011**, *32*, 814–817.
14. Li, C.X.; Li, H.; Zhou, B. Experimental study on steelmaking using limestone instead of lime as the slagging material in 100t converter. *China Metall.* **2015**, *25*, 22–26.
15. Liang, Y.J.; Che, C.Y. *Handbook for Inorganic Thermodynamic Calculation;* Northeast University Press: Shenyang, China, 2001.
16. Li, G.Q.; Zhang, F.Q.; Zhang, L. Recycle of converter thermal reducction. *J. Mater. Metall.* **2003**, *2*, 167.
17. Jia, J.R.; Ai, L.Q. Reduction dephosphorization behavior of steelmaking slag in microwave heating field. *Iron Steel* **2012**, *47*, 70.

18. Dong, J.S.; Xu, G.; Shigeru, U. Separation of phosphorus and manganese from steelmaking slag by selective reduction. *Metall. Mater. Trans. B* **2019**, *50*, 1248–1259.
19. Xue, Y.K.; Tian, P.; Li, C.X. Reduction mechanism of $P_2O_5$ insteel slag. *Trans. Indian Inst. Met.* **2020**, *73*, 251–258. [CrossRef]
20. Tong, S.; Li, C.X.; Wang, S.H. Effect of temperature on the reuse of dephosphorization slag from converter gasification. *Iron Steel Vanadium Titan.* **2021**, *42*, 109–114.
21. Li, C.X.; Wang, S.H.; Xue, Y.K. Experiment on gasification dephosphorization of converter slag with coke powder reduction. *Iron Steel* **2018**, *53*, 20–24.
22. Zhou, C.G.; Ai, L.Q.; Wang, S.H. Study on the mechanism of gasification dephosphorization and its influence on the next smelting proces. *Vanadium Titan. Iron Steel* **2018**, *39*, 129–136.
23. Guo, R.H.; Wang, S.H.; Li, C.X. Gasification dephosphorization of converter slag by coke reduction. *Iron Steel* **2020**, *55*, 118–124.
24. Wang, S.H.; Tong, S.; Li, C.X. Phosphorus migration behaviorin the process of converter slag gasification dephosphorization. *Metalurgija* **2022**, *61*, 149–152.
25. Li, C.X.; Zhao, D.G.; Xue, Y.K. Experimental study on gasification and dephosphorization of semi-steel slag in Chenggang. *Steelmaking* **2019**, *35*, 19–22.
26. Yang, Y.; Ma, L.; Yu, J. Life cycle assessment introduced by using nanorefrigerant of organic rankine cycle system for waste heat recovery. *J. Renew. Mater.* **2023**, *11*, 27. [CrossRef]
27. Zhang, H.; Sun, W.; Li, W. A carbon flow tracing and carbon accounting method for exploring $CO_2$ emissions of the iron and steel industry: An integrated material–energy–carbon hub. *Appl. Energy* **2022**, *309*, 118485. [CrossRef]
28. Paustian, K.; Ravindranath, N.; Amstel, H. *2006 IPCC Guidelines for National Greenhouse gas Inventories*; International Panel on Climate Change: Geneva, Switzerland, 2006.
29. Gomez, D.R.; Wattersson, J.; Americano, B. *Chapter 2, Energy, 2006 IPCC Guidelines for National Greenhouse Gas Emissions Inventories*; Intergovernmental Panel on Climate Change: Geneva, Switzerland, 2008.
30. Li, C.X.; Li, H.; Zhu, S.N. Estimation of energy saving and emission reduction of limestone replacing lime as raw material for steelmaking and slagging. *China Metall.* **2015**, *25*, 66–69+72.
31. Liu, Y.Q.; Lin, L.; He, S. Research status of dephosphorization and resource utilization of phosphorus-containing steel slag by carbothermal reduction. *J. Iron Steel Res.* **2022**, *34*, 1034–1046.
32. Beijing Peking University Pioneer Technology Co., Ltd. The World's First Set of Sulphuric Acid Plant for Purifying CO from Converter Gas [EB/OL]. 2021. Available online: https://www.pioneer-pku.com/show-312.html (accessed on 15 June 2021).
33. News Center. Beijing's First Central Enterprise Moved Abroad Settled in Handan. [EB/OL]. 2014. Available online: http://news.sina.com.cn/o/2014-05-16/015930145049.shtml (accessed on 16 May 2014).
34. Guo, H.Y.; Wang, G.X.; Jia, L.C. Progress of the synthesis of energy materials using ferrophosphorus. *Phosphate Compd. Fertil.* **2012**.

*processes*

MDPI

*Article*

# Effect of Powder Formulation and Energy Density on the Nitrogen Content, Microstructure, and Mechanical Properties of SLMed High-Nitrogen Steel

**Xin Sun, Jianbiao Ren, Shuhuan Wang and Dingguo Zhao ***

School of Metallurgy and Energy, North China University of Science and Technology, Tangshan 063009, China; sunxin8835@163.com (X.S.); rjianbiao@163.com (J.R.); wshh88@163.com (S.W.)
* Correspondence: gyyzhao@163.com; Tel.: +86-315-8805020

**Abstract:** The effects of powder formulation, including elemental mixed powder (EMP) and alloy mixed powder (AMP), and energy density on the nitrogen content and microstructural characteristics of high-nitrogen steel prepared by selective laser melting were investigated. The results reveal that the samples prepared with EMP had more nonfusion flaws and a relatively low density, with a maximum of only 92.36%, while samples prepared with AMP had fewer defects and a relative density of up to 97.21%. The nitrogen content and microstructural characteristics were significantly influenced by the laser energy density. The relative density of the EMP samples increased from 88.29% to 92.36% as the laser energy density increased from 83.3 $J/mm^3$ to 125 $J/mm^3$, while the relative density of the AMP samples rose from 93.31% to 97.21%, and the number of defects and the nitrogen content decreased. The mechanical properties of the AMP samples were superior to those of the EMP samples when the energy density rose, and the strength of the high-nitrogen steel first rose and then fell. The AMP samples showed the best mechanical properties when the energy density was 104.2 $J/mm^3$, which corresponds to a laser power of 250 W, a scanning speed of 1000 mm/s, and a layer thickness of 30 μm. The corresponding values of yield strength, ultimate tensile strength, and elongation were 958.8 MPa, 1189.2 MPa, and 30.66%, respectively.

**Keywords:** mixed powder; selective laser melting; high-nitrogen steel; mechanical properties

**Citation:** Sun, X.; Ren, J.; Wang, S.; Zhao, D. Effect of Powder Formulation and Energy Density on the Nitrogen Content, Microstructure, and Mechanical Properties of SLMed High-Nitrogen Steel. *Processes* **2023**, *11*, 1937. https://doi.org/10.3390/pr11071937

Academic Editor: Carlos Sierra Fernández

Received: 7 June 2023
Revised: 24 June 2023
Accepted: 25 June 2023
Published: 27 June 2023



## 1. Introduction

Nitrogen has long been regarded as one of the harmful impurities in traditional steel. However, in the early 20th century, researchers discovered that the addition of a certain amount of nitrogen could influence the mechanical properties of steel and the stability of the austenite phase region. This finding sparked a wave of research on nitrogen-containing stainless steels, and it gradually gained recognition in the field [1–3]. Despite the growing acceptance of high-nitrogen stainless steel, there is no unified standard for its definition. Speidel [4] considers "high-nitrogen steel" to be steel with a nitrogen content exceeding the limit achievable under normal pressure (0.1 MPa). However, the classification of nitrogen-containing stainless steel depends on the mass fraction of nitrogen in austenitic stainless steel. It can be divided into three types: the nitrogen-controlled type (0.05 wt% to 0.10 wt%), the medium-nitrogen type (0.10 wt% to 0.40 wt%), and the high-nitrogen type (above 0.40 wt%). For ferritic stainless steel and martensitic-matrix stainless steel, a nitrogen content exceeding 0.08% can be referred to as high-nitrogen steel. Furthermore, based on their microstructures, high-nitrogen steels can be categorized into five main types: nitrogen-containing austenitic stainless steel, nitrogen-containing duplex stainless steel, nitrogen-containing ferritic stainless steel, nitrogen-containing martensitic stainless steel, and nitrogen-containing precipitation-hardened stainless steel [5]. High-nitrogen steel is an innovative type of stainless steel that offers significant advantages, such as a low cost and high strength, by utilizing inexpensive nitrogen elements to replace expensive nickel.

By adding nitrogen as interstitial atoms to the stainless-steel matrix, the solid solution is strengthened, resulting in improved strength, toughness, and corrosion resistance [6,7]. These outstanding properties have made high-nitrogen steel increasingly popular in various industries, including the bioenergy industry, aerospace industry, petrochemical industry, marine engineering, biomedical applications, and others [8,9]. The utilization of high-nitrogen steel has revolutionized these industries, offering enhanced performance and cost-effectiveness as compared to traditional stainless steels. In the bioenergy industry, for example, high-nitrogen steel components exhibit excellent resistance to corrosion and fatigue, making them ideal for critical applications in biofuel production and processing. In the aerospace sector, the high strength of high-nitrogen steel enables the construction of lightweight structures without compromising safety and reliability. Similarly, in the petrochemical industry, the superior corrosion resistance of high-nitrogen steel ensures the durability and longevity of equipment exposed to harsh chemical environments. Moreover, high-nitrogen steel finds extensive use in marine engineering, where its enhanced strength and corrosion resistance make it suitable for shipbuilding, offshore platforms, and underwater structures. In the biomedical field, high-nitrogen steel's biocompatibility and corrosion resistance make it an excellent choice for medical devices, implants, and surgical instruments, where durability and long-term performance are paramount [9–13].

Supersaturated nitrogen can be added to steel to significantly increase its strength, according to the research [8]. The insufficient solubility of nitrogen in iron-based materials at atmospheric pressure makes the production of high-nitrogen steel challenging. At 1 standard atmospheric pressure and 1600 °C, the solubility is only 0.045 wt% [14]. The solubility and distribution uniformity of nitrogen during the steel solidification process are therefore improved by applying pressure, which is typically done using high-pressure technology [15]. High-nitrogen steel can only be produced on a large scale because the equipment and method required to prepare it under high pressure are complicated, dangerous, and expensive. In addition, strain strengthening happens due to cast high-nitrogen steel's relatively high hardness. While strain strengthening occurs throughout the manufacturing process, the hardness of cast high-nitrogen steel is also rather high, making typical machining more challenging [16].

Metal additive manufacturing (MAM), also known as metal 3D printing, is a revolutionary manufacturing technology that emerged in the 1980s. It uses digital model files and metal powder as the processing material, and it prints layer by layer using a high-energy heat source, enabling the fast and moldless fabrication of complex components [17]. Its low material consumption, high processing efficiency, and capacity for achieving integrated structures have drawn significant attention from researchers, making it one of the most popular topics in the manufacturing industry [18]. Metal additive manufacturing has developed into various shaping techniques, classified by different heat sources. These include laser engineered net shaping (LENS), electron beam selective melting (EBSM), selective laser sintering (SLS), and selective laser melting (SLM). These techniques have been widely applied and advanced abroad, forming mature and large-scale systems [19]. Metal materials for additive manufacturing have been explored both domestically and internationally, covering a broad spectrum of alloys, such as titanium-based alloys [20,21], nickel-based alloys [22,23], iron-based alloys [24], aluminum alloys [25], and high-entropy alloys [26]. These materials have unique properties and performance features that suit different applications. Metal additive manufacturing is a significant breakthrough in the field of manufacturing technology in the past 30 years and is considered to be a key driver of the third industrial revolution. It has advantages, such as producing complex geometries, customizing designs, and integrating functions, that have changed traditional manufacturing processes and opened up new possibilities for product development and production [27]. With continuous improvement and increasing research and development efforts, metal additive manufacturing is expected to play an increasingly important role in various industries, such as aerospace, automotive, medical, and consumer goods, revolutionizing the way we design and manufacture components [28–30]. The ability to net-form parts using metal additive manufacturing is possible [31], greatly reducing the difficulty of subsequent processing and enabling the processing of high-nitrogen steel. By simultaneously feeding wire powder and a TIG arc additive, Wu et al. [32] formed a straight wall made of high-nitrogen steel, and the nitrogen content was greatly raised by adding $Cr_2N$ powder, yielding a nitrogen content of 1.04%. Unfortunately, the droplets react with the air during the arc additive process, which lends a susceptibility to flaws, such as inclusions and porosity.

In order to effectively remove the defects, such as oxidation inclusions and pores during the AM process, selected laser melting has a protective gas compartment that can be filled with nitrogen or argon gas with a purity value of 99.999%. In the selective laser melting (SLM) process, metal powder is the most basic and crucial raw material. In theory, any powder that can be melted and metallurgically bonded using a laser heat source can be used for SLM forming. However, in practice, SLM differs from traditional powder metallurgy as the morphology, sphericity, particle size distribution, and flowability of metal powders can all affect the formability of SLM parts. This requires the metal powder to have excellent properties, such as particle size, sphericity, and purity [33,34]. Lv et al. [35,36] investigated the advantages of selective laser melting, which may be used to create complicated parts, including high forming precision and precise positioning. Using home-made, aerosolized, high-nitrogen steel powder is one way to conduct the selective laser melting process of high-nitrogen steel. However, the preparation of high-nitrogen steel powder is expensive and complicated. Thus, it is urgent to find a new method for preparing high-nitrogen steel.

In order to obtain powders with a nitrogen content much higher than the permissible nitrogen content (0.4 wt%), and to effectively reduce the production cost of the powder, our research method adds chromium nitride powder to the premixed powder. The effects of powder formulation, including elemental mixed powder (EMP) and alloy mixed powder (AMP), and energy density on the nitrogen content and microstructural characteristics of high-nitrogen steel prepared by selective laser melting were investigated. The relevant research results can provide guidance for the selective laser melting of high-nitrogen steel in engineering.

## 2. Materials and Methods

The distribution of the particle sizes was between 15 and 53 μm. To create the mixed powder, the elemental powder and alloy powder were first mixed separately. Mixed powder is defined as a powder with a nitrogen content significantly higher than the maximum nitrogen content of 0.4 wt% in steel under atmospheric pressure. Iron powder, chromium nitride, chromium powder, manganese powder, and molybdenum powder with a purity of higher than 99.0% were used as raw ingredients for the elemental mixed powders (EMPs). While CrN and Mn are asymmetric powders; Fe, Cr, and Mo are spherical powders. To calculate the weight of each metal powder, we used the following formula: Fe: Cr: Mn: Mo: CrN = 63:12:12:3:10 per 100 g. At a ratio of 98:2 per 100 g, alloy mixed powders (AMP) were prepared from spherical stainless-steel powder with a nitrogen content of 0.29 wt% and a nonspherical chromium nitride powder with a purity of 99.0%. Table 1 provides the chemical composition of the mixed powders.

**Table 1.** Chemical composition of the mixed powders (wt%).

| Elements | Cr | Mn | Mo | N | C | Fe |
|---|---|---|---|---|---|---|
| EMP | 20.53 | 10.61 | 3.23 | 1.472 | 0.038 | Bal |
| AMP | 21.14 | 11.21 | 3.04 | 1.01 | 0.036 | Bal |

Following weighing, the powder was first extensively combined in a planetary ball mill for 4 h at a speed of 400 rpm. The ball mill tank was sealed and vacuum treated for 10 min before being mixed to lessen oxidation. Two different types of mixed powders were evaluated for flowability, apparent density, and compacted density, and the corresponding findings are displayed in Table 2. The particle size distribution of the mixed powder was measured using a laser diffraction particle size distribution meter, and the results are displayed in Figure 1. The D10, D50, and D90 of the EMP were 20.31, 36.57, and 58.35 μm, respectively, and the average particle size was 36.57 μm. The D10, D50, and D90 of the AMP were 21.91, 34.23, and 53.25 μm, respectively. Both kinds of powders met the specifications for particle size distribution for selective laser melting, with an average particle size of 34.23 μm.

**Table 2.** Comparison of the characteristics of the mixed powders.

| Mixed Powder | Size (μm) | Flowability (s/50 g) | Apparent Density (g/cm$^3$) | Compacted Density (g/cm$^3$) |
|---|---|---|---|---|
| EMP | 15-53 | 40.25 | 3.44 | 3.85 |
| AMP | 15-53 | 29.81 | 4.15 | 4.36 |

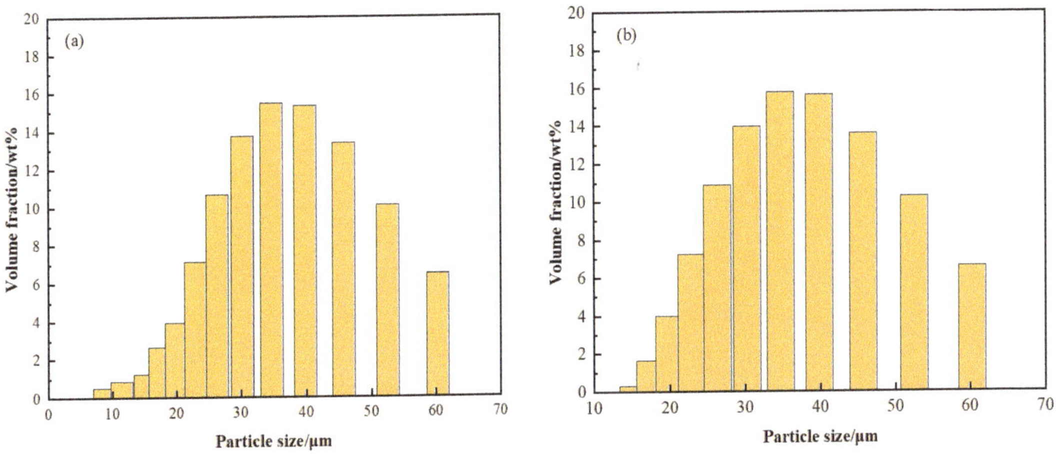


**Figure 1.** Particle size distribution of the mixed powders: (**a**) EMP; (**b**) AMP.

DiMetal-280 equipment with a substrate preheating temperature of 150 °C and a forming chamber covered by nitrogen gas was used for selective laser melting. The powder layer thickness was 30 μm for the selective laser melting. The scan interval was 80 μm, the scanning speed was 1000 mm/s, and the laser power was 200, 225, 250, 275, and 300 W, respectively. The corresponding laser energy densities were 83.3, 93.8, 104.1, 114.6, and 125.0 J/mm$^3$, respectively. The tensile specimen is shown in Figure 2. The block sample had dimensions of 5 × 5 × 5 mm. It was intended for room-temperature tensile testing according to GB/T 228.1-2021.

The morphology and elemental distribution of the powder, as well as the pore flaws, microstructure, and tensile fracture surface of the samples, were examined using optical microscopy and a Czech TESCAN MIRA LMS SEM. The Archimedes drainage method was used to calculate the samples' relative density. The nitrogen content was measured by using a LECO oxygen, nitrogen, and hydrogen analyzer. Tensile testing was performed at room temperature using CMT4204 static hydraulic universal testing equipment to measure tensile strength, yield strength, and elongation.

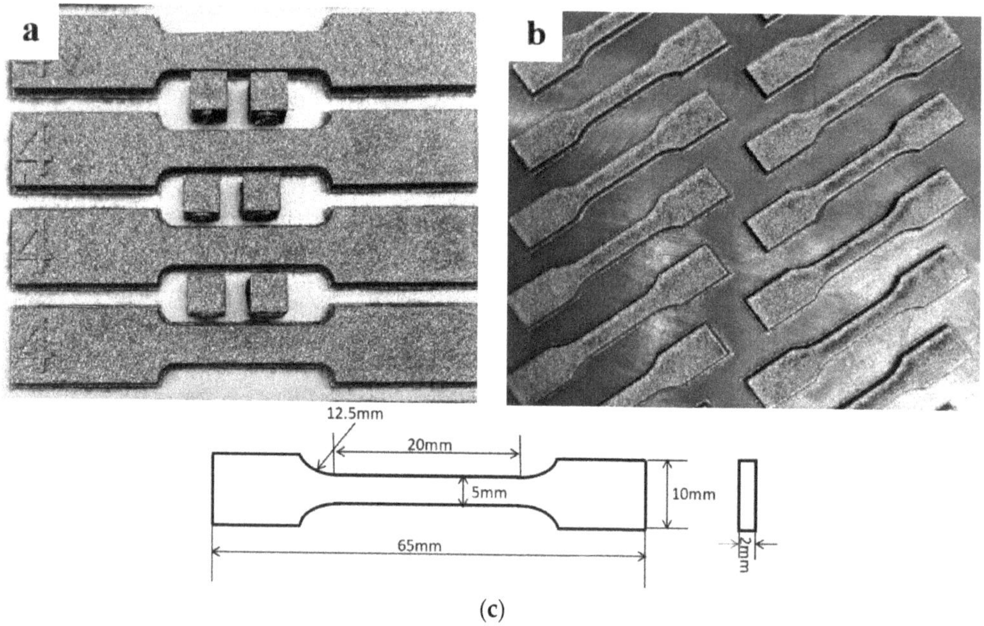


**Figure 2.** The SLMed samples prepared by: (**a**) EMP and (**b**) AMP; (**c**) Schematic diagram of tensile specimen dimensions.

## 3. Results and Discussion

### *3.1. Morphological Characteristics and Elemental Distribution of the Mixed Powder*

In Figure 3, the two powders' morphologies and elemental distributions are depicted. The results demonstrate that the sphericity of the EMP was poor, and there were numerous nonspherical powder particles, while there was only a very small amount of nonspherical powder in the AMP. There was a small amount of aggregation in the elemental distribution of the EMP, while the elemental energy spectrum scanning results of the AMP were more uniform. This might be a result of the alloy mixed powder having greater flowability, apparent density, and compacted density due to the use of more irregular powders, specifically CrN and Mn, which accounted for up to 22% of the total, and which have uneven particle sizes distributed throughout, as shown in Table 2.

### *3.2. The Effect of Powder Formulation and Energy Density on the Relative Density of High-Nitrogen Steel*

Figure 4 shows the relative density of the EMP and AMP samples under different energy densities. The results demonstrate that, as the energy density increased, the relative density of the two samples increased, and that the density of the AMP samples was superior to that of the EMP samples. When the energy density was 104.2 J/mm$^3$, the relative density of the AMP samples reached 97.21%, while it was only 92.36% for the EMP samples. Theoretical studies have demonstrated that, when the laser energy density is low, the actual printing temperature is effectively improved as the laser energy density rises, greatly improving the density of the printed samples [37]. The behavior of the powder and the relative density can both be impacted by the flowability and apparent density of the powder [38,39]. With an increase in energy density, the relative density of the two samples rose. This happened as the density of the samples grew; both the energy density and the energy absorbed by the powder increased. As a result, the powder totally melted, and the fluidity in the micro melt pool increased. However, since the gas could not be entirely expelled during printing, the density was significantly reduced, and more gas was created in the micro melt pool due to the large amount of fine powder, poor fluidity, and low apparent density of the EMP.

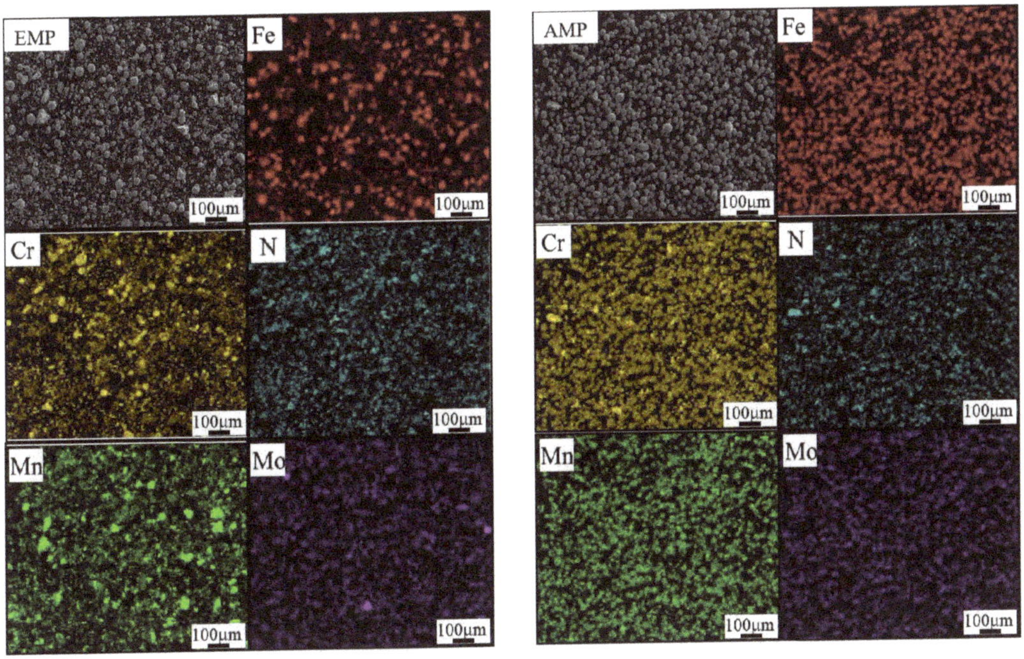


(a) (b)

**Figure 3.** Comparison of the morphology and element distribution of the mixed powders: (**a**) EMP and (**b**) AMP.

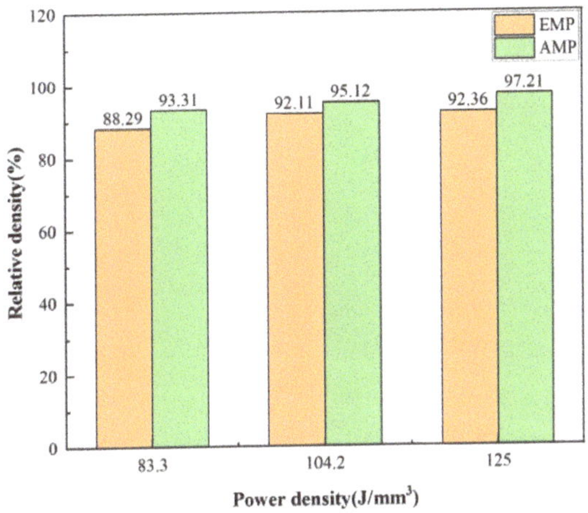


**Figure 4.** Relative density [40] of the EMP and AMP samples under different energy density conditions.

*3.3. The Effect of Powder Formulation and Energy Density on the Microstructure of the High-Nitrogen Steel*

Figure 5 displays the microstructure of the samples prepared with the EMP and AMP. Both types of samples had numerous pore flaws, as shown in Figure 5a–f. While there were few flaws and a smooth edge to the faults in the AMP samples, there were more nonfusion defects and spheroidized particles in the EMP samples. Meanwhile, nonfusion defects also appeared when the energy density was low. When the melt pool's temperature is low, the liquid-phase dynamic viscosity is high; thus, the melt's fluidity is low and the wettability between the melt and the matrix is low [41]. When the energy density is low, as demonstrated by [42], the metal in the micro melt pool cools quickly. As liquid metal solidifies before the bubbles have a chance to leave the surface of the micro melt pool, they can shape themselves into virtually circular pores. With an increase in energy density, the phenomenon of nonfusion in the two samples improves since increasing energy density improves the melt's fluidity and wettability with the substrate. Yet, the morphology of the two types of samples differed at the same power because of the two powders' inconsistent sphericity, fluidity, and elemental homogeneity.

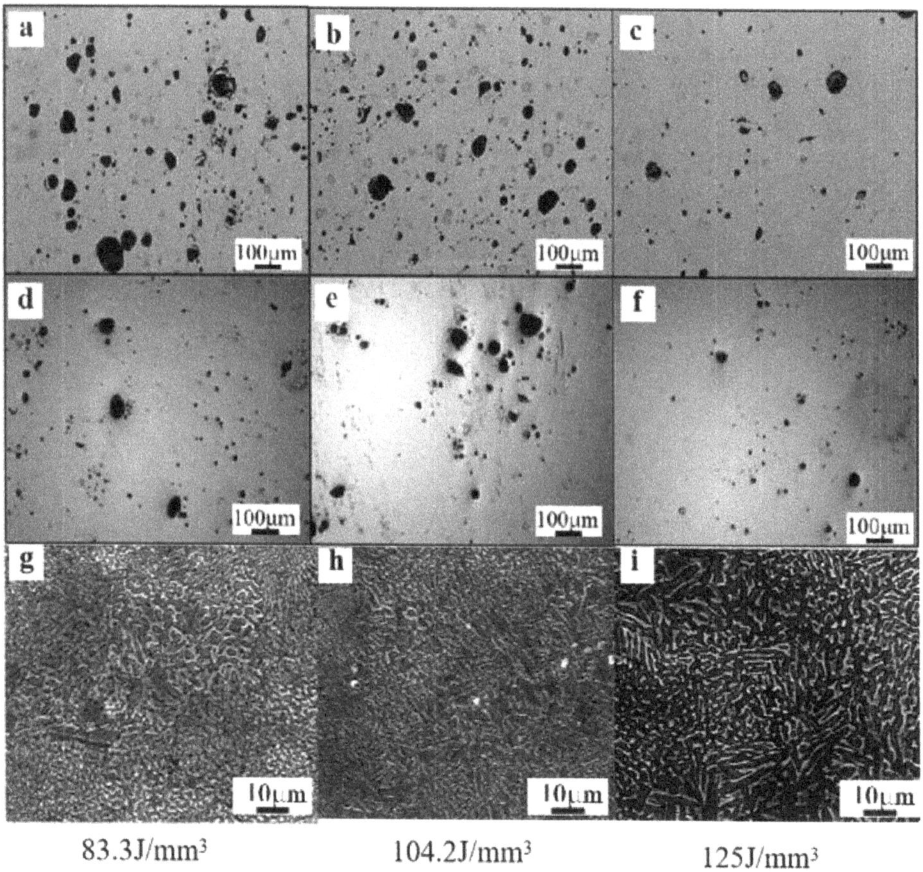


**Figure 5.** Comparison of the defects in samples prepared with: (**a**–**c**) EMP and (**d**–**i**) AMP.

In order to obtain a clearer observation of the morphological structure of the samples under three different energy densities, scanning electron microscopy (SEM) was utilized to scan the samples. The columnar and cellular crystals made up the microstructure of the AMP samples, as can be seen in Figure 5g–i. The cellular and columnar crystals gradually became coarser as the energy density rose. Theoretical studies have demonstrated that increasing both the volume energy density [43] and the surface energy density inputs into the molten pool [44] will result in an increase in the size of the cellular structure [45]. Due to the relatively high temperature gradient and rapid solidification of the alloy mixed powder during the SLM process, cellular dendrites, instead of planar crystals, preferentially developed in the SLM material. The microstructure was therefore found to have cellular structures [46]. The molten pool's temperature rises as the energy density does as well, and as the laser strength rises, the temperature of the molten pool rises and prolongs the existence of the molten pool, which favors crystal growth. As a result, the cellular crystals tend to grow in size, while the columnar crystals progressively become coarser.

Figure 6 depicts the results of the composition analysis of the EMP samples. It was discovered that the spheroidized particles were made up of elemental elements including Mo and Cr in the matrix. Unfused particles and element aggregation in the EMP samples were the product of the uneven distribution of elemental elements in the powder. When comparing the microstructure and morphology of the two samples, it can be seen that both improved with increases in the energy density. However, the morphology of the EMP samples was less complete than that of the AMP samples, and the improvement in morphology was greater with increases in the energy density. In order to produce samples with a higher nitrogen content and better morphology, it is necessary to fully consider the quality of the powder while considering the nitrogen content when configuring two different types of mixed powder samples.

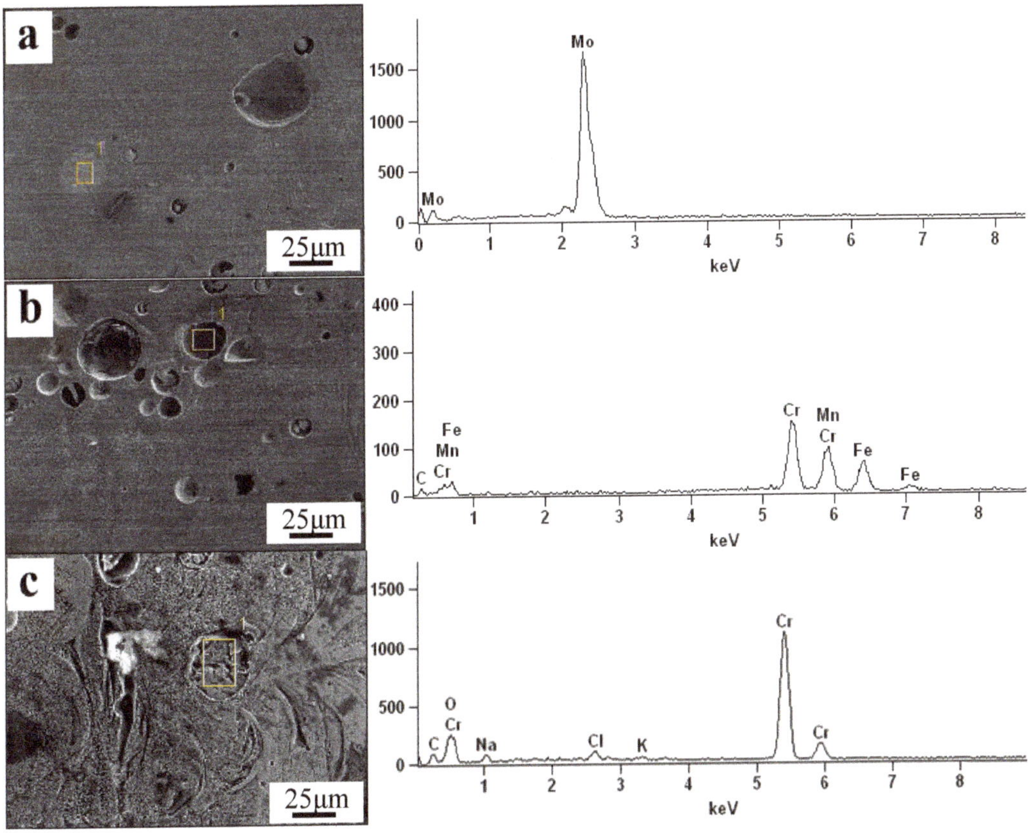


**Figure 6.** Composition analysis of the samples prepared with EMP: (**a**) unmelted particles of Mo; (**b**) unmelted particles of Fe, Cr, and Mn compounds; (**c**) unmelted particles of Cr.

### 3.4. The Effect of Powder Formulation and Energy Density on the Microstructure of the High-Nitrogen Steel

Figure 7 depicts the relationship between the nitrogen content in the EMP and AMP samples and the energy density. It can be seen that the nitrogen content of the EMP samples decreased from 0.996 wt% to 0.96 wt%, and the nitrogen loss rate rose from 32.34% to 34.78% when the energy density increased from 83.3 J/mm$^3$ to 125 J/mm$^3$. The nitrogen content of the AMP samples decreased from 0.836 wt% to 0.768 wt%, and the nitrogen loss rate rose from 17.23% to 23.96%. When the energy density was increased, the nitrogen contents of the two types of samples dropped by 0.036 wt% and 0.068 wt%, respectively, and the nitrogen loss rate rose by 2.44% and 6.73%, respectively. As the energy density rose, the nitrogen content gradually declined, and the nitrogen loss rate gradually rose. Although the nitrogen loss rate increased with the increase of initial nitrogen content in the powder, it is possible to obtain high-nitrogen steel samples above 0.4 wt% through powder blending under atmospheric-pressure printing conditions.

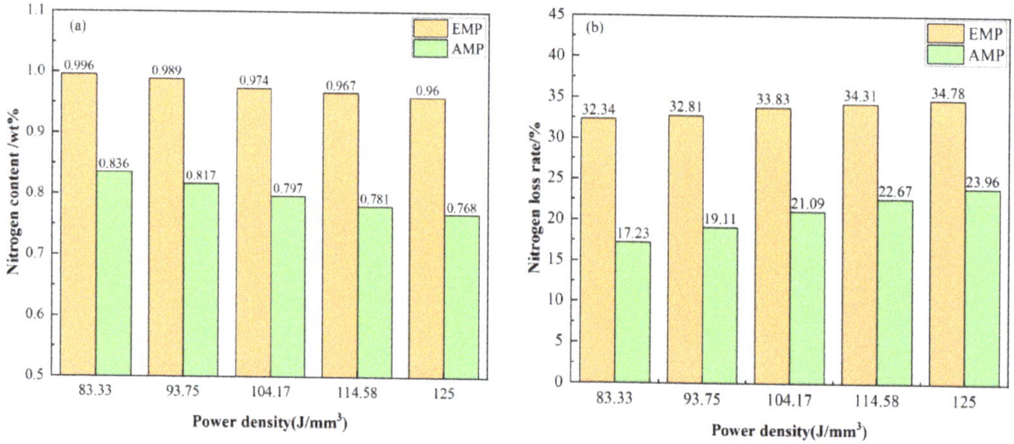


**Figure 7.** Comparison of nitrogen content of the formed samples under different energy densities: (**a**) Nitrogen content and (**b**) Nitrogen loss rate.

In conclusion, it is clear that a lower laser energy density can successfully prevent nitrogen escape. It is hypothesized that the cause is that, when the molten pool's nitrogen content is supersaturated, the lower energy density accelerates the micro molten pool's solidification process. As a result, the gas cannot escape before the molten pool solidifies, effectively reducing nitrogen loss. Under the same printing conditions, EMP with a higher initial nitrogen content can produce a sample with a higher nitrogen content, but it will also result in a higher nitrogen loss, suggesting that a sample's nitrogen content can be increased by increasing the initial nitrogen content of the mixed powder.

### 3.5. Mechanical Properties of the High-Nitrogen Steel

The mechanical properties of the EMP and AMP samples were examined under various energy density conditions. The typical tensile curves are displayed in Figure 8. The results show that the AMP samples had mechanical properties (yield strength, tensile strength, and elongation) superior to those of the EMP samples. The equivalent comparison of yield strength and tensile strength is shown in Figure 9. When the energy density was 114.6 J/mm$^3$, the EMP samples had the highest ultimate tensile strength, at 1005.74 MPa; when the energy density was 105 J/mm$^3$, the AMP samples had the highest ultimate tensile strength, at 1189.2 MPa. As the energy density increased, the tensile strength of the two samples first increased and then decreased. At each energy density, the strength of the AMP samples was greater than that of the EMP samples. The equivalent elongation comparison is shown in Figure 10. The EMP samples had the greatest elongation, 13.07%, at the energy density of 300 W. The elongation of the AMP samples was above 24%, which was much better than that of the EMP samples.

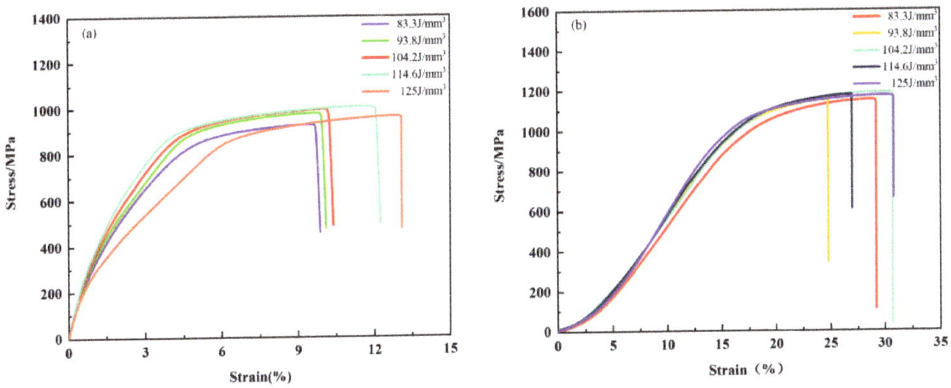


**Figure 8.** Typical stress–strain curves under different energy density conditions: (**a**) sample formed from EMP and (**b**) sample formed from AMP.

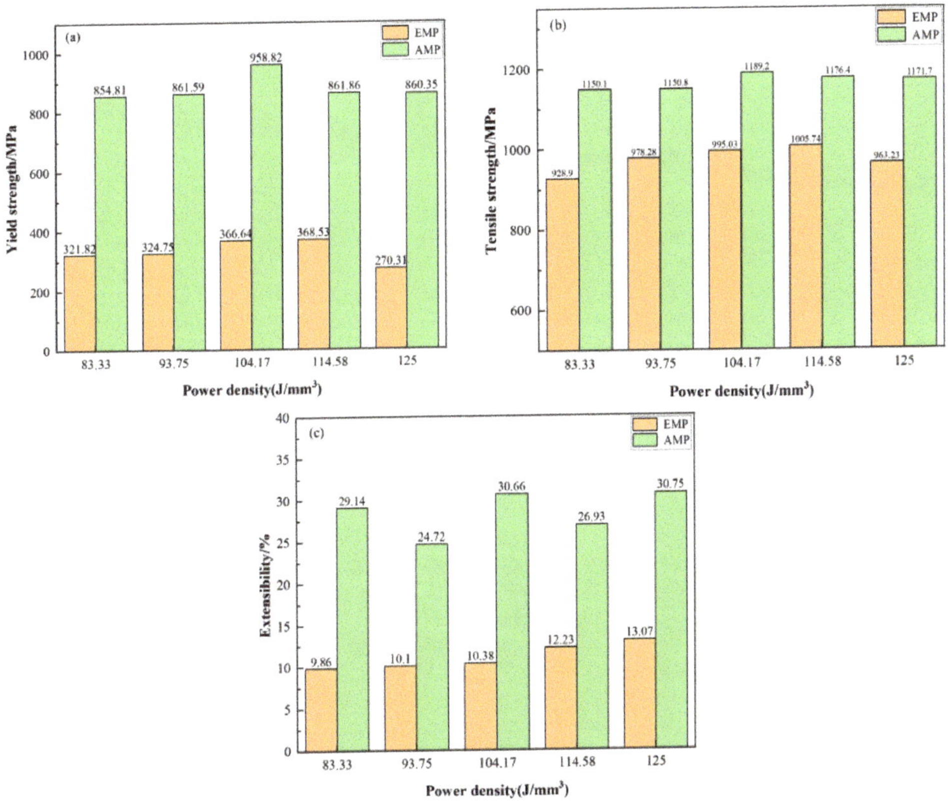


**Figure 9.** Comparison of YS, UTS, and elongation of the formed samples under different energy densities: (**a**) yield strength; (**b**) tensile strength; (**c**) extensibility.

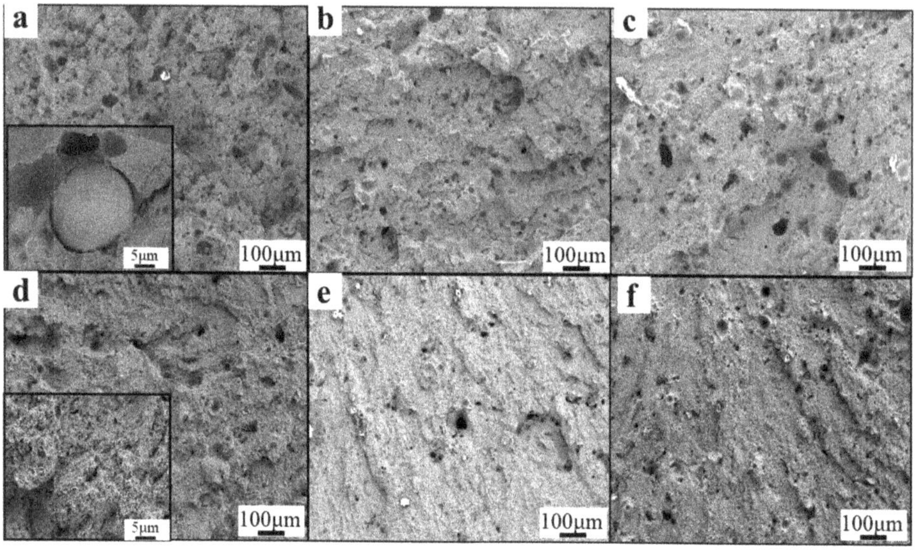


**Figure 10.** Fracture morphology of tensile specimens prepared with: (**a**–**c**) EMP and with (**d**–**f**) AMP.

The above-described research indicates that the steel's strength can be greatly increased by the nitrogen in high-nitrogen steel by acting as solid solution strengthening [47]. The Hall–Petch relationship [48] states that the strength of the metal will diminish as the grain size increases, and that the absence of fusion (pores) will diminish the sample's tensile properties [49]. A positive correlation exists between strength and high or low density [50]. As the energy density rose, the nitrogen content decreased, while the size of the samples' grains grew. The samples' unfused defects progressively worsened as the energy density was increased. When a number of variables are coupled together, the strength is low. The samples' absence of fusion flaws gradually declined as the energy density rose, and the strength gradually rose. An increase in pores inside the samples and the disruption of the matrix continuity occur when the power is too high because these factors cause the micro melt pool to evaporate more quickly and the keyhole changes considerably in size. In the end, the energy density does not continue to grow. The EMP samples exhibited numerous incomplete fusion defects and spheroidized particles, as shown in Figure 5. Despite the higher nitrogen content of the EMP samples, the strength was significantly lower than that of the AMP samples.

Figure 10 displays the fracture morphology of the two types of samples under various energy density conditions. On the fracture surface of the EMP samples, there were numerous spheroidized substances with a size of 25 μm and pore defects. There were only a few dimples, and the predominant fracture mode was that of brittle fracture. Consequently, the ductility was not very high. On the other hand, the fracture morphology of the AMP samples (Figure 10d–f) contained numerous dimples, a negligible number of pore flaws, and no signs of spheroidization, indicating ductile fractures. Compared to that of the EMP samples, the ductility of the AMP samples was markedly enhanced because the EMP samples had a high percentage of nonspherical powder, poor flowability, and an uneven element distribution.

## 4. Conclusions

The nitrogen content, microstructure, and mechanical properties of high-nitrogen steel prepared using selective laser melting were investigated in this study, along with the impacts of powder formulation (elemental mixed powder and alloy mixed powder) and energy density. The main conclusions are as follow:

(1) The formulation of mixed powders significantly affects the quality of high-nitrogen steel. The samples prepared with EMP had more nonfusion flaws and a relatively low density, with a maximum of only 92.36%, while the samples prepared with AMP had fewer defects and a density of up to 97.21%. This is due to the uneven distribution of elements, poor sphericity, and poor flowability of the elemental mixed powder.
(2) The nitrogen content and microstructural characteristics are significantly influenced by the laser energy density. The relative density of the EMP samples increased from 88.29% to 92.36% with the increase in laser energy density from 83.3 J/mm$^3$ to 125.0 J/mm$^3$, while the relative density of the AMP samples increased from 93.31% to 97.21%, and it contained fewer defects and a lower nitrogen content.
(3) The mechanical properties of the AMP samples were superior to those of the EMP samples when the energy density rose, and the strength of the high-nitrogen steel first rose and then fell. The AMP samples possessed the best mechanical properties when the energy density was 104.2 J/mm$^3$, which corresponds to a laser power of 250 W, a scanning speed of 1000 mm/s, and layer thickness of a 30 μm. These values of yield strength, tensile strength, and elongation were 958.8 MPa, 1189.2 MPa, and 30.66%, respectively.

**Author Contributions:** Conceptualization, D.Z.; methodology, S.W.; formal analysis, X.S.; investigation, J.R.; writing—original draft preparation, X.S.; writing—review and editing, X.S. All authors have read and agreed to the published version of the manuscript.

**Funding:** The current work was supported by the Central Guidance on Local Science and Technology Development Fund of Hebei Province (226Z1008G), the National Natural Science Foundation of China (No. 52274336), and the Research Project of Basic Scientific Research Expenses of Hebei Province (No. JYG2021004).

**Data Availability Statement:** No new data were created or analyzed in this study. Data sharing is not applicable to this article.

**Conflicts of Interest:** The authors declare no conflict of interest.

## References

1. Degallaix, S.; Foct, J.; Hendry, A. Mechanical Behaviour of High-Nitrogen Stainless Steels. *Mater. Sci. Technol.* **1986**, *2*, 946–950. [CrossRef]
2. Li, H.; Jiang, Z.; Zhang, Z.; Xu, B.; Liu, F. Mechanical Properties of Nickel Free High Nitrogen Austenitic Stainless Steels. *J. Iron Steel Res. Int.* **2007**, *14*, 330–334. [CrossRef]
3. Chai, G.; Siriki, R.; Nordström, J.; Dong, Z.; Vitos, L. Roles of Nitrogen on TWIP in Advanced Austenitic Stainless Steels. *Steel Res. Int.* **2022**, 2200359. [CrossRef]
4. Speidel, H.J.; Speidel, M.O. Nickel and Chromium-Based High Nitrogen Alloys. *Mater. Manuf. Process.* **2004**, *19*, 95–109. [CrossRef]
5. Rashev, T.V.; Eliseev, A.V.; Zhekova, L.T.; Bogev, P.V. High-Nitrogen Steel. *Steel Transl.* **2019**, *49*, 433–439. [CrossRef]
6. Li, S.; Zhang, C.; Lu, J.; Chen, R.; Chen, D.; Cui, G. A Review of Progress on High Nitrogen Austenitic Stainless-Steel: Ingenta Connect. *Mater. Express* **2021**, *11*, 1901–1925. [CrossRef]
7. Di Schino, A.; Kenny, J.M. Grain Refinement Strengthening of a Micro-Crystalline High Nitrogen Austenitic Stainless Steel. *Mater. Lett.* **2003**, *57*, 1830–1834. [CrossRef]
8. Li, H.; Jiang, Z.; Feng, H.; Ma, Q.; Zhan, D. Aging Precipitation Behavior of 18Cr-16Mn-2Mo-1. 1N High Nitrogen Austenitic Stainless Steel and Its Influences on Mechanical Properties. *J. Iron Steel Res. Int.* **2012**, *19*, 43–51. [CrossRef]
9. Xia, L.; Li, H.; Feng, H.; Jiang, Z.; Zhu, H.; Zhang, S.; Wang, X. Enhanced Strength and Toughness of High Nitrogen Stainless Bearing Steel by Controlling Interstitial Partitioning via V-Microalloying. *J. Mater. Sci. Technol.* **2023**, *151*, 204–218. [CrossRef]
10. Anderson, M.C.; Shin, Y.C. Laser-Assisted Machining of an Austenitic Stainless Steel: P550. *Proc. Inst. Mech. Eng. Part B J. Eng. Manuf.* **2006**, *220*, 2055–2067. [CrossRef]
11. Kuznetsov, Y.V.; Muradyan, O.S.; Muradyan, S.O. Development of High-Strength Corrosion Resistant Austenitic Steel for Oil Equipment Shafts. *Metallurgist* **2022**, *65*, 1095–1099. [CrossRef]
12. High-Manganese and Nitrogen Stabilized Austenitic Stainless Steel (Fe–18Cr–22Mn–0.65N): A Material with a Bright Future for Orthopedic Implant Devices—IOPscience. Available online: https://iopscience.iop.org/article/10.1088/1748-605X/ac265e/meta (accessed on 21 June 2023).
13. Radice, S.; Impergre, A.; Fischer, A.; Wimmer, M.A. Corrosion Resistance of the Nickel-free High-nitrogen Steel FeCrMnMoN0.9 under Simulated Inflammatory Conditions. *J. Biomed. Mater. Res. Part B Appl. Biomater.* **2021**, *109*, 902–910. [CrossRef] [PubMed]

14. Simmons, J.W. Overview: High-Nitrogen Alloying of Stainless Steels. *Mater. Sci. Eng. A* **1996**, *207*, 159–169. [CrossRef]
15. Preparation of Cr17Mn11Mo3N Powders by High-Pressure Gas Atomization and the Nitrogen Increasing Mechanism—ScienceDirect. Available online: https://www.sciencedirect.com/science/article/abs/pii/S0032591021002163 (accessed on 21 June 2023).
16. Cheng, B.; Wei, F.; Teh, W.H.; Lee, J.J.; Meng, T.L.; Lau, K.B.; Chew, L.T.; Zhang, Z.; Cheong, K.H.; Ng, C.K.; et al. Ambient Pressure Fabrication of Ni-Free High Nitrogen Austenitic Stainless Steel Using Laser Powder Bed Fusion Method. *Addit. Manuf.* **2022**, *55*, 102810. [CrossRef]
17. Armstrong, M.; Mehrabi, H.; Naveed, N. An Overview of Modern Metal Additive Manufacturing Technology. *J. Manuf. Process.* **2022**, *84*, 1001–1029. [CrossRef]
18. Full Article: Topology Optimization for Metal Additive Manufacturing: Current Trends, Challenges, and Future Outlook. Available online: https://www.tandfonline.com/doi/full/10.1080/17452759.2023.2181192 (accessed on 21 June 2023).
19. Frazier, W.E. Metal Additive Manufacturing: A Review. *J. Mater. Eng. Perform.* **2014**, *23*, 1917–1928. [CrossRef]
20. Attar, H.; Ehtemam-Haghighi, S.; Kent, D.; Dargusch, M.S. Recent Developments and Opportunities in Additive Manufacturing of Titanium-Based Matrix Composites: A Review. *Int. J. Mach. Tools Manuf.* **2018**, *133*, 85–102. [CrossRef]
21. Tong, J.; Bowen, C.R.; Persson, J.; Plummer, A. Mechanical Properties of Titanium-Based Ti–6Al–4V Alloys Manufactured by Powder Bed Additive Manufacture. *Mater. Sci. Technol.* **2017**, *33*, 138–148. [CrossRef]
22. Ming, Q.; Lim, L.C.; Chen, Z.D. Laser Cladding of Nickel-Based Hardfacing Alloys. *Surf. Coat. Technol.* **1998**, *106*, 174–182. [CrossRef]
23. Zhang, M.; Zhang, B.; Wen, Y.; Qu, X. Research Progress on Selective Laser Melting Processing for Nickel-Based Superalloy. *Int. J. Miner. Metall. Mater.* **2022**, *29*, 369–388. [CrossRef]
24. Chen, H.; Gu, D.; Kosiba, K.; Lu, T.; Deng, L.; Xi, L.; Kühn, U. Achieving High Strength and High Ductility in WC-Reinforced Iron-Based Composites by Laser Additive Manufacturing—ScienceDirect. *Addit. Manuf.* **2020**, *35*, 101195.
25. Huang, W.; Chen, S.; Xiao, J.; Jiang, X.; Jia, Y. Laser Wire-Feed Metal Additive Manufacturing of the Al Alloy. *Opt. Laser Technol.* **2021**, *134*, 106627. [CrossRef]
26. Ostovari Moghaddam, A.; Shaburova, N.A.; Samodurova, M.N.; Abdollahzadeh, A.; Trofimov, E.A. Additive Manufacturing of High Entropy Alloys: A Practical Review. *J. Mater. Sci. Technol.* **2021**, *77*, 131–162. [CrossRef]
27. Gong, G.; Ye, J.; Chi, Y.; Zhao, Z.; Wang, Z.; Xia, G.; Du, X.; Tian, H.; Yu, H.; Chen, C. Research Status of Laser Additive Manufacturing for Metal: A Review. *J. Mater. Res. Technol.* **2021**, *15*, 855–884. [CrossRef]
28. Blakey-Milner, B.; Gradl, P.; Snedden, G.; Brooks, M.; Pitot, J.; Lopez, E.; Leary, M.; Berto, F.; du Plessis, A. Metal Additive Manufacturing in Aerospace: A Review. *Mater. Des.* **2021**, *209*, 110008. [CrossRef]
29. Naseer, M.U.; Kallaste, A.; Asad, B.; Vaimann, T.; Rassõlkin, A. A Review on Additive Manufacturing Possibilities for Electrical Machines. *Energies* **2021**, *14*, 1940. [CrossRef]
30. Selema, A.; Ibrahim, M.N.; Sergeant, P. Electrical Machines Winding Technology: Latest Advancements for Transportation Electrification. *Machines* **2022**, *10*, 563. [CrossRef]
31. Hu, G.; Huang, B.; Xie, H.; Zuo, L.; Wang, N.; Li, D. Ultrasonic-Assisted Direct Writing Metal Additive Manufacturing Technique—ScienceDirect. *J. Mater. Process. Technol.* **2023**, *312*, 117830. [CrossRef]
32. Wu, T.; Liu, J.; Wang, K.; Wang, L.; Zhang, X. Microstructure and Mechanical Properties of Wire-Powder Hybrid Additive Manufacturing for High Nitrogen Steel. *J. Manuf. Process.* **2021**, *70*, 248–258. [CrossRef]
33. Yasa, E.; Kruth, J.-P.; Deckers, J. Manufacturing by Combining Selective Laser Melting and Selective Laser Erosion/Laser Re-Melting. *CIRP Ann.* **2011**, *60*, 263–266. [CrossRef]
34. Gao, B.; Zhao, H.; Peng, L.; Sun, Z. A Review of Research Progress in Selective Laser Melting (SLM). *Micromachines* **2023**, *14*, 57. [CrossRef] [PubMed]
35. Guo, S.; Li, Y.; Gu, J.; Liu, J.; Peng, Y.; Wang, P.; Zhou, Q.; Wang, K. Microstructure and Mechanical Properties of Ti6Al4V/B4C Titanium Matrix Composite Fabricated by Selective Laser Melting (SLM)—ScienceDirect. *J. Mater. Res. Technol.* **2023**, *23*, 1934–1946. [CrossRef]
36. Li, Y.; Liu, S.; Zhi, H.; Yang, X.; Zhang, J.; Wang, Y. In-Situ Generation of High-Strength AISI 1045 Steel with SiO2 Nano-Precipitation by Selective Laser Melting (SLM). *J. Manuf. Process.* **2023**, *94*, 374–386. [CrossRef]
37. Tonelli, L.; Fortunato, A.; Ceschini, L. CoCr Alloy Processed by Selective Laser Melting (SLM): Effect of Laser Energy Density on Microstructure, Surface Morphology, and Hardness. *J. Manuf. Process.* **2020**, *52*, 106–119. [CrossRef]
38. Lan, L.; Yang, Z.; Wang, W.; Cui, Z.; Hao, X. Effect of Initial Powder Particle Size on Densification Behavior and Mechanical Properties of Laser Additive Manufacturing of AlCoCrFeNi2.1 Eutectic High-Entropy Alloy. *Powder Technol.* **2023**, *420*, 118379. [CrossRef]
39. Měsíček, J.; Čegan, T.; Ma, Q.-P.; Halama, R.; Skotnicová, K.; Hajnyš, J.; Juřica, J.; Krpec, P.; Pagáč, M. Effect of Artificial Aging on the Strength, Hardness, and Residual Stress of SLM AlSi10Mg Parts Prepared from the Recycled Powder. *Mater. Sci. Eng. A* **2022**, *855*, 143900. [CrossRef]
40. Xiang, H.; Chen, G.; Zhao, W.; Wu, C. Densification Behavior and Build Quality of Duplex Stainless Steel Fabricated by Laser Powder Bed Fusion. *Metals* **2013**, *13*, 741. [CrossRef]
41. Dwivedi, S.; Rai Dixit, A.; Kumar Das, A. Wetting Behavior of Selective Laser Melted (SLM) Bio-Medical Grade Stainless Steel 316L. *Mater. Today Proc.* **2022**, *56*, 46–50. [CrossRef]

42. Weingarten, C.; Buchbinder, D.; Pirch, N.; Meiners, W.; Wissenbach, K.; Poprawe, R. Formation and Reduction of Hydrogen Porosity during Selective Laser Melting of AlSi10Mg. *J. Mater. Process. Technol.* **2015**, *221*, 112–120. [CrossRef]
43. Wang, L.; Wang, S.; Hong, X. Pulsed SLM-Manufactured AlSi10Mg Alloy: Mechanical Properties and Microstructural Effects of Designed Laser Energy Densities. *J. Manuf. Process.* **2018**, *35*, 492–499. [CrossRef]
44. Ma, M.; Wang, Z.; Zeng, X. A Comparison on Metallurgical Behaviors of 316L Stainless Steel by Selective Laser Melting and Laser Cladding Deposition. *Mater. Sci. Eng. A* **2017**, *685*, 265–273. [CrossRef]
45. Zhong, Y.; Liu, L.; Wikman, S.; Cui, D.; Shen, Z. Intragranular Cellular Segregation Network Structure Strengthening 316L Stainless Steel Prepared by Selective Laser Melting. *J. Nucl. Mater.* **2016**, *470*, 170–178. [CrossRef]
46. Wang, D.; Song, C.; Yang, Y.; Bai, Y. Investigation of Crystal Growth Mechanism during Selective Laser Melting and Mechanical Property Characterization of 316L Stainless Steel Parts. *Mater. Des.* **2016**, *100*, 291–299. [CrossRef]
47. Kim, S.-T.; Kim, S.-Y.; Lee, I.-S.; Park, Y.-S.; Shin, M.-C.; Kim, Y.-S. Effects of Shielding Gases on the Microstructure and Localized Corrosion of Tube-to-Tube Sheet Welds of Super Austenitic Stainless Steel for Seawater Cooled Condenser. *Corros. Sci.* **2011**, *53*, 2611–2618. [CrossRef]
48. Li, H.; Jiang, Z.; Zhang, Z.; Yang, Y. Effect of Grain Size on Mechanical Properties of Nickel-Free High Nitrogen Austenitic Stainless Steel. *J. Iron Steel Res. Int.* **2009**, *16*, 58–61. [CrossRef]
49. Sohrabi, M.J.; Mirzadeh, H.; Sadeghpour, S.; Mahmudi, R. Grain Size Dependent Mechanical Behavior and TRIP Effect in a Metastable Austenitic Stainless Steel. *Int. J. Plast.* **2023**, *160*, 103502. [CrossRef]
50. Liang, Z.; Chen, X.; Sun, Z.; Guo, Y.; Li, Y.; Chang, H.; Zhou, L. A Study on the Compressive Mechanical Properties of 316L Diamond Lattice Structures Manufactured by Laser Powder Bed Fusion Based on Actual Relative Density. *J. Manuf. Process.* **2022**, *84*, 414–423. [CrossRef]

MDPI
St. Alban-Anlage 66
4052 Basel
Switzerland
www.mdpi.com

*Processes* Editorial Office
E-mail: processes@mdpi.com
www.mdpi.com/journal/processes

www.ingramcontent.com/pod-product-compliance
Lightning Source LLC
Chambersburg PA
CBHW061455210926
51850CB00008B/1721
* 9 7 8 3 7 2 5 8 0 6 3 9 3 *